STP 992

Masonry: Materials, Design, Construction, and Maintenance

Harry A. Harris, editor

ASTM
1916 Race Street
Philadelphia, PA 19103

Library of Congress Cataloging-in-Publication Data

Masonry: materials, design, construction, and maintenance/Harry A. Harris, editor.
(STP; 992)
Papers from a symposium held in New Orleans, LA, 2 Dec. 1986 and sponsored by ASTM Committees C-7 on Lime, C-12 on Mortars for Unit Masonry, and C-15 on Manufactured Masonry Units.
Includes bibliographies and indexes.
"ASTM publication code number (PCN) 04-992000-07."
ISBN 0-8031-1168-1
1. Masonry—Congresses. I. Harris, Harry A. II. ASTM Committee C-7 on Lime. III. ASTM Committee C-12 on Mortars for Unit Masonry. IV. ASTM Committee C-15 on Manufactured Masonry Units. V. Series: ASTM special technical publication; 992.
TA670.M378 1988 88-15446
693.1—dc19 CIP

NOTE

The Society is not responsible, as a body, for the statements and opinions advanced in this publication.

Peer Review Policy

Each paper published in this volume was evaluated by three peer reviewers. The authors addressed all of the reviewers' comments to the satisfaction of both the technical editor(s) and the ASTM Committee on Publications.

The quality of the papers in this publication reflects not only the obvious efforts of the authors and the technical editor(s), but also the work of these peer reviewers. The ASTM Committee on Publications acknowledges with appreciation their dedication and contribution of time and effort on behalf of ASTM.

Printed in Baltimore, MD
July 1988

Foreword

This publication, *Masonry: Materials, Design, Construction, and Maintenance,* contains papers presented at the symposium of the same name held in New Orleans, LA on 2 Dec. 1986. The symposium was sponsored by ASTM Committees C-7 on Lime, C-12 on Mortars for Unit Masonry, and C-15 on Manufactured Masonry Units. Harry A. Harris, Ash Grove Cement Co., presided as symposium chairman and was editor of this publication.

Contents

MAINTENANCE

Overview

This book stems from the fifth in a series of symposia on masonry sponsored by ASTM Committees C-7 on Lime, C-12 on Mortars for Unit Masonry, and C-15 on Manufactured Masonry Units. Like those that have preceded it, this symposium provided a forum for the dissemination and exchange of information and experiences related to masonry construction.

Three of the four preceding symposia were also published by ASTM:

STP 589—Masonry: Past and Present, published in August 1975, was the first in this series. It provided the basis for future symposia by reviewing the specifications and test methods from a historical perspective. Research and new developments in the field of masonry construction were also covered.

STP 778—Masonry: Materials, Properties, and Performance, published in September 1982, covered the third symposium on masonry. This publication presented a forum for research on masonry units, mortar and grout (including their components), and masonry assemblages.

STP 871—Masonry: Research, Application, and Problems, published in April 1985, covered the fourth symposium. The objective of this symposium was to cover field applications, end-use problems, and research.

The second symposium in this series was not published except for several papers appearing in ASTM's *Journal of Testing and Evaluation.* The scope of this symposium was similar to that of the first.

For the current symposium, Masonry: Materials, Design, Construction, and Maintenance, papers dealing with current technology in each of these four major areas of masonry were requested. The areas were selected to provide general coverage of current developments in the industry and thus provide an update to previous publications on this subject.

Materials

The first three papers relate to testing procedures and properties of masonry materials and assemblages. New test procedures are described and data presented on bond and shear strengths of masonry assemblages. These papers will provide new guidelines for writing future specifications and codes for masonry materials.

Robinson and Brown examine the existing C 270 requirements (ASTM Specification for Mortar for Unit Masonry) for mortar and its shortcomings. Test data on the bond strengths of several C 270 mortars are offered as a basis of writing a new performance specification.

Johal and Anderson have evaluated masonry cement mortars when used in the construction of shear-wall specimens. Both static and cyclic load tests were conducted in this investigation of concrete masonry block and clay brick walls.

Ribar and Dubovoy have explored the surface characteristics of brick. A new technique for measuring surface characteristics provides important new insights into factors controlling bond and shear strengths. Additional research on surface characteristics may provide a means of evaluating the performance of each material in a masonry assemblage.

Design

Design of masonry construction is dealt with in the next five papers. Subjects range from the detailing of tie systems to the testing of assemblages under various types of simulated loading.

This important work will provide the means for better and safer construction even under the extreme conditions found during an earthquake.

Raths has presented a preconstruction testing program for the selection of materials used in a brick-veneer cavity wall. The program uses ASTM standards to evaluate materials and their compatibility as a means of preventing both construction problems and unsatisfactory performance.

Gensert and Bretnall have documented the construction and performance of a masonry structure by means of photography and computer graphics. Methods of analyzing architectural details and the interactions between masonry and structural frame works are shown.

Chin et al. have examined the relative stiffness between brick veneer and metal studs and its effects upon wall design. This paper shows that, by using shorter length metal studs, critical flexural bond stresses do not exist on typical brick veneer/metal stud walls under designed loading conditions.

Arnold et al. compare two methods of analysis for the design of brick spandrel panels. Torsion stresses during placement were of particular interest in this study.

Chen and Shah have studied methods of improving seismic design of masonry structures by testing masonry single pier models. The behavior of these piers was studied under dynamic shaking and slowly applied cyclic loading.

Construction

The section on construction provides a direct link between the researcher and field application. Three papers are offered in this category, each dealing with a different subject but all of value to both the university laboratory as well as the masonry contractor.

Grimm's first paper is a review of methods for sampling and statistical data reduction for brick masonry. Based on the techniques described, a concept of structural reliability is introduced.

Matthys et al. present data on extended life or ready-mix mortars. The paper describes a study in optimizing a mix to achieve specific strength and setting characteristics. The optimized mix is then compared to standard portland cement lime Type N mortar for mechanical properties.

Coney and Stockbridge address water leakage problems through a study of waterproofing coatings, surface grouting, and tuckpointing. Field studies were conducted prior to and during the repairs of a building.

Maintenance

The preservation of our national heritage is gaining greater interest year by year, and the most outstanding symbols of our past are found in masonry structures. To this end we devoted the final section of this symposia to maintenance and rehabilitation. The four papers presented here cover unique reviews of major rehabilitation projects. Details are presented on the materials and methods used, where and how failures in the original construction occurred, and how failures were corrected. These papers will be of great interest and value to those involved in this area of masonry work.

Thomasen and Searls describe the deterioration of terra-cotta claddings and their repair. Special emphasis is given to the cause of glaze spalling as related to repair and prevention measures.

Manmohan et al. have studied the compressive stresses in terra-cotta cladding due to frame shorting. Methods of relieving these stresses by cutting the bed and head joints are described.

Sourlis presents a detailed review of the restoration of a 100-year-old historical masonry structure. All aspects of the project from bidding the job to final cleanup and landscape repairs are described.

Grimm's second paper is a study of masonry cracks and how they affect the performance of a masonry structure. The various types of cracks are described and their causes and method for repair are given.

This publication is the result of the combined efforts of many people. I want to thank the members of my subcommittee and those assisting with the presentation and review of papers. A special thanks to those on the ASTM staff who helped guide me through the many stages of this process from conception to final publication.

Harry A. Harris

Ash Grove Cement Co., Kansas City, KS 66103; symposium chairman and editor.

Materials

Gilbert C. Robinson[1] *and Russell H. Brown*[2]

Inadequacy of Property Specifications in ASTM C 270

REFERENCE: Robinson, G. C. and Brown, R. H., "**Inadequacy of Property Specifications in ASTM C 270,**" *Masonry: Materials, Design, Construction, and Maintenance, ASTM STP 992,* H. A. Harris, Ed., American Society for Testing and Materials, Philadelphia, 1988, pp. 7–17.

ABSTRACT: Bond strength and water leakage were determined for Types M, S, and N mortars joined to six different brick types. The air content, water retention, and flow rate were varied for each mortar-brick combination. It was found that air content, water retention, and compressive strength were inadequate indices of bond strength and water leakage. Yet these are the only properties specified in ASTM Specification for Mortar for Unit Masonry (C 270-84). A plea is made to add a bond strength requirement to present property requirements. The initial rate of absorption (IRA) of brick was found to be significant in predicting water leakage. Low IRA brick provided the highest probability of zero leakage.

KEY WORDS: masonry, mortar, bond strength, water retention, air content, compressive strength, water leakage

Consumers of mortars have supported the concept of a performance standard for mortar. As a consequence, ASTM Specification for Mortar for Unit Masonry (C 270-84) includes a property specification as well as a proportion specification. Inclusion of the property specification option is a move in the right direction, but the designated qualities are inadequate for assuring quality mortar. It is intended to plea for the addition of a bond strength requirement to present property requirements and possibly the addition of a water leakage specification.

Property requirements should relate to key performance qualities desired in a mortar. The most rigorous mortar requirements are to provide adequate and uniform bond strength and to prevent water leakage. All other mortar functions are easy to fulfill. ASTM C 270 fails to address the most demanding mortar qualities and instead specifies only three properties: compressive strength, water retention, and air content. It is intended to present test results that will question the ability of these three properties to assure quality mortar performance.

Compressive Strength

ASTM C 270 prescribes a minimum compressive strength for each type of mortar. Does this relate to bond strength or water leakage? Figure 1 shows the insignificance of compressive strength to bond strength. Three different types of mortar and four to six different types of brick for each mortar type were evaluated for bond strength. A skilled brick mason constructed a brick prism consisting of four bricks and three mortar joints for each combination for mortar type and brick type. The mason proportioned the mortars by volume as prescribed by ASTM C 270. He used conventional field procedures for mixing the mortar and adjusted the mixing water until he judged the mixture to be of desired consistency. The prisms were constructed and

[1]Professor of Ceramic Engineering, Clemson University, Clemson, SC 29634-0907.
[2]Professor and Head of Civil Engineering, Clemson University, Clemson, SC 29634-0911.

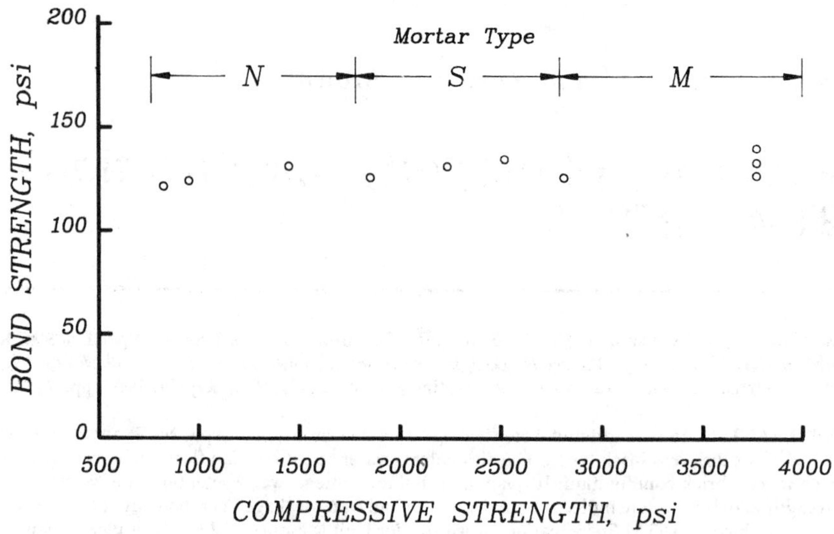

FIG. 1—*The bond strengths of mortars of various compressive strengths.*

then cured in a room environment for 28 days under a cover of plastic sheeting. The bond strength was determined by the bond wrench method [ASTM Standard Method for Measurement of Masonry Flexural Bond Strength (C 1072-86)]. Each point of Fig. 1 represents the average results from several brick types as shown in Table 1. The compressive strengths of the mortars were determined on mortar cubes as instructed by ASTM C 270 and ASTM Test Method for Compressive Strength of Hydraulic Cement Mortars (Using 2-in. or 50-mm Cube Specimens) (C 109).

All the bond strengths were within plus or minus 10 psi (69 kPa) of 135 psi (931 kPa). At the same time, the compressive strength of the mortars varied from 800 to 3800 psi (5.5 to 26.2 MPa). This suggests that compressive strength is valueless as an index of bond strength.

Water Retention

ASTM C 270 specifies that mortars of all types shall have a minimum water retention of 75%. This seems to make sense because a mortar should hold water and remain fluid against the suction of the masonry unit, but does it work to assure quality mortar? The results of Fig. 2 suggest that this property too provides little assurance of quality mortar. The figure shows the relation between water retention and bond strength. Each point is an average of several brick

TABLE 1—*The number of brick types and joints tested for each datum point of Figs. 1 and 2.*

Type Mortar	Number of Brick Types	Number of Joints Tested Per Datum Point
Type M	6	18
Type S	5	15
Type N	4	12

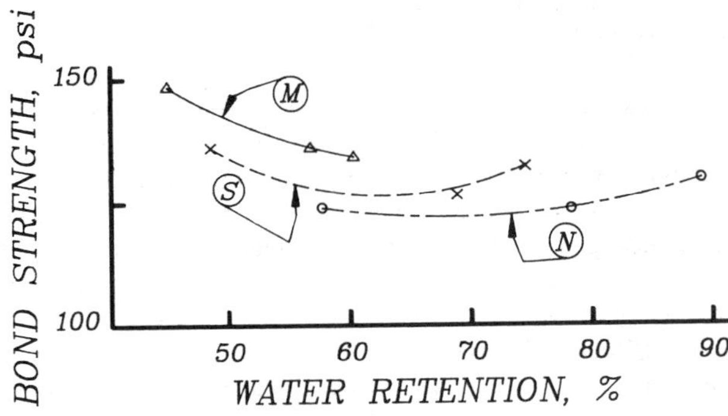

FIG. 2—*The influence of water retention on bond strength.*

types as shown in Table 1. The water retention follows the expected pattern of being higher with Type N and lowest with Type M. The unexpected result was the little change in bond strength with large changes in water retention. All of the combinations showed good bond strength, and yet all of the M mortars were below the prescribed water retention, one being only 45%. Two out of three of the Type S mortars failed to meet the water retention standard while one of the Type N mortars was below the standard. Nevertheless, they all gave approximately the same level of bond strength as the mortars that did meet the specified 75% water retention. There is even a suggestion with the Type M mortars that substandard water retention provides a slight improvement in bond strength. This supports the work by Hogberg as cited by Goodwin and West [*1*]. Hogberg demonstrated that mortars with low water retentivity gave better bond strength on absorbent brick. Low water retention coupled with extra water in the mortar quenches the thirst of high IRA brick and improves bonding. This was demonstrated by Robinson and Salmond [*2*]. It is suggested that three mechanisms may explain the observed results on water retention.

The first is that the mortar should give up easily its water to high IRA brick in order to nullify the deleterious effect of high suction rate. Low water retention coupled with high water content mortars provides for satisfaction of the suction demand of the brick while retaining sufficient fluidity of the mortar for penetration into the surface structure of the brick. In contrast, increasing mortar water retention through additives can retain the water against the suction of the brick and retard filtration of the mortar paste into the brick surface.

The second consideration is for low IRA brick. Their limited water suction suggests the desirability of a mortar that would stiffen with very little water removal.

The third consideration applies to all units. Dewatering of the mortar at the brick mortar interface provides a more favorable water-to-cement ratio for strength development.

Once again the inclusion of the property "water retention" bears little relation to mortar performance and is harmfully restrictive in designing a mortar-brick combination for optimum performance.

Air Content

The last quality specified is the air content of the mortar. A maximum air content of 12% is specified for portland cement lime mortars—Types M and S—and 14% for Type N. In contrast, masonry cements are limited to 18% maximum air content only if structural reinforcement is incorporated. Otherwise no maximum is listed, although C 270 does state that the masonry cement should conform to the requirements of ASTM Specification for Masonry Cement

(C 91). C 91 lists minimum air contents of 12% by volume and maximum contents of 20% for Types M and S and 22% for Type N.

The influence of air content on bond strength is shown in Fig. 3. The same procedure was used as described in the section entitled "Compressive Strength." The results of Fig. 1 were limited to PCL mortars with less than 5.3% air content. These specimens were included in Fig. 3 together with additional specimens with higher air quantities and mortars made with masonry cements. There is considerable scatter in results, but there appears to be a trend of decreasing bond strength with increasing air content. In fact, it appears that the air content is more significant than the type of mortar. Thus Types N, S, and M almost overlap at 3% air, and Types N and M and masonry cement overlap at 12% air. The masonry cement seems to form a continuation of the curve to lower levels of bond strength because of the higher permissible levels of air content. Thus Type N masonry cement with 18% air (near its maximum limit) showed a bond strength of about 35 psi (241 kPa), whereas Type N-PCL (portland cement-lime) mortar at 10% air content (near its maximum limit) showed bond strengths of 75 psi (517 kPa). However, the masonry cement compared to the PCL at the same air content showed about the same bond strength. It would seem that masonry cements do themselves a disservice by specifying a minimum air content. This is further illustrated by the bond strength designated as M-MC in Fig. 3. These results are from a different investigator and were prepared with a Type M masonry cement in which the air content was controlled by mixing procedure. It will be observed that this produced somewhat higher bond strengths than the Type M-PCL mortar when compared at equivalent air content. Furthermore, it seems to follow a similar trend of decreasing bond strength with increasing air content.

These results suggest that air content is a more meaningful quality indicator than the other property requirements. However, air content alone is not sufficient to assure mortar quality, as shown in Fig. 4. This shows the results on seven different mortar batches of similar air content.

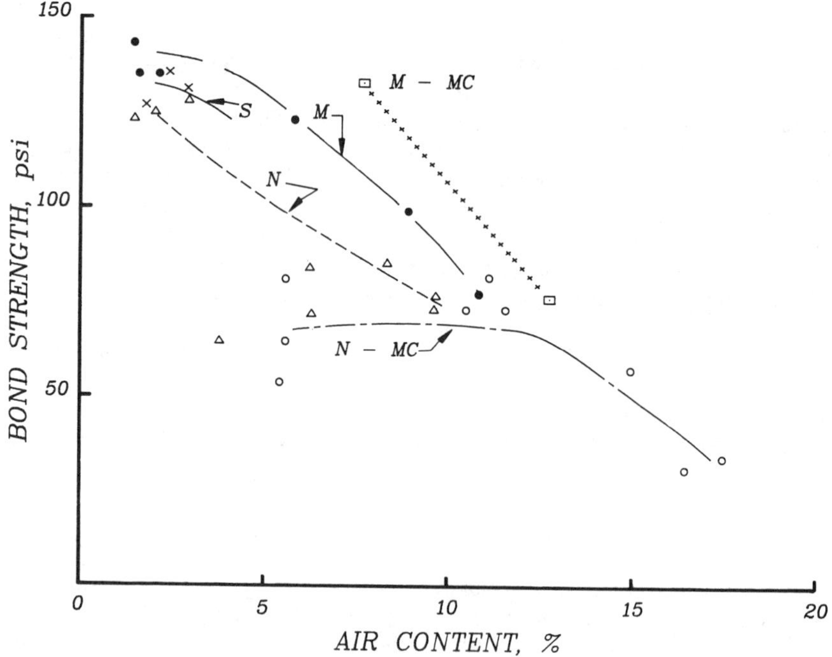

FIG. 3—*The influence of mortar air content on bond strength.*

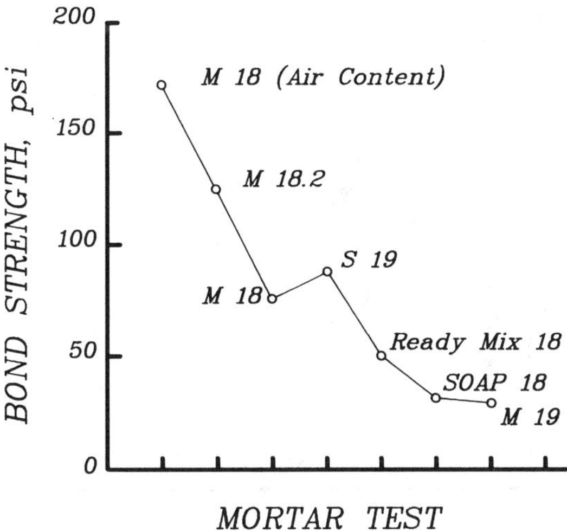

FIG. 4—*Variation in bond strength of mortars with similar air content.*

However, the air was introduced by use of different air entraining agents or different procedures. This shows a change from 179 to 30 psi (1234 to 207 kPa) bond strengths for the various trials. All of the compositions were proportioned to be Type M except for the one labeled S and Ready-Mix. The Type S was a Type S masonry cement and the Ready-Mix was also a Type S masonry cement. These results point out that although air content may be significant to bond strength, it is not sufficient to assure bond strength. Instead, the type of entraining agent and the pore structure will influence bonding. A fine, uniform bubble structure favors strength while an irregular structure with large bubbles harms strength. Schmidt [*3*] reports a similar conclusion of European investigators of bond strength.

Once again the property specification of C 270 is inadequate. Replacement of the air requirement with a bond strength requirement would assure adequate performance and permit masonry cements and PCL mortars to compete without prejudice.

Brick IRA

The properties of masonry units are not part of the scope of C 270. Nevertheless, bond strength is influenced by the masonry unit, and a bond strength requirement would evaluate the compatibility of a particular mortar-brick pair. The initial rate of absorption (IRA) has been suggested as a key quality of brick. Tests were made to evaluate the bond strength of brick of different IRAs and Figs. 5 and 6 show typical results. It is interesting that the IRA appears to have little influence on bond strength.

Other investigators have come to a similar conclusion. Gazzola [*4*] presented a similar result in his paper at the last masonry symposium presented in Florida in 1983. Additional results available from the Holmes Laboratory [*5*] in another mortar investigation had a similar finding. The reason for the contradiction in results between these investigations and earlier ones is being explored and will be the subject of another report.

Water Leakage

The topic of water leakage has been avoided because of its complexity and the wide scattering of results. The same mortar-brick combinations previously presented under "Compressive

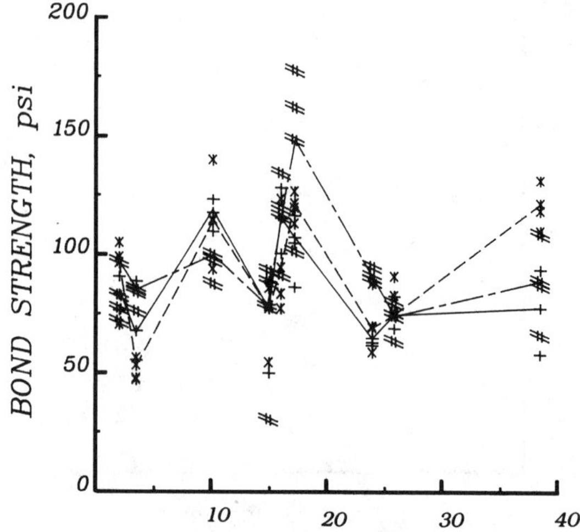

FIG. 5—*The relationship between bond strength and IRA for a Type N PCL mortar.*

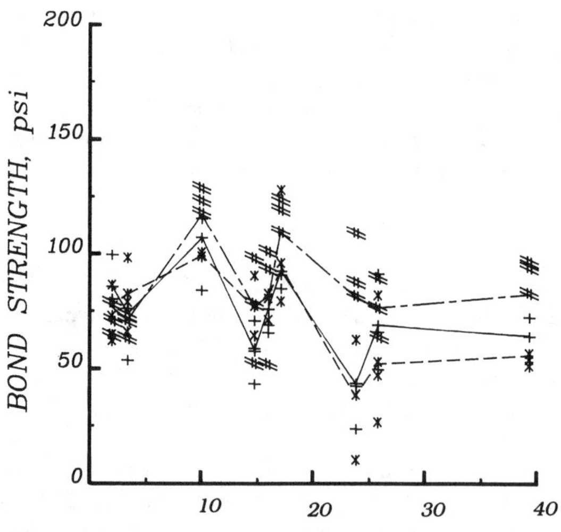

FIG. 6—*The relationship between bond strength and IRA for a Type N masonry cement mortar.*

Strength" were evaluated for their tendency towards leakage of water. This was done by pooling water to a depth of 2 in. on top of horizontal prisms, then measuring the rate in drop of water level. Each type of brick was tested with Types N, S, and M mortar and each combination was tested at low, medium, and high air content and low, medium, and high flow rate.

The average water leakage was calculated for each brick type by combining the results for all mortar types, flow rates, and air contents. The results are shown in Fig. 7 and the number at each point is the number of prisms tested for each point. The solid curve indicates the average water leakage. An attempt was made to assign a number to each trial which would indicate the inches of water drop occurring per hour. There were two problems with this. First, there would be an initial period where the water would soak into the assembly but not contribute to leakage through the assembly. The higher the absorption of the brick, the greater the amount of this initial water level drop. At the other extreme, trials that approached 2 in. of water leakage in the 6-h test period showed a tailing off of the rate of drop when the water level became 0.5 in. or less. It would appear that the slope of the curve between these extremes better illustrates the tendency to water leakage. As a consequence, a numerical value expressing water leakage was obtained by measuring the drop in water level occurring between 1-h elapsed time and 6-h elapsed time and dividing by five. Furthermore, if the water level dropped to 0.5 in. prior to 6-h elapsed time, the reading was taken at this point and the time duration adjusted accordingly. In the case of severe leakage the water would drop 2 in. in less than an hour, and in this event the slope was evaluated in the period from zero to 1 h.

In addition to a numerical ranking, the specimens were also given a letter grade with A awarded to those prisms that showed leakages of less than 0.05 in./h, a B for leakages of 0.05 to 0.10 in./h, etc. The broken line curve of Fig. 7 is the inverse of the first and obtained by a different method. This curve shows the percentage of nonleakers or A grade trials for each type of brick.

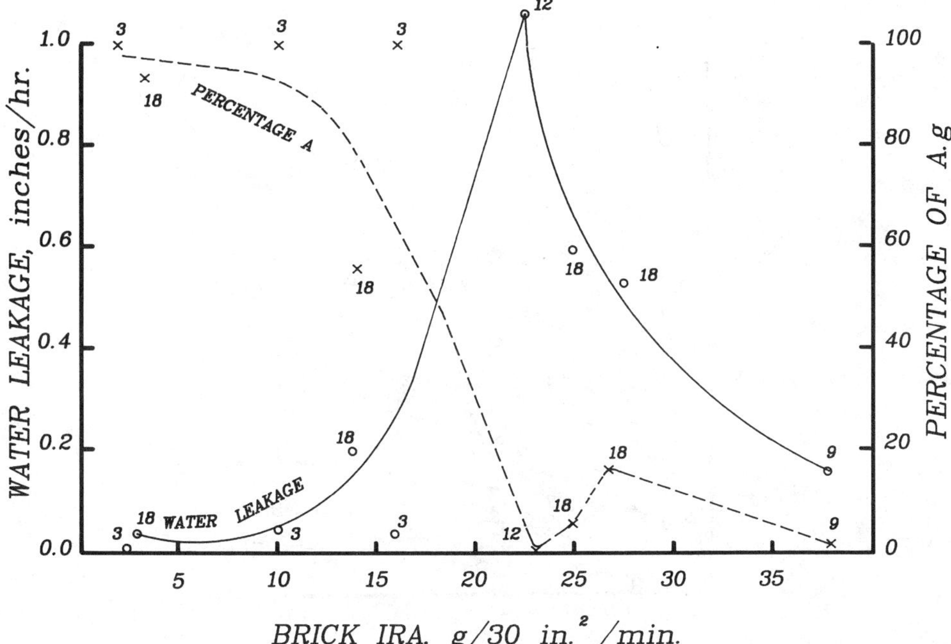

FIG. 7—*The relationship between brick IRA and rate of water leakage or percentage of nonleaking units (A grade).*

Both methods of expressing results agree that water leakage is lowest with low IRA brick. It would appear that IRAs of 10 or less are preferable.

The results of the two methods are distinct for higher IRA brick. The water leakage curve indicates maximum leakage at 23 g IRA and then decreasing leakage as the IRA increases to 38 g. In contrast, the percentage of A prisms decreases to below 12% for all IRAs of 23 and above. This would suggest that the rate of water leakage may decrease at higher IRA but that the probability of a joint showing leakage is high for any value of 23 or above. These results were obtained for brick without prewetting prior to application of mortar.

The scatter in results at higher IRAs precludes definite conclusions. The results do seem to support the conviction that brick type and IRA have a major influence on water leakage.

Brick continue to imbibe water from mortar beyond the 1-min limit, and it was wondered how an extended IRA test would rank the brick type. This is shown in Fig. 8, which plots the grams

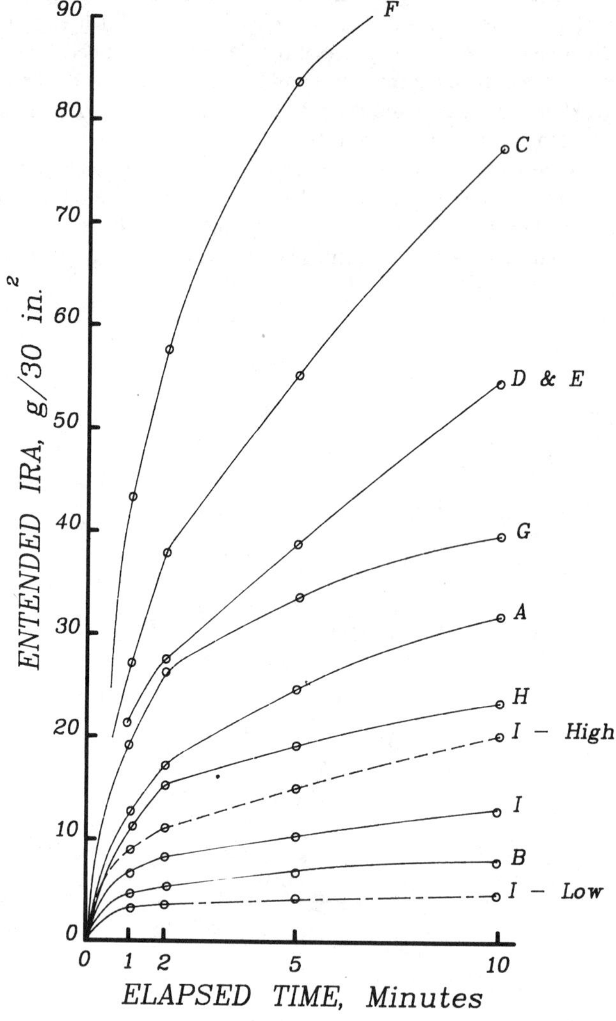

FIG. 8—*The waterpickup versus elapsed time for the types of brick used in bond strength tests.*

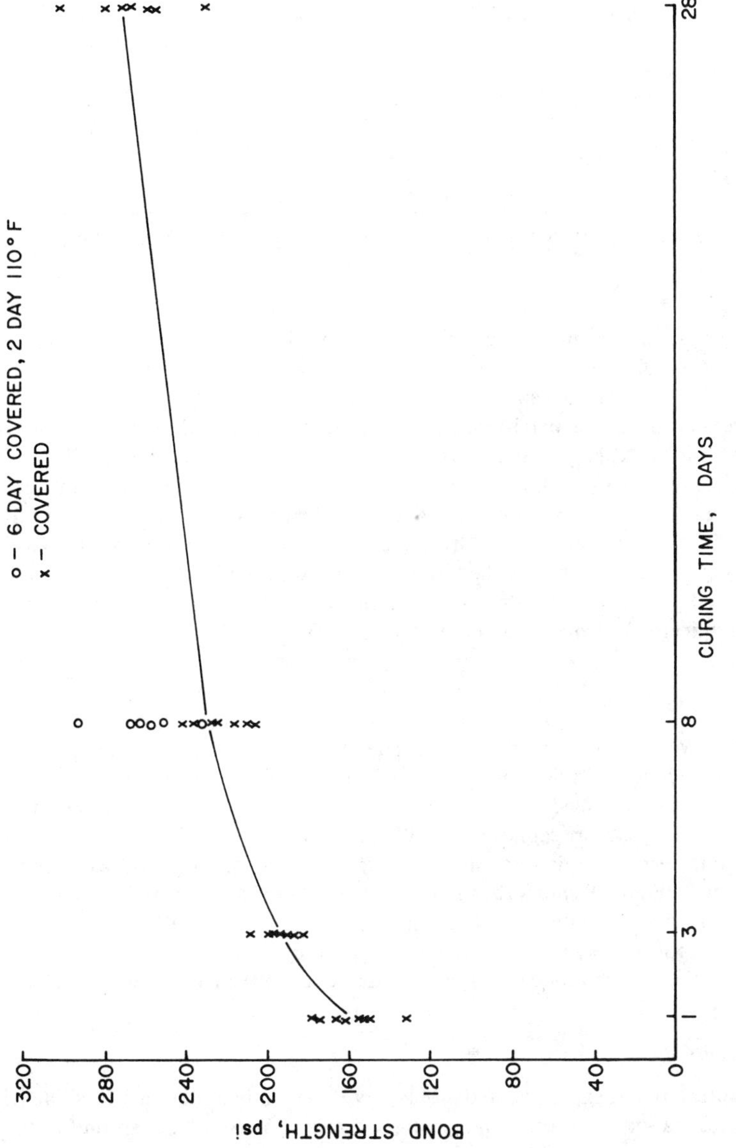

FIG. 9—*The influence of curing time on bond strength.*

TABLE 2—*The influence of air content and mortar flow on resistance to water leakage.*

Brick Type	IRA	Air Content	Type N Mortar by Flow			Type M Mortar by Flow		
			Low	Medium	High	Low	Medium	High
		Low	$\bar{A}$	A	F	A	B	A
A	18	Medium	C	C	F			
		High	$\underline{B}$	A	A	C	A	A
		Low	D	F	$\bar{A}$	D	D	$\bar{C}$
C	27	Medium	C	F	A			
		High	D	C	$\underline{D}$	F	F	$\underline{A}$
		Low	F	B	$\bar{A}$	F	C	$\bar{C}$
D	25	Medium	F	D	C			
		High	F	F	$\underline{D}$	F	F	$\underline{F}$

NOTE: A = no leakage (less than 0.05 in./h); F = severe leakage (greater than 0.39 in./h).

for water pickup per net 30 in.^{2} of surface area against minutes of exposure. It will be noticed that the bad leaking brick (C, D, E, and F) show an increased gap from the other brick at 10 min. Good Brick G and bad Brick D and E gave similar results at 1 and 2 min but a widening gap with increased time. This persistent suction appears detrimental to leakage resistance.

The results on water leakage cause a suspicion that mortar flow may be significant. Table 2 abstracts leakage rankings (A, B, C, D, or F) from a larger table. The numerical value of leakage was obtained by averaging all the values for each brick type. It will be noticed that high flows gave the best results for problem Brick C and D; however, a reverse result was obtained with Brick A of lower IRA. Perhaps the flow should be adjusted from low to high as the brick IRA increases. Also, this table illustrates the ranking of brick in Fig. 7. Note that the percentage of A grades decreases from brick Types A to C to D.

Proposed Change to C 270

Test results have been presented that indicate that the present property specification of ASTM C 270 is inadequate. It does not provide the consumer with assurance of quality mortar. Such assurance could be provided if the present requirement for water retention, air content, and compressive strength were replaced with a bond strength test of the intended mortar-brick combination. Furthermore, it is suggested that such a test could be performed within 24 h of forming specimens. Figure 9 shows the relation between bond strength and curing time. It will be noticed that the majority of bond strength develops within 24 h. Bond wrench testing procedures are easy to apply to field testing as well as to the laboratory. The inclusion of a bond test in C 270 would be a major step forward in quality assurance to the consumer.

Acknowledgments

Grateful appreciation is expressed to the Masonry Research Foundation for its financial support of this investigation. Appreciation is expressed also to Wilson U. Johnson for his performance of most of the test procedures of this project.

References

[*1*] Goodwin, J. F. and West, H. W. H., "A Review of Literature on Brick/Mortar Bond," British Ceramic Research Association, Stoke-on-Trent, ST47LQ, England, December 1980.

[*2*] Robinson, G. C. and Salmond, W. H., work in process, Ceramic Engineering Department, Clemson University, Clemson, SC, 1986.

[*3*] Schmidt, S. B., president, Addiment, Inc., Doraville, GA, personal communication.

[*4*] Gazzola, E., Bagnariol, D., Toneff, J., and Drysdale, R. G., "Influence of Mortar Materials on the Flexural Tensile Bond Strength of Block and Brick Masonry," *Masonry: Research, Application and Problems, ASTM STP 871*, American Society for Testing and Materials, Philadelphia, PA, 1985, pp. 15–25.

[*5*] "Final Test Results Mortar Life Project," H. H. Holmes Testing Laboratories, Inc., Wheeling, IL, June 1983.

L. S. Paul Johal[1] *and Eric D. Anderson*[1]

Shear Strength of Masonry Piers Under Cyclic Loading

REFERENCE: Johal, L. S. P. and Anderson, E. D., "**Shear Strength of Masonry Piers Under Cyclic Loading,**" *Masonry: Materials, Design, Construction, and Maintenance, ASTM STP 992,* H. A. Harris, Ed., American Society for Testing and Materials, Philadelphia, 1988, pp. 18–32.

ABSTRACT: This paper describes a test program to evaluate shear behavior of masonry piers constructed with portland cement and masonry cement-based mortars and tested under simulated seismic loading. Both clay brick and concrete masonry block pier specimens were subjected to in-plane cyclic shear stresses. Several brands of masonry cement and portland cement were used in fabrication of 32 specimens. The pier portion of specimens was minimally reinforced for flexure and partially grouted. No shear reinforcement was provided. The specimens were loaded in a manner to eliminate any axial force in the pier cross section. Results indicate that the shear strength of masonry piers constructed with masonry cement-based mortars compared favorably with those constructed with portland cement-based mortars. Values of maximum shear stress obtained with portland cement and masonry cement-based mortars were close and within the range of scatter normally expected in experimental data. The shear strength was not influenced significantly by the use of Types M or S mortars.

KEY WORDS: masonry, masonry piers, shear, shear stress, shear strength, cyclic loading, masonry cement-based mortar, portland cement-based mortar, Type M mortar, Type S mortar

In masonry construction, shear walls are often used to resist lateral loads through in-plane stresses. Shear behavior of masonry, therefore, is a significant factor influencing the overall response of masonry structures. During the last three decades, a limited number of experimental programs have attempted to identify and investigate parameters affecting the shear strength of masonry. One of the parameters that has not been adequately investigated is the type of cement used in the mortar. Both portland cement and masonry cement can be used to achieve a selected mortar compressive strength. Almost all experimental programs on shear strength, however, have been based on mortars containing a combination of portland cement and lime. A shortage of test data on the behavior of masonry walls, using masonry cement-based mortars, has led to regulatory constraints on the use of masonry cement. The current edition of the Uniform Building Code (UBC) does not allow the use of masonry cement in structural frames for Seismic Zone Nos. 2, 3, and 4.

A test program was undertaken to expand the data base on the behavior of masonry walls. Both clay brick and concrete masonry block pier specimens were subjected to in-plane cyclic shear stresses. The overall objective of the program was to evaluate and compare shear behavior of masonry piers constructed with portland cement and masonry cement-based mortars and tested under simulated seismic loading. Several brands of masonry cement and portland cement were used in the fabrication of 32 pier-type masonry shear wall specimens. Sixteen of these specimens were constructed with hollow clay brick and 16 were constructed with concrete block.

[1]Senior structural engineer and associate structural engineer, respectively, Structural Experimental Section, Construction Technology Laboratories, Skokie, IL.

The test region (pier portion) of specimens was minimally reinforced and partially grouted so that the shear strength of masonry is primarily influenced by the mortar. Only the outer pier cells were reinforced with a single vertical bar and grouted. In order to determine the shear strength of masonry only, no horizontal shear reinforcement was provided in the pier portion. In addition, loads were applied in a manner to eliminate the application of axial load to the pier portion. This provided lower bound values of shear strength. Shear behavior of the specimens was then compared to evaluate the performance of different cement types.

Experimental Program

The experimental program included two categories of mortars. The first category included mortars obtained from Types M and S masonry cements. The second category had Types M and S mortars prepared from a combination of portland cement and hydrated lime, with no masonry cement included. Three representative brands each of Types M and S masonry cement were used to prepare three mixes each of M and S mortar. One mix each of Types M and S mortar was prepared by blending three brands of portland cement with lime. Both hollow clay bricks and hollow concrete blocks were included. Running bond construction pattern was used throughout this program.

Test Matrix

The test matrix is presented in Table 1. Three representative brands of Type M masonry cement designated MC1, MC2, and MC3 were used to prepare three mixes of Type M mortar. Three representative brands of Type S masonry cement designated SC1, SC2, and SC3 were used to prepare three mixes of Type S mortar. Three representative brands of portland cement were used to form a blended sample designated PC1. One mix each of Type M and Type S mortar was obtained by mixing blended portland cement with hydrated lime.

Specimens

Thirty-two I-shaped pier type specimens were tested. This included 16 specimens constructed with hollow clay brick and 16 with hollow concrete block. Specimens were cured for at least 28

TABLE 1—*Test matrix.*

Mortar Cementitious Materials	Cement Brand Designation	Mortar Type	Mix Proportions[a]	Specimen Designation	
				Hollow Clay Brick Pier	Hollow Concrete Block Pier
Masonry cement, Type M	MC1	M	0:1:0:2 1/2[b]	CM1, CM2	DM1, DM2
	MC2	M	0:1:0:2 1/2[b]	CM3, CM4	DM3, DM4
	MC3	M	0:1:0:3	CM5, CM6	DM5, DM6
Masonry cement, Type S	SC1	S		CS1, CS2	DS1, DS2
	SC2	S	0:1:0:3	CS3, CS4	DS3, DS4
	SC3	S		CS5, CS6	DS5, DS6
Blended portland cement and lime	PC1	M	1:0:1/4:3 3/4	CP1, CP2	DP1, DP2
		S	1:0:1/2:4 1/2	CP3, CP4	DP3, DP4

[a]Mix proportions are expressed as proportions of portland cement : masonry cement : lime: fine aggregate (by volume).

[b]Mix proportions used as specified by the manufacturer.

days in a laboratory maintained at 23° ± 3°C (73° ± 5°F) and 50 ± 5% relative humidity prior to testing. Specimen designation, type of mortar used, and mix proportions are listed in Table 1.

Brick Pier Specimens. Sixteen specimens constructed with two-core nominal 200 by 100 by 300-mm (8 by 4 by 12-in.) hollow brick were tested. The specimen included 2.1-m by 0.6-m by 200-mm (7-ft by 2-ft by 8-in.) top and bottom spandrels as shown in Fig. 1. A 0.9-m^2 (3-ft^2) pier formed the test portion of the specimen. The spandrels were used to apply horizontal shear forces that were transferred across the pier portion of the specimens. The pier portion was reinforced with one Grade 60 No. 5 bar in each of the two outer cells. These two cells were filled with portland cement-based grout. No horizontal shear reinforcement was provided in the pier region. Thickness of mortar joints was maintained at approximately 13 mm (1/2 in.) Full mortar bedding was used. Mortar joints on both faces were cut flush and then tooled. The spandrels were overreinforced to force damage into the pier portion. Both spandrels were completely grouted.

Block Pier Specimens. Sixteen specimens constructed with two-core nominal 200 by 200 by 400-mm (8 by 8 by 16-in.) concrete block were tested. The specimen included 2.0-m by 0.6-m by 200-mm (6-ft 8-in. by 2-ft by 8-in.) top and bottom spandrels as shown in Fig. 2. The test portion consisted of a 0.6-m^2 (2-ft 8-in.^{2}) pier. Thickness of the mortar joint in the pier region was approximately 10 mm (3/8 in.). Mortar was applied only to the face shells. Other details were similar to those given for brick pier specimens.

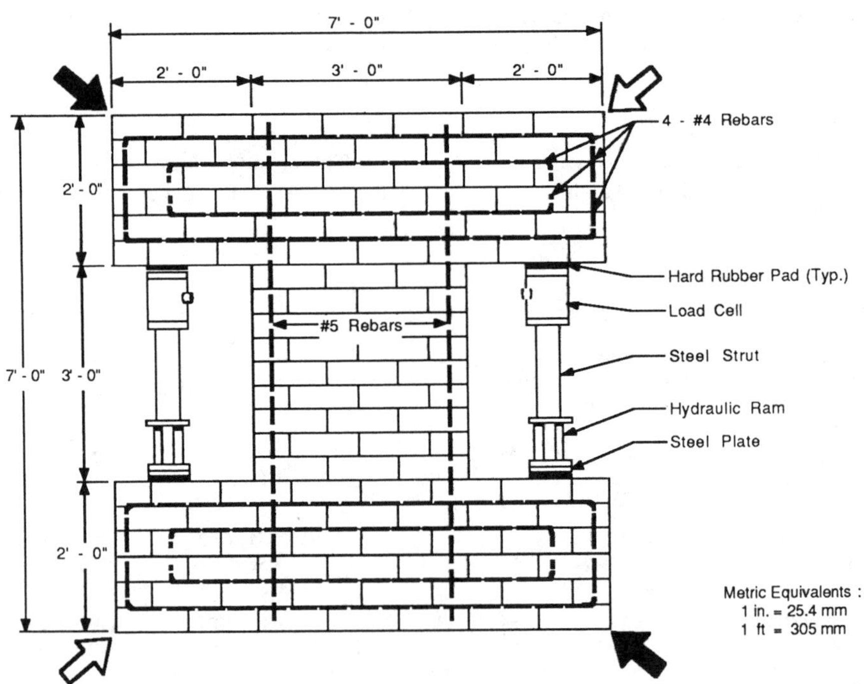

FIG. 1—*Brick pier specimen.*

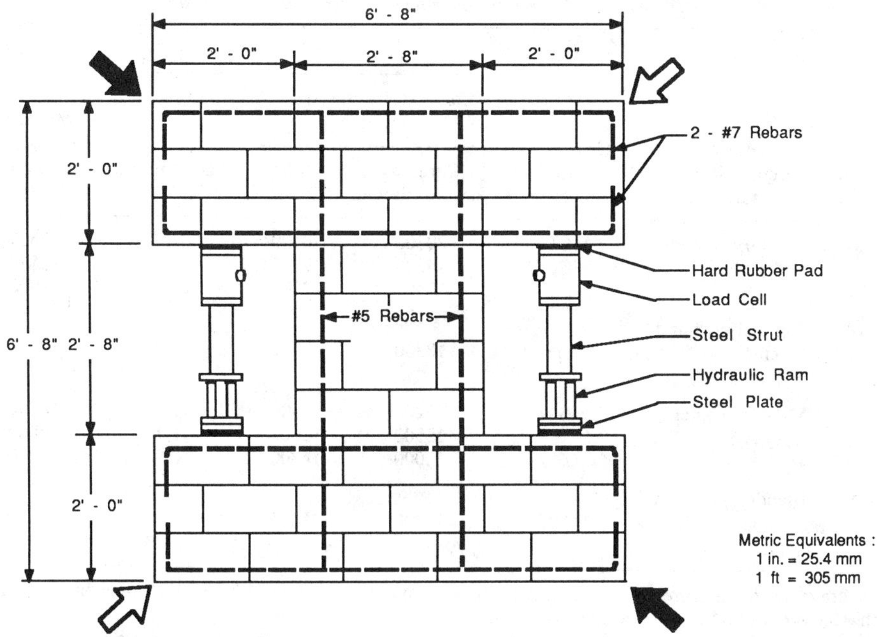

FIG. 2—*Block pier specimen.*

Materials

Masonry Units. Average compressive strengths of hollow clay brick based on gross and net areas were 22.1 and 32.8 MN/m^2 (3210 and 4750 psi), respectively. Average compressive strengths of concrete masonry block based on gross and net areas were 10.1 and 19.3 MN/m^2 (1470 and 2800 psi), respectively. Net areas of brick and block units were determined in accordance with the ASTM Method of Sampling and Testing Concrete Masonry Units (C 140-75).

Mortar. Types M and S mortar mixes were proportioned in accordance with the ASTM Specifications for Mortar for Unit Masonry (C 270-84). Materials were proportioned as shown in Table 1. At least six 50 by 50-mm (2 by 2-in.) mortar cubes were prepared for each mortar mix. Three of these cubes were cured under the same laboratory conditions as the test specimens and three were cured in a moist room maintained at 23° ± 2°C (73° ± 3°F) and 100% relative humidity for comparison. Air content of wet mortar at desired consistency was also determined for each mortar mix. Compressive strengths and air content levels of mortar for brick and block pier specimens are listed in Tables 2 and 3, respectively.

Grout. Grout used to fill the outer cells of pier portions of the specimens was prepared in accordance with the ASTM Specifications for Grout for Masonry (C 476-83). The grout mixture included by volume one part of Type I/II portland cement, one tenth part of hydrated lime, three parts of sand, and two parts of pea gravel with a maximum size of 10 mm (3/8 in.). In addition, a grout aid admixture, amounting to approximately 1% of cement by weight, was added. Water-to-cement ratio was approximately 0.80. Average 28-day grout compressive strengths measured on 75 by 150-mm (3 by 6-in.) cylinders were 23.0 and 19.3 MN/m^2 (3330 and 2800 psi), respectively, for brick and block pier specimens.

Masonry Prisms. At least three brick and three block prisms were constructed and tested for

TABLE 2—*Material properties for brick pier specimens.*

Mortar Type	Mortar Cementitious Materials	Cement Brand Designation	Mortar Properties: Compressive Strength, kN/m²[a] Cured as Pier Specimens	Mortar Properties: Compressive Strength, kN/m²[a] Moist Cured	Mortar Properties: Air Content, %	Masonry Compressive Strength[b], kN/m²
M	Masonry cement, Type M	MC1	20300	20300	7.3	17500
		MC2	14200	17400	12.4	15800
		MC3	15700	17300	13.5	11900
M	Blended portland cement and lime	PC1	18300	31700	1.0	18000
S	Masonry cement, Type S	SC1	15200	14800	11.4	18300
		SC2	15500	15600	11.1	18000
		SC3	15000	16300	10.9	17200
S	Blended portland cement and lime	PC1	15400	21400	3.5	14900

[a]Average mortar compressive strength determined from at least three 51 by 51 mm cubes in accordance with ASTM C 270-84 (UBC Standard No. 24-20).
[b]Average masonry compressive strength determined from at least three masonry prisms in accordance with ASTM E 447-84 (UBC Section 2405 C).

TABLE 3—*Material properties for block pier specimens.*

Mortar Type	Mortar Cementitious Materials	Cement Brand Designation	Mortar Properties: Compressive Strength, kN/m²[a] Cured as Pier Specimens	Mortar Properties: Compressive Strength, kN/m²[a] Moist Cured	Mortar Properties: Air Content, %	Masonry Compressive Strength[b], kN/m²
M	Masonry cement, Type M	MC1	15800	19200	13.4	8610
		MC2	17200	28600	14.4	9230
		MC3	19600	22800	18.0	8410
M	Blended portland cement and lime	PC1	22300	38700	1.0	11200
S	Masonry cement, Type S	SC1	17300	15400	11.2	9230
		SC2	18100	16500	13.3	11000
		SC3	19600	23000	13.6	8410
S	Blended portland cement and lime	PC1	17900	22000	4.0	8340

[a]Average mortar compressive strength determined from at least three 51 by 51 mm cubes in accordance with ASTM C 270-84 (UBC Standard No. 24-20).
[b]Average masonry compressive strength determined from at least three masonry prisms in accordance with ASTM E 447-84 (UBC Section 2405 C).

each type of mortar mix. Construction and testing of prisms conformed to the requirements of the ASTM Test for Compressive Strength of Masonry Prisms (E 447-84). The height of the brick prisms was approximately 400 mm (16 in.) obtained by stacking four units. The height of the block prisms was approximately 600 mm (24 in.) obtained by stacking three units. Test results are given in Tables 2 and 3.

Test Equipment

Loads were applied diagonally to the top and bottom spandrel beams as illustrated in Fig. 3. The loads were applied by 445-kN (100-kip) capacity hydraulic rams using special loading shoes and transferred through four 16-mm-diameter (5/8-in.) high-strength rods. The horizontal component of the applied load represented the shear load transferred across the pier section. Two struts, one on each side of the pier as shown in Figs. 1 and 2, were provided to reduce rotation of the spandrel beams. Each strut included a 155-kN (35-kip) capacity hydraulic ram and a 222-kN (50-kip) capacity load cell. Vertical strut force was applied simultaneously on each side of the pier. The compression induced by each strut was held to one half of the applied shear force to prevent application of axial load to the pier section. Thus, the pier portion of the specimen was only subjected to a combination of shear and flexural loads. A photograph of the representative test setup is shown in Fig. 4.

Test Procedure

After placing the specimen in the loading rig and attaching all instrumentation, the test was started by applying load through diagonally opposite corners of the top and bottom spandrel beams. The load was applied in a series of increments alternating from one direction to the

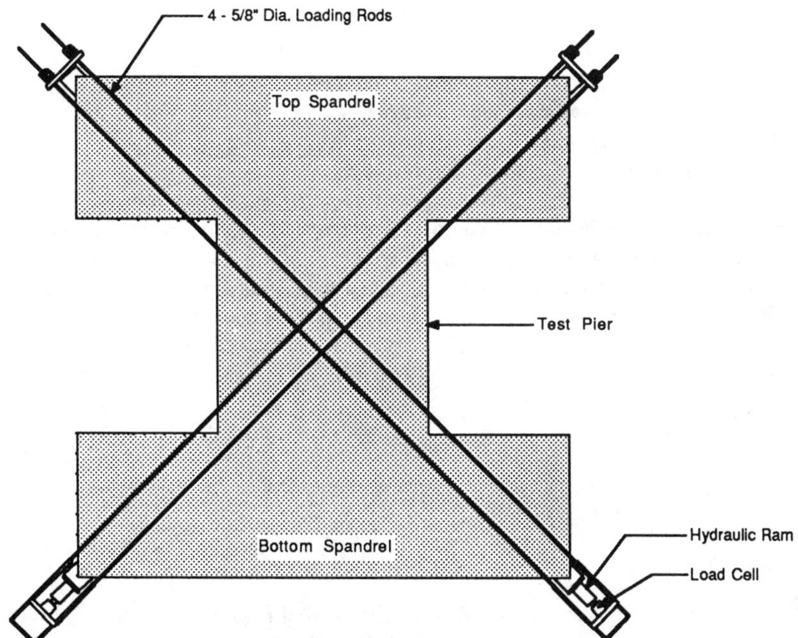

FIG. 3—*Loading arrangement.*

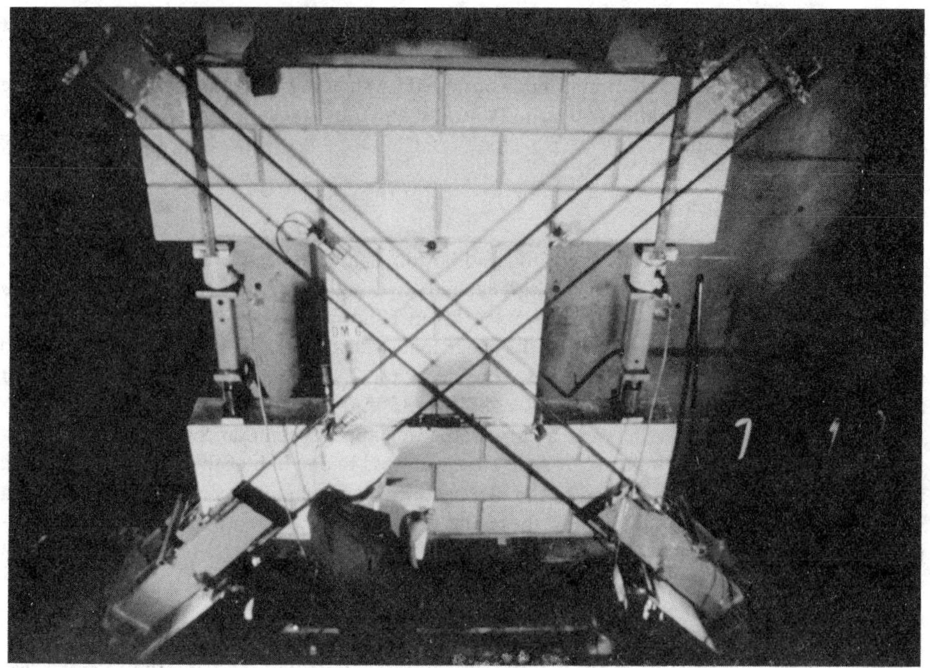

FIG. 4—*Representative test setup.*

opposite direction. When the specimen exhibited large shear deformations at a specific load level, the specimen was assumed to be approaching its load capacity. The load was then continuously increased to obtain maximum shear stress. Testing was stopped at a stage when the load dropped under increasing shear deformation.

Specimens were instrumented to measure applied loads and induced deformations. A layout of instrumentation is given in Fig. 5. The diagonally applied loads were measured by a pair of 445-kN (100-kip) capacity load cells located behind the loading shoes at the corners of the bottom spandrel. Vertical strut forces on each side of the pier were measured by two 222-kN (50-kip) capacity load cells. Deformations along the diagonals of the pier were measured by two linear potentiometers.

For the square pier section, the shear strain, γ, was calculated as follows

$$\gamma = \frac{(d_1 - d_2)d}{2ab}$$

where

a, b = height and width of the instrumented region ($a = b$ in this case),
d = original diagonal length of the instrumented region, and
$d_1 - d_2$ = change in diagonal lengths of the instrumented region at a given load stage.

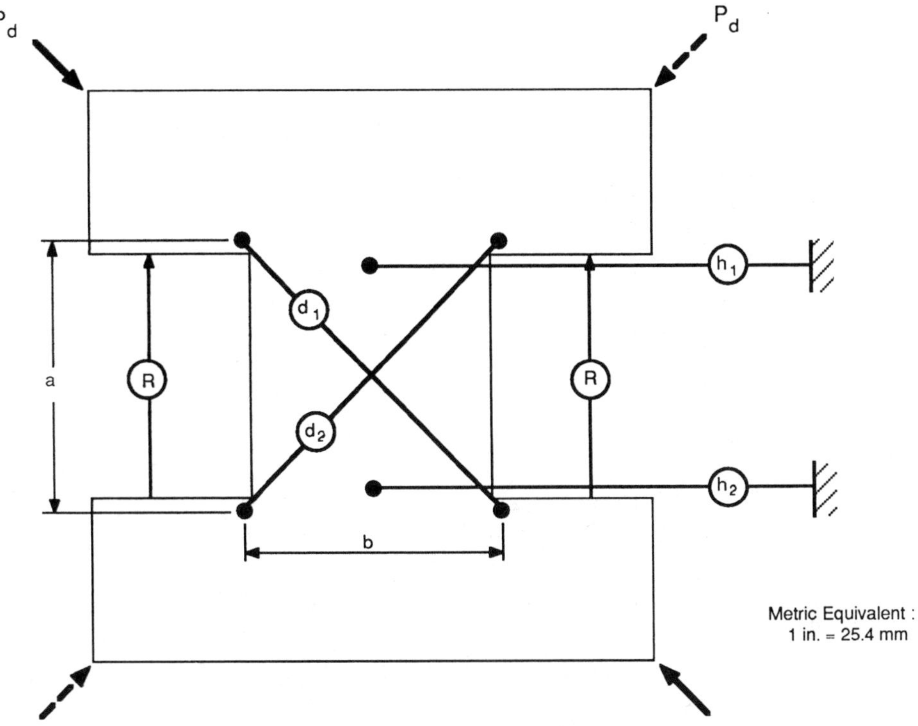

Note :

P_d = Diagonally Applied Load

R = Vertical Strut Force

d_1, d_2 = Change in diagonal lengths at a given Load Stage

a, b = Height and Width of Instrumented Region

(38 in. for Brick and 34 in. for Block Pier Specimens)

h_1, h_2 = Transverse Displacement Measurements

FIG. 5—*Layout of instrumentation.*

Test Results

Shear Strength

A summary of test results including maximum shear stress, corresponding shear strain, and shear modulus is listed in Tables 4 and 5 for brick and block pier specimens, respectively. Maximum shear stress was determined from the measured maximum shear force divided by the net shear area of the pier section. Shear strain was calculated as described in the previous section. The values reported corresponded to measured maximum shear forces. As judged from the slopes of shear stress versus shear strain plots, the shear modulus was determined at shear stresses of approximately 517 kN/m^2 (75 psi) and 345 kN/m^2 (50 psi) for brick and block pier specimens, respectively. Values of average maximum shear stress and coefficient of variation

TABLE 4—*Test results for brick pier specimens.*

Mortar Type	Mortar Cementitious Materials	Specimen Designation	Max Shear Stress, kN/m^2	Shear Strain[a], %	Shear Modulus[b], MN/m^2	Max Shear Stress, kN/m^2	
						Average	C.O.V.[c], %
M	Masonry cement, Type M	CM1	806	0.821	2790	946	8.99
		CM2	992	0.129	2670		
		CM3	985	0.335	2670		
		CM4	999	0.651	3130		
		CM5	1070	0.588	2780		
		CM6	985	0.569	2750		
	Portland cement and lime	CP1	944	0.591	2110	955	0.82
		CP2	965	0.300	3020		
S	Masonry cement, Type S	CS1	1070	0.884	2610	794	14.69
		CS2	861	0.739	1890		
		CS3	834	...	2940		
		CS4	744	0.125	3410		
		CS5	758	0.968	2490		
		CS6	772	0.430	2550		
	Portland cement and lime	CP3	868	0.568	3690	868	7.97
		CP4	1020	0.888	3090		

[a]Shear strain indicated corresponds to the maximum shear stress.
[b]The modulus of rigidity (shear modulus) was calculated at a shear stress of approximately 520 kN/m^2.
[c]C.O.V. = coefficient of variation.

TABLE 5—*Test results for block pier specimens.*

Mortar Type	Mortar Cementitious Materials	Specimen Designation	Max Shear Stress, kN/m^2	Shear Strain[a], %	Shear Modulus[b], MN/m^2	Max Shear Stress, kN/m^2	
						Average	C.O.V.[c], %
M	Masonry cement, Type M	DM1	930	0.630	7540	1040	6.08
		DM2	1050	0.389	765		
		DM3	1130	0.559	3350		
		DM4	1080	0.574	2620		
		DM5	999	0.585	3050		
		DM6	1030	0.700	2400		
	Portland cement, and lime	DP1	1120	0.314	4180	1060	5.18
		DP2	1010	0.468	3280		
S	Masonry cement, Type S	DS1	1100	0.416	3600	1060	2.25
		DS2	1040	0.390	2670		
		DS3	1070	0.355	5090		
		DS4	1050	0.546	1760		
		DS5	1050	0.333	2990		
		DS6	1030	0.595	4000		
	Portland cement and lime	DP3	1010	0.426	2580		
		DP4	978	0.551	4710	1010	1.82

[a]Shear strain indicated corresponds to the maximum shear stress.
[b]The modulus of rigidity (shear modulus) was calculated at a shear stress of approximately 340 kN/m^2.
[c]C.O.V. = coefficient of variation.

for all specimens are also listed in Tables 4 and 5. Representative plots of shear stress versus shear strain are shown in Figs. 6 and 7.

Modes of Behavior

All specimens exhibited a shear mode of failure in the pier portion. Photographs of representative brick and block specimens after completion of tests are provided in Figs. 8 and 9, respectively. Cracking in most cases initiated at one of the pier corners. These were usually small cracks. In some cases, cracking appeared along the entire length of a diagonal instantaneously. In these cases, the mode of behavior was brittle and the failure occurred at loads only 10 to 15% above the cracking loads. Cracking loads ranged from 40 to 90% of the maximum loads. Early cracking (below 60% of maximum load) usually resulted in subsequent behavior that was more ductile. Initiation of cracking was also observed in some specimens in the interior portion of the pier. As the applied load increased, the diagonal tensile stress reached the tensile capacity of masonry resulting in large diagonal cracks. Cracks generally extended through horizontal and vertical mortar joints along the pier diagonals prior to reaching maximum load. At failure loads, however, the cracks extended through some masonry units also.

Discussion of Test Results

Brick Pier Specimens. As listed in Table 4, average maximum shear stress was 972 kN/m^2 (141 psi) for the six brick pier specimens constructed with Type M masonry cement-based mortar. The two corresponding specimens constructed with portland cement and lime reached an average maximum shear stress of 958 kN/m^2 (139 psi). Thus, the maximum shear stress values in these two series were very close. Also, the average cracking loads and the behavior at maximum load were similar.

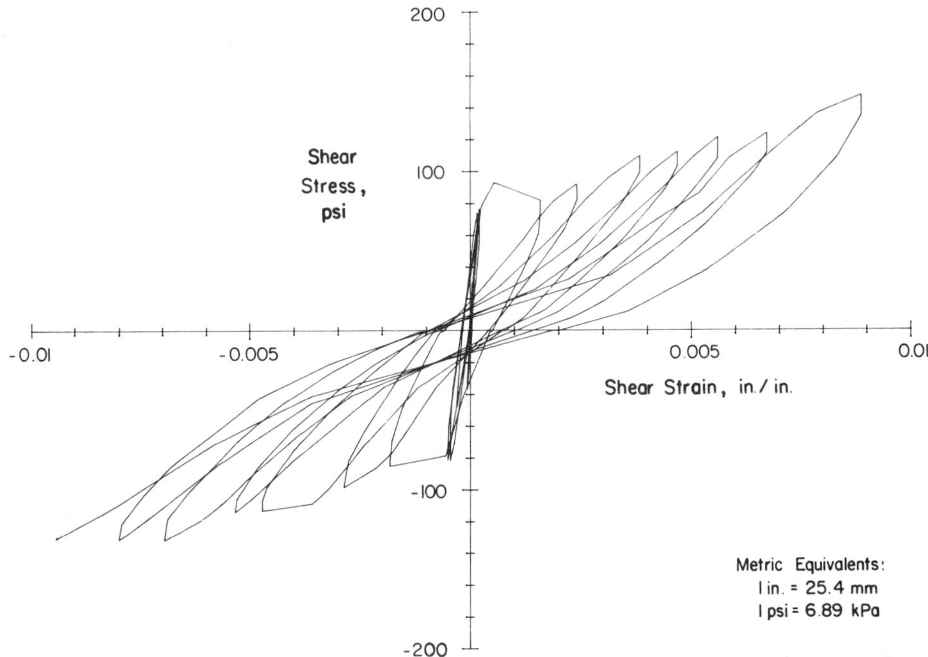

FIG. 6—*Shear stress versus strain for brick pier specimen CP4.*

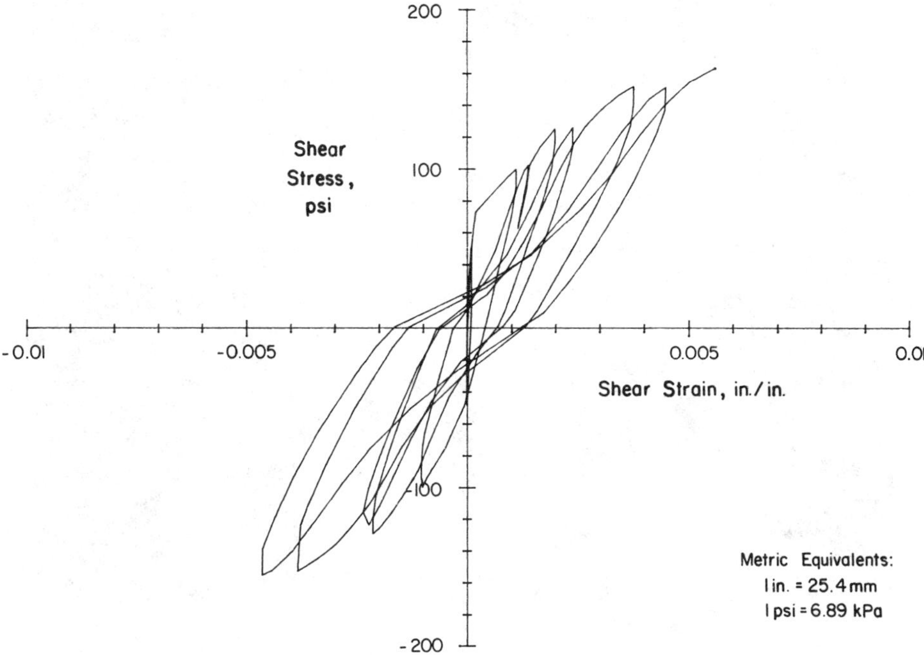

FIG. 7—*Shear stress versus strain for block pier specimen DM3.*

For the six brick specimens constructed with Type S masonry cement-based mortar, average maximum shear stress was 841 kN/m^2 (122 psi). The two corresponding specimens constructed with portland cement and lime reached an average maximum shear stress of 945 kN/m^2 (137 psi). The shear stress values obtained in these two series are within 11% of each other, indicating similar strengths for masonry cement-based and portland cement-based mortars. The overall behavior and cracking pattern of specimens were also similar.

As can be seen in Table 4, the shear strains and shear moduli for all specimens fall within acceptable range of scatter expected in this type of experimental data. Also, there is no appreciable difference between the results of specimens constructed with Type M and Type S mortars. Within the scope of this test program and within the data scatter normally expected, the type of cement used in the mortar had no influence on the shear strength of brick pier specimens.

Block Pier Specimens. As listed in Table 5, the six block specimens constructed with Type M masonry cement-based mortar reached an average maximum shear stress of 1034 kN/m^2 (150 psi). This compares very favorably with the two corresponding specimens constructed with portland cement-based Type M mortar. These specimens reached an average maximum shear stress of 1062 kN/m^2 (154 psi). Overall behavior and cracking pattern in the two series were similar.

For the six block pier specimens constructed with Type S masonry cement-based mortars, average maximum shear stress was 1062 kN/m^2 (154 psi). This is greater than the average maximum shear stress of 993 kN/m^2 (144 psi) obtained in the corresponding portland cement-based specimens.

There is no appreciable difference between the results of specimens constructed with Type M and Type S mortars. Thus, within the scope of the test program and the data scatter normally

(a) Specimen CM5

(b) Specimen CS5

FIG. 8—*Representative brick pier specimens after test:* (a) *Specimen CM5;* (b) *Specimen CS5.*

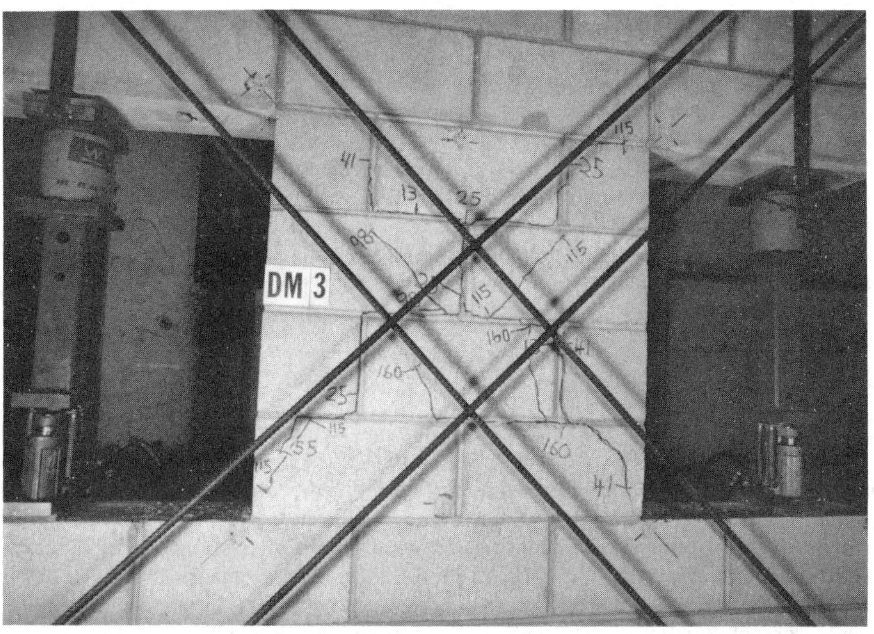

(a) Specimen DM3

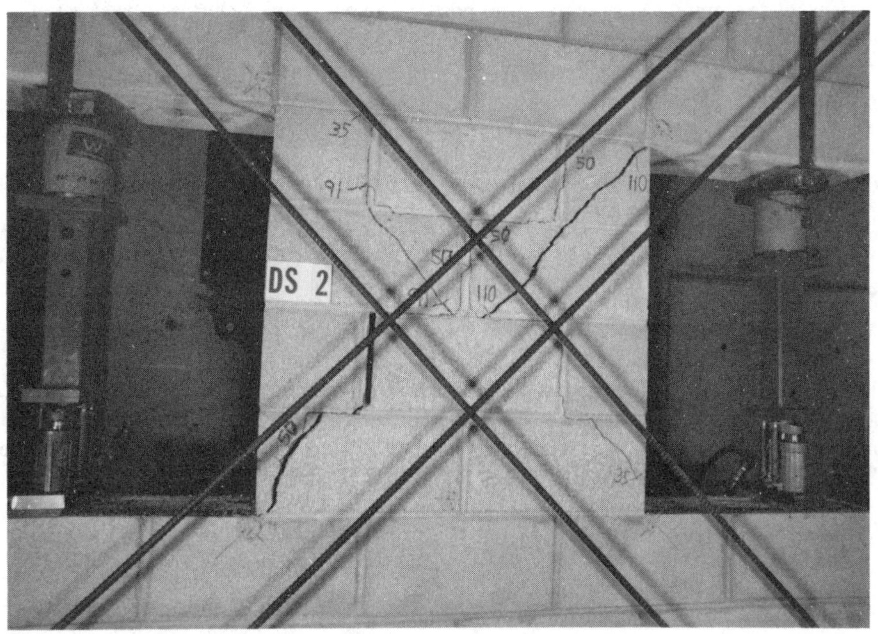

(b) Specimen DS2

FIG. 9—*Representative block pier specimens after test:* (a) *Specimen DM3;* (b) *Specimen DS2.*

expected, the type of cement used in mortar did not affect the shear strength of block pier specimens.

Comparison with Uniform Building Code Allowable Shear Stresses. Based on the measured average compressive strength of masonry used in this program, the UBC allowable shear stress varies from 276 to 379 kN/m^2 (40 to 55 psi). This was determined according to UBC Section 2406(c) for a shear wall with in-plane flexural reinforcement present, and "M/Vd" ratio less than 1. The maximum measured shear stress ranged from approximately two to four times these values for individual pier specimens. This is quite significant because the pier portion of each specimen was only minimally reinforced for flexure, was partially grouted, had no shear reinforcement, and had no axial force transferred through the pier section. All these factors combined to provide a lower bound value of the measured shear strength.

Conclusions and Recommendations

Comparisons of test results obtained from specimens constructed with portland cement-based and masonry cement-based mortars indicate that the shear strength of masonry piers constructed with masonry cement-based mortars compared favorably with those constructed with portland cement-based mortars. Values of maximum shear stress obtained with portland cement and masonry cement-based mortars were close and within the range of scatter normally expected in experimental data. Overall behavior including cracking loads and damage patterns at maximum loads were similar.

The air contents for the masonry cement-based mortars, shown in Tables 2 and 3, were less than or equal to the 18% maximum air content specified in ASTM C 270-82. The air contents for the portland cement-based mortars, also shown in Tables 2 and 3, were significantly less than the 12% maximum specified in ASTM C 270-82. In view of the similar performance for specimens made from masonry cement and portland cement-based mortars, the effect of substantial variations in the air content was apparently insignificant for these laboratory-prepared mortars. In addition, the shear strength was not influenced significantly by the use of Type M or Type S mortar. The measured maximum shear stresses ranged from approximately two to four times those allowed by the Uniform Building Code.

Based on the results of the test program described in this paper, it is recommended that restrictions on the use of masonry cement in the structural frames of buildings in Seismic Zone Nos. 2, 3, and 4 be eliminated from the Uniform Building Code.

Acknowledgments

Work described in this paper was sponsored by the Portland Cement Association under the direction of George B. Barney, Divisional Director, Engineering Services Division. The investigation was carried out at the Construction Technology Laboratories of the Portland Cement Association under the direction of Henry G. Russell, Director, Structural Development Department. The authors acknowledge review of a report on this work by J. Huisman, R. W. Kriner, C. W. Miller, W. C. Reed, J. A. Rhineberger, and D. L. Rydquist. Suggestions and comments during this investigation by members of an unofficial ad hoc committee are greatly appreciated.

Jacob W. Ribar[1] *and Val S. Dubovoy*[1]

Investigation of Masonry Bond and Surface Profile of Brick

REFERENCE: Ribar, J. W. and Dubovoy, V. S., "**Investigation of Masonry Bond and Surface Profile of Brick,**" *Masonry: Materials, Design, Construction, and Maintenance, ASTM STP 992,* H. A. Harris, Ed., American Society for Testing and Materials, Philadelphia, 1988, pp. 33–37.

ABSTRACT: The original objective of this research was to investigate the influence of masonry cements on flexural bond of mortar to clay brick units and on the water penetration of masonry assemblies fabricated using these materials. The test variables included 20 masonry cements and 11 types of brick, each representing materials produced in different areas of North America. Results of flexural bond strength tests on masonry assemblies indicated that factors other than mortar composition affected bond strength. Accordingly, an investigation was made to evaluate the effect of surface texture of clay brick units on bond strength. A surface profilometer, from which results could be quantified, was used for this work. Results suggest that surface texture of the brick is a primary factor in the development of bond strengths.

KEY WORDS: bond, bond wrench, brick, initial rate of absorption, mortar-brick prisms, brick surface texture, surface profilometer

The masonry research community has, in recent years, expressed considerable interest and concern regarding bond between mortar and masonry units. It has enthusiastically accepted the bond wrench testing technique introduced by Hughes and Zsembery [*1*] as a mean of measuring bond strength because of its versatility and ability to produce data with less effort and cost. This capability has generated a flood of information that is frequently contradictory and more questions remain unanswered than are resolved.

Industry research efforts on masonry bond have become restricted due to the multitude of combinations of materials available and the great variety of individual physical characteristics of these materials. Research is often limited to mortar types and composition and to not more than a few brick types and their physical properties. The characteristic of the bricks used in these studies is almost exclusively the initial rate of absorption. Characteristics of the mortar include composition, air content, water retention, and compressive strength. When combined with frequently overlooked variables that exist among aggregates, cementitious materials, and masonry units, these can produce an infinite number of combinations. Therefore, research confined to a narrow spectrum of materials reflects the performance of the tested materials only. Masonry bond is an intricate process not completely understood. It is obvious that those parameters currently being used to predetermine masonry bond are inadequate.

Construction Technology Laboratories (CTL) has been involved in a research program to evaluate masonry mortars prepared with 20 masonry cements representing a cross-section of manufacturers in North America. These cements were combined first with one locally produced solid brick and later with ten additional bricks representative of brick manufactured in the eastern two thirds of the United States. This investigation concentrated on testing assemblies

[1]Principal masonry research engineer and research engineer, respectively, Concrete Materials/Technical Services Department, Construction Technology Laboratories, Skokie, IL 60077.

for flexural bond strength with tension normal to the bed joint, and for water penetration. The primary variables in the bond strength study included masonry cement formulation, brick and mortar properties, and the sources of brick and cement. A total of 454 clay-masonry prisms were tested for bond strength, yielding 2844 data points. The water penetration study evaluated 20 test assemblies prepared with each of the 20 masonry cement mortars and the one solid brick unit.

Results of this research disclosed several definite trends. Characteristics appearing to have favorable influence on bond with all units tested were higher water content of the mortar and greater fineness of the cement. A poor correlation was found between bond values and water penetration and leakage, and between mortar compressive strength and bond strength. Also, bond tests on seven-unit-high specimens made with Type S mortar that were removed from wall test assemblies had consistently higher bond values than individually constructed prisms. Half of the Type N mortars also showed this tendency.

A wide range of bond strengths was measured when each of the 20 cements was coupled with the ten types of clay brick units. None of the traditional parameters mentioned above was useful in explaining these variations. The least explainable circumstance occurred when two different bricks with similar physical characteristics yielded widely contrasting bond data when combined with the same mortar. The physical properties of these two bricks, A and H, are given in Table 1. Even though the two bricks had identical initial rates of absorption, Brick H produced significantly higher bond strengths with each of the 20 cements. Attempts to use the traditional parameters, including those given in Table 1, to explain these differences in bond strength proved to be inadequate.

The bedding surfaces of Bricks A and H revealed a significant difference in surface texture, as seen in Figs. 1 and 2. An attempt was then made to quantify this difference by examining the surface texture using a surface profilometer.

Resulting plots of the surface profiles, with pertinent surface texture parameters for these bricks, are presented in Figs. 1b and 2b. Some of the parameters in Figs. 1b and 2b that best describe and quantify the surface texture are defined as follows:

1. *Ra* is the most widely used parameter of roughness. It is the arithmetic mean of the departure of the profile from the mean line.
2. *Rt* is the maximum peak-to-valley height of the profile in the assessment traverse lengths.
3. *Rv* is the maximum depth of the profile below the mean line within the assessment traverse length.
4. *Rp* is the maximum height of the profile above the mean line within the assessment traverse length.
5. *Pz* is the average height difference between the five highest peaks and the five lowest valleys within the traverse length.

Comparing these parameters, it can be seen that Brick H has a much rougher surface texture than Brick A, thus providing for greater absolute surface area contacted by mortar. It is the

TABLE 1—*Data on brick.*

Brick Designation	Initial Rate of Absorption, g	Absorption, %	Compressive Strength, psi
A	17	4.0	14,160
H	17	2.2	13,990

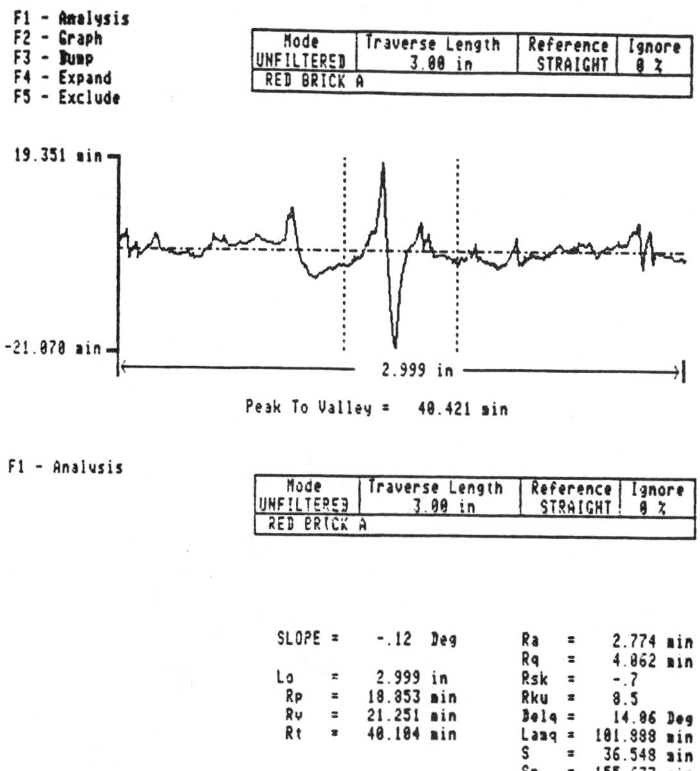

FIG. 1—(a) *Close-up photograph of bedding surface of Brick A (X10);* (b) *the surface profile of Brick A.*

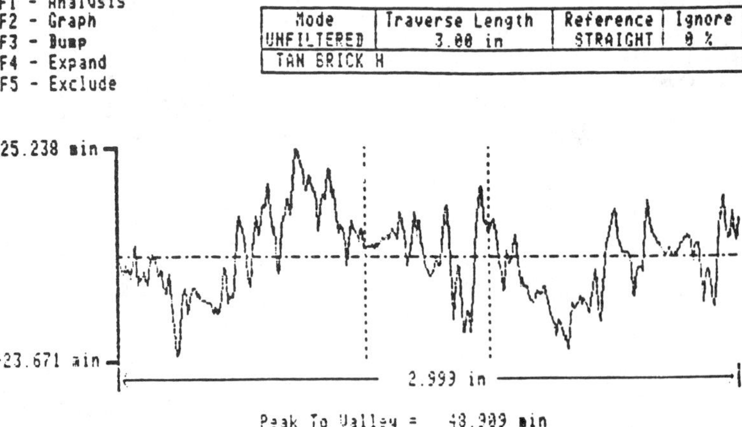

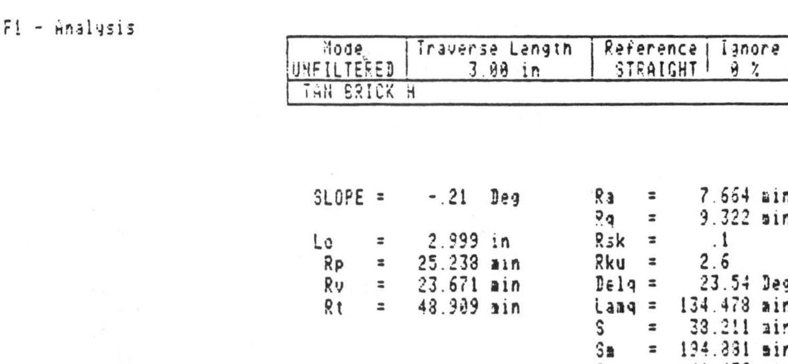

FIG. 2—(a) *Close-up photograph of bedding surface of Brick H (X10);* (b) *the surface profile of Brick H.*

judgment of the investigators that this greater surface area is a primary factor in the greater bond development measured for Brick H.

It is planned to further evaluate the effect of surface texture of the masonry units on bond strength in future work at CTL.

Conclusions

Concepts within the masonry community that masonry performance can be predetermined by the characteristics of individual components is not justified. The authors agree with Yorkdale [2] that bond between clay masonry units and mortar is not well understood. Additional research is needed to understand the complex mechanism of masonry bond. This research has identified a useful tool that may assist research community.

References

[1] Hughes, D. M. and Zsembery, S., "A Method of Determining the Flexural Bond Strength of Brickwork at Right Angles to the Bed Joint," Second Canadian Masonry Symposium, Ottawa, June 1980.

[2] Yorkdale, A. H., "Initial Rate of Absorption and Mortar Bond," *Masonry: Materials, Properties, and Performance, ASTM STP 778,* J. G. Borchelt, Ed., American Society for Testing and Materials, Philadelphia, 1982, pp. 91-98.

Design

Donald C. Raths[1]

Preconstruction Brick Veneer Evaluation Testing

REFERENCE: Raths, D. C., "**Preconstruction Brick Veneer Evaluation Testing,**" *Masonry: Materials, Design, Construction, and Maintenance, ASTM STP 992,* H. A. Harris, Ed., American Society for Testing and Materials, Philadelphia, 1988, pp. 41–56.

ABSTRACT: This paper discusses and presents a preconstruction testing program for a brick veneer cavity wall. It presents the testing program and the application of ASTM and industry standards to the evaluation testing of the brick masonry materials. It provides findings on brick and mortar characteristics, steps taken to improve the performance of the selected brick wall materials, and engineering procedures for determining material compatibility and performance.

KEY WORDS: bond, bond wrench test, brick masonry, cavity wall, preconstruction, cope, mortar performance, veneer, water permeability

Contract documents typically present the masonry requirements that must be complied with during the construction of a project. However, they generally do not address preconstruction testing that should be conducted to ensure proper selection and compatibility of masonry materials. This paper explores some of the methods and procedures that may be utilized in making these often very necessary preconstruction evaluations.

Preconstruction testing and evaluation of the masonry materials that comprise the brick veneer of a typical cavity wall can be an important factor in the prevention of construction problems and/or unsatisfactory performance of exterior masonry walls. A proper program ensures the compatibility of materials and establishes construction procedures relating to brick prewetting, tooling, and mortar ingredients. Also, a successful preconstruction testing program establishes a clear understanding of construction requirements, which should lead to a completed exterior veneer wall with satisfactory service performance.

To illustrate the benefits of a preconstruction testing program, a recent project is presented as an example. The project can be described as a complex of several low-rise office structures with a nominal 100-mm (4-in.) brick veneer and 60-mm (2$^3/_8$-in.) cavity. The exterior brick veneer was anchored to the backup system with flexible wire anchors. The typical brick unit was a standard size and shape with a special feature created by a continuous 16-mm ($^5/_8$-in.)-radius quarter circle (cope) removed from the top exterior edge of the brick (Figs. 1 and 2).

Preliminary architectural drawings and masonry specifications were reviewed at the start of the evaluation testing program to ensure a complete understanding of the project requirements and typical construction details. Suggested changes were offered to the architect in order to clarify ASTM testing requirements, material selections, and field workmanship practices during construction. Also, two brick types (A and B) were evaluated during the evaluation testing program.

Based on a review of the project requirements and the general nature of the selected brick properties, a masonry testing and evaluation program was developed that would include testing of brick, mortar, and fabricated masonry assemblies.

[1]Principal, Raths, Raths & Johnson, Inc., Willowbrook, IL 60521.

FIG. 1—*Brick A with cope feature and 25-mm (1-in.) exposed mortar bed joint.*

Program Criteria

Brick A had been initially selected by the architect based on its color, texture, and its aesthetic appearance on the building in combination with a colored mortar. The selected brick was a sand-molded clay product manufactured by the wet-mud process. The project specifications required the brick to conform to ASTM Specification for Facing Brick (Solid Masonry Units Made from Clay or Shale) (C 216), Grade SW, Type FBS [*7*], and mortar to be ASTM Specification for Unit Masonry (C 270), Type N [*8*]. The special feature of a continuous cope along the top exposed edge of the brick results in a nominal 25-mm (1-in.) mortar bed joint on the exposed masonry wall (Figs. 1 and 2). Concerns were raised by the architect regarding the performance and durability of the large mortar bed joint at the exposed face of the brick veneer wall and in part was the impetus for the evaluation testing program.

The author's firm was requested to develop a masonry testing program to determine compatibility of materials and to measure conformance to ASTM requirements. This program consisted of testing the brick for compliance with ASTM C 216 requirements, compressive strength of

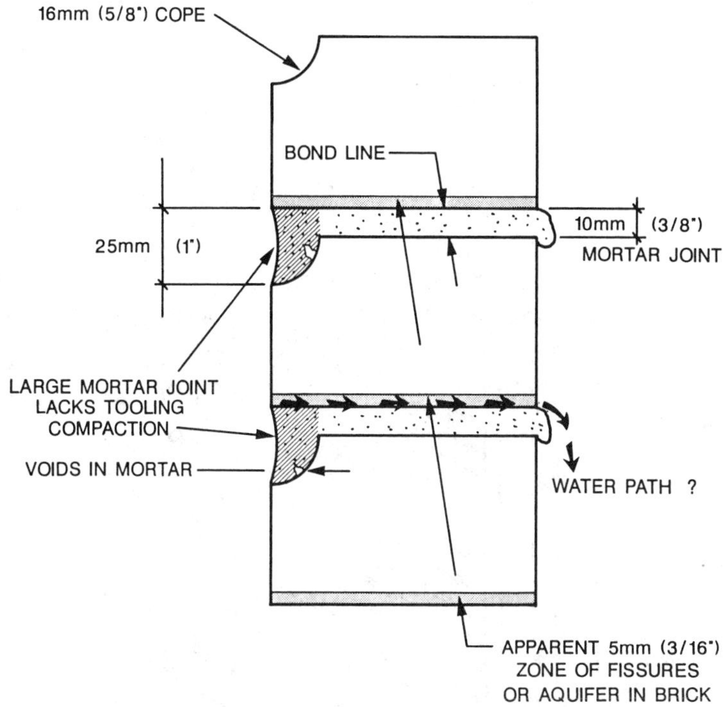

FIG. 2—*Sketch of possible water paths at bed joint.*

mortar per ASTM C 270, water permeance rating of fabricated assemblies in accordance with field-adapted ASTM Test Method for Water Permeance of Masonry (E 514) [*1,2*] (Fig. 3), flexural bond strength of wall prisms (cut from test panels) by the bond wrench method, ASTM Method for Measurement of Masonry Flexural Bond Strength (C 1072) [*4*] (Fig. 4), and freeze/thaw testing of a masonry specimen (brick and mortar) for a minimum of 150 cycles in accordance with ASTM Method of Sampling and Testing Brick and Structural Tile (C 67) [*6*]. Also, due to the high initial rate of absorption nature of this sand-molded brick, brick wetting procedures were to be evaluated to establish requirements and procedures for field construction.

Phase I—Brick A Testing and Evaluation

This phase of testing resulted in the fabrication of four laboratory masonry test panels. Approximately 600 bricks (Type A) were received from the brick manufacturer for the testing work. To ensure meaningful testing data, the actual masonry sand and colored portland cement to be used in the construction were used in the panel construction.

Due to the high initial rate of absorption (IRA) of the brick, approximately 32 g/min/30 in.[2] (Table 1), it was decided to fabricate the four masonry test panels with varying wetting procedures. One panel was fabricated with no prewetting of the brick (as delivered brick), two panels had water sprayed on the brick to represent a typical field wetting application, and a fourth panel was constructed with the bricks in a saturated condition (bricks kept moist for several days and covered).

[2]All water permeance values reported in this paper are based on the area of ASTM E 514 test chamber.

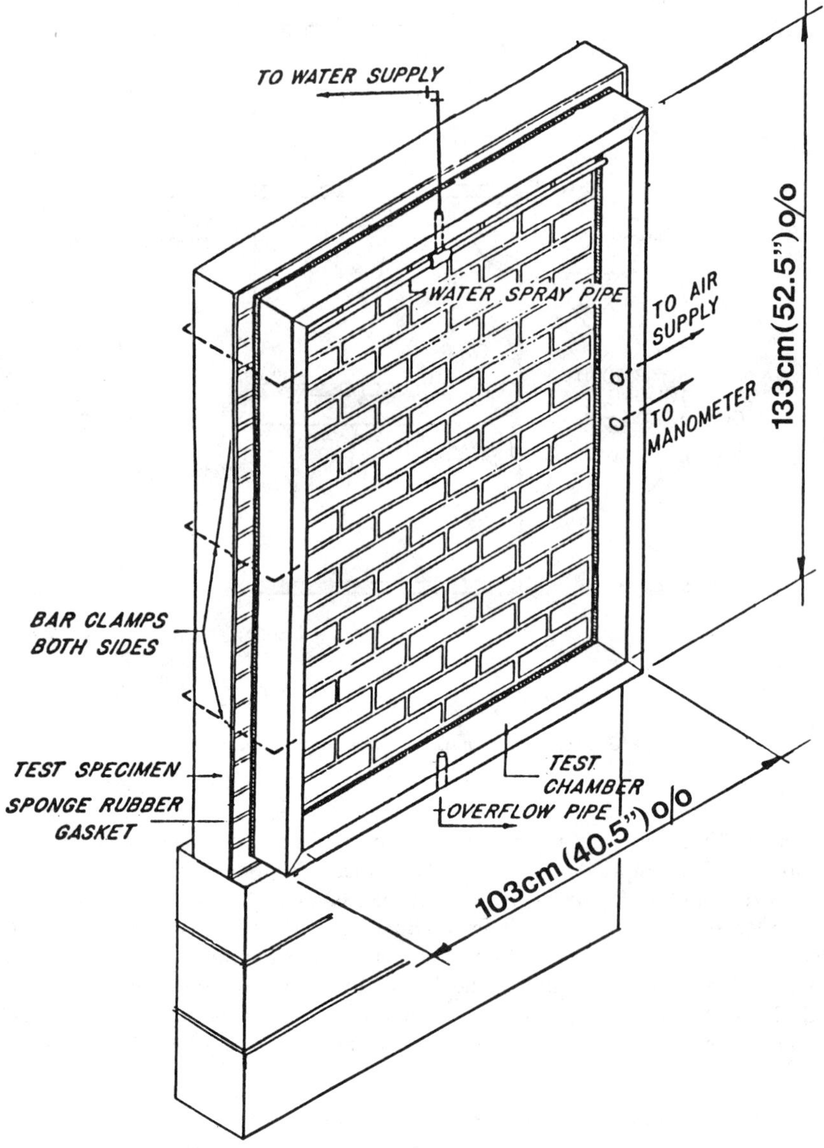

FIG. 3—*Field-adapted ASTM E 514 water permeability test. Figure taken from ASTM E 514* [3].

The four masonry test panels, 137 cm (54 in.) wide by 142 cm (56 in.) high were constructed outdoors by a journeyman mason employed by the masonry subcontractor for this project (Fig. 5). The panels were constructed of one brick wythe (one vertical section of wall one unit in thickness) [5] in running bond on a pedestal about 60 cm (24 in.) above grade and were completed in two days. The mortar ingredients were proportioned by volume in accordance with ASTM C 270, Type N mortar (1 : 1 : 5½ to 6 proportions) and mixed for 5 min in a paddle-type mixer. The head and bed joints were completely filled with mortar as called for in the specifications. The completed panels were representative of a high level of workmanship (Fig. 6). The

FIG. 4a—*Brick prisms removed from test wall.*

FIG. 4b—*Bond wrench test on brick prism.*

large 25-mm (1-in.) mortar bed joint was tooled using a section of metal pipe with an approximate diameter of 100 mm (4 in.) in accordance with the specified architectural requirements.

Brick A test results are presented in Table 1. A sieve analysis of the mortar sand was made and it was determined that the sand conformed to ASTM Specification for Aggregate for Masonry Mortar (C 144) requirements. After a period of 28 days, field-adapted ASTM E 514 water permeance [*2*] and flexural bond strength tests, using the bond wrench method, ASTM C 1072 [*4*], were executed on the four masonry test panels. Samples were removed from test panels by a masonry saw for the bond wrench tests (see Fig. 4). The Phase I test results are presented in Table 2 and Fig. 7.

TABLE 1—*ASTM C 67 Tests, Brick A.*

Compressive strength	Average five bricks = 36.0 MPa (5224 psi) Range 31.8 to 41.1 MPa (4615 to 5958 psi)
Initial rate of absorption	Upper surface 24.5 g/min/30 in.^{2} Lower surface 32.0 g/min/30 in.^{2}
24-h cold soak	Average = 6.50%
5-h boil	Average = 11.30%
Coefficient of saturation	Cold soak/5-h boil = 0.58 ratio
Freeze/thaw test	150 cycles completed on small brick prism specimen (brick and mortar) with negligible weight loss and no visual deterioration
Comments	Fine, powdery residue on upper surfaces of brick may affect bond strengths Lower surface exhibited a fissured open surface that yields higher initial rate of absorption

Additional bond wrench tests were conducted on a series of brick prisms (stack bond) to make a preliminary evaluation of the significance of the fine powdery residue (common to sand-molded bricks) which adhered to the upper surfaces of the brick. Four brick specimens were fabricated by a laboratory technician using a hand-mixed Type N mortar (1:1:5½ proportions) and bricks uniformly prewetted using a hand sprayer. The specimens were fabricated and cured in the laboratory at about 75°F (24°C) and then tested at 14 days. The test results are shown in Table 3.

The Phase I masonry testing results indicated the brick met ASTM C 216 requirements [*7*], including the freeze/thaw test on a brick and mortar assemblage for 150 cycles. Because Brick A exhibited a very high initial rate of absorption [*6*], they were determined to require prewetting during construction. The ASTM E 514 water permeance tests indicated water leakage rates well beyond a 1.89 L/h (½ gal/h)[2] criterion established for a typical good wall [*1*] on Test Panels 1 and 3. The bond strength values were satisfactory for bricks properly prewetted. The Phase I testing program results lead to the following observations:

1. Field-adapted ASTM E 514 water leakage rates varied widely according to the degree of prewetting of the brick.
2. Test Panel No. 3 construction was most representative of field workmanship and its water leakage rate of 8.50 L/h (2¼ gal/h) was deemed unacceptable.
3. Test Panel No. 4 exhibited the lowest water leakage rate, but was deemed to be impractical to construct due to the "floating" of the brick and "bleeding" of mortar during fabrication. This was caused by the saturated condition of the brick at the time of fabrication.
4. Bond strength test values provided an adequate factor of safety compared to recommended practice for brick masonry which allows flexural tension of 193 kPa (28 psi) for Type N mortar [*5*].
5. The large volume of mortar at the 25-cm (1-in.) exposed bed joint created by the cope along the top edge of the brick was difficult to properly compact. Small horizontal voids were visually detected along the brick surface at the cope area when specimens were broken apart for examination (Fig. 2).
6. The fine powdery residue on the top surfaces of the brick, resulting from the brick manufacturer's production process, reduced the bond strengths approximately 12 to 15% (see Table 3).
7. The texture and density of the bricks' bottom surface were suspected to be partly a cause of high water leakage rates (see Fig. 2).

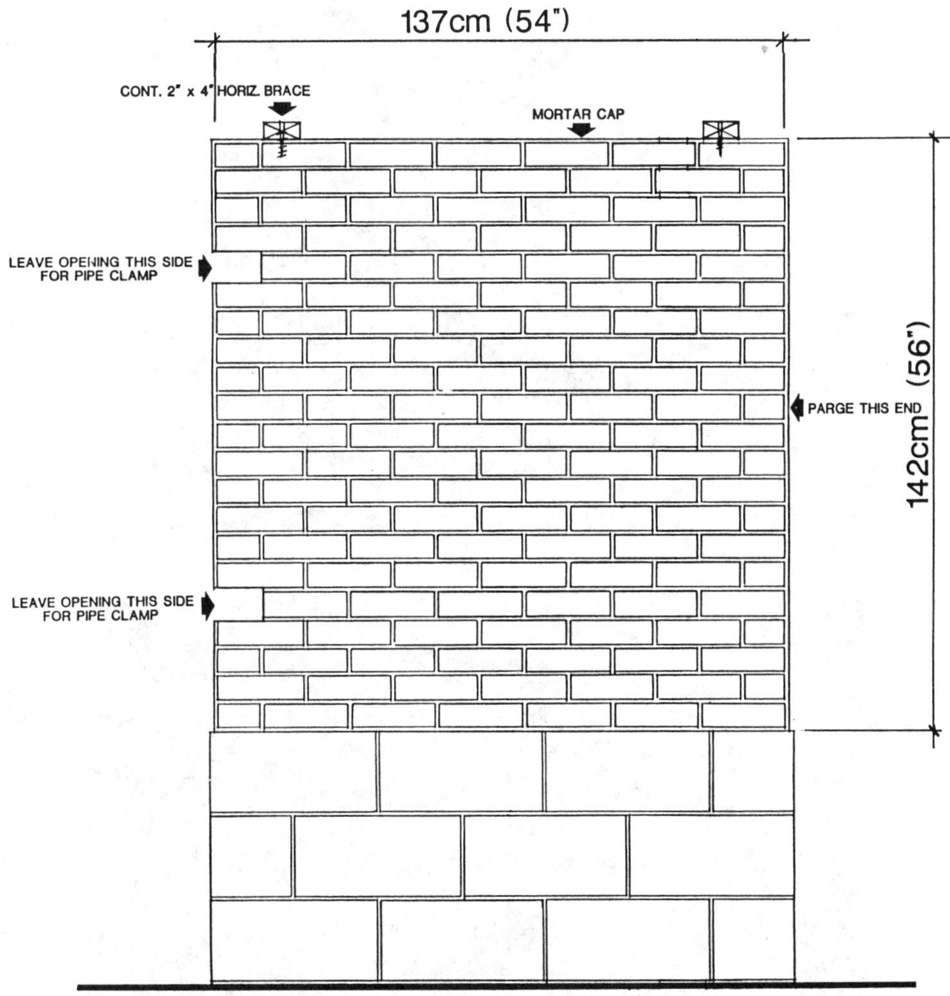

FIG. 5—*Typical brick masonry test panel.*

Due to the high water leakage rates, the use of Brick A was not recommended. It was recommended to evaluate an alternate Brick B as well as regular stretcher bricks (solid—no cores) without the cope feature in the next testing phase. Also, the next test phase needed to more fully address brick prewetting requirements considering the extreme sensitivity of water leakage to prewetting moisture contents.

Phase IA—Additional Testing/Evaluation (Brick A)

Based upon concerns raised regarding the fissured texture and zone of lesser density near the bottom surface of Brick A, it was decided to examine the surfaces of the brick more closely and

FIG. 6a—*Test panel workmanship at head and bed joints, interior surface.*

FIG. 6b—*Close up of head and bed joints, interior surface.*

to perform a special water penetration test using a dye to allow identification of actual water leakage paths through the brick masonry assembly.

A typical brick was sectioned vertically along its long axis to allow a close examination of the cross section. This examination revealed the presence of a less dense zone of material, about 5 mm (3/16 in.) thick, at the entire bottom surface of the brick. This led to removal of a large masonry prism section, approximately 46 cm (18 in.) by 76 cm (30 in.) in size from one of the completed test walls. It was placed in a horizontal position with the exposed face upward, and ponded water with a green dye was allowed to penetrate the wall for about 12 h. Upon selective

TABLE 2—*Phase I test results of masonry wall panels, Brick A.*

Test Panel	Brick Prewetting	Field Adapted E 514 Rate, L/h (gal/h)	Average Flexural Tension, kPa (psi)	Comments
Panel 1	Dry (as delivered)	22.9 (6.0)	365 (53)	Leakage very high, poor bond
Panel 2	Sprayed	...	...	Tests not run due to cracked bed joint in test panel
Panel 3	Sprayed	8.5 (2.25)	1531 (222)	Typical of field production, good bond, bricks brushed to remove powder residue
Panel 4	Saturated (surface dry)	3.2 (0.85)	1420 (206)	Bricks too wet, wall difficult to lay up—floating

NOTES:

1. Type N Mortar (1 : 1 : 6 proportions Panel 1, 1 : 1 : 5½ proportions Panels 2 through 4).
2. Compressive strength of mortar cubes, ASTM C 270: 6123 kPa (888 psi), 6206 kPa (900 psi), and 6295 kPa (913 psi).
3. Flexural tension determined by bond wrench tests, ASTM C 1072, on brick prism specimens (removed from test panel) [*4*]. Tooled face in tension.

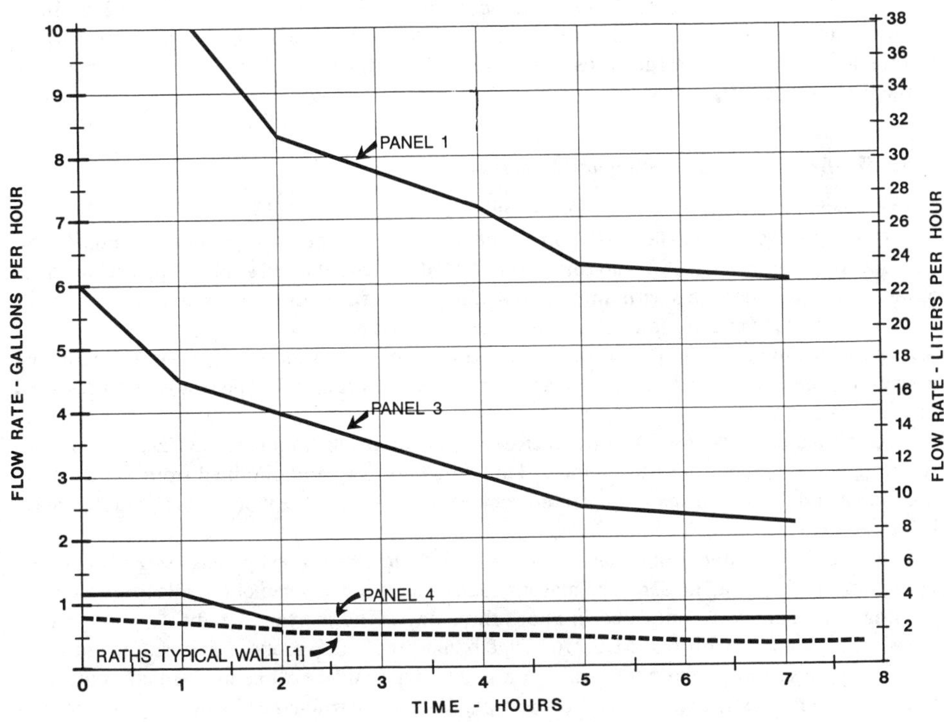

FIG. 7—*Phase I test results, modified ASTM E 514 water permeance test.*

TABLE 3—*Bond wrench stack bond tests, brushed versus normal surfaces, Brick A.*

Specimen	Brick Surface	Average Flexural Tension, kPa (psi)
1	Normal	869 (126)
2	Normal	979 (142)
3	Brushed	1083 (157)
4	Brushed	1007 (146)

NOTES:

1. Type N Mortar (1:1:5½ proportions).
2. Compressive strength of mortar cubes, ASTM C 270: 5116 kPa (742 psi) and 5461 kPa (792 psi).
3. Flexural tension determined by bond wrench tests, ASTM C 1072, on brick prism specimens [*4*]. Tooled face in tension.
4. All bricks uniformly wetted by a hand sprayer prior to fabrication of samples.

demolition of the masonry prism, a detailed examination was made of the mortar/brick interface at the head joints, the bed joints, and in the cope area (top front portion of the brick). This examination resulted in the conclusion that the water paths were occurring in the bed joint behind the mortar bond line through the less dense zone at the brick bottom surface, in addition to the expected water paths at the head joints. Also, it was noted that water appeared to migrate laterally at the cope area through the veins (voids) in the uncompacted mortar (see Fig. 2).

Additional testing was deemed necessary to further study the effects of prewetting, surface texture, and tooling of the large exposed bed joint to determine their effects on water permeance of the brick veneer wall.

Phase II—Brick A and B Testing and Evaluation

The second phase of testing consisted of comparative evaluations of Bricks A and B. Bricks of both types with the special 16-mm (⅝-in.) cope feature along the top exterior edge, Fig. 1, and a regular stretcher brick (solid—no cores) were included. This allowed study of the performance of identical brick materials with and without the cope feature, and comparison of test results with two different brick bottom surface textures, fissured and smooth.

As a prelude to fabrication of the test panels, a series of brick saturation tests were conducted to establish the wetting and drying characteristics of Bricks A and B. The results are presented in Tables 4 and 5.

An additional supply of bricks was received from each Brick A and B manufacturer for constructing additional masonry test panels. The same masonry sand obtained from the local job site source and the actual colored portland cement were used for the mortar mix as in the Phase I tests.

A series of five masonry test panels, similar in size to the Phase I test panels, were constructed by the same journeyman mason. The masonry panels were constructed of a single brick wythe in running bond and were completed in about three days. The mortar ingredients were proportioned in accordance with ASTM C 270, Type N mortar (1:1:5 proportions) with slightly less sand than Phase I and mixed for 5 min in a paddle-type mixer. All head and bed joints were filled with mortar as specified. The fifth test panel was constructed utilizing a Type N mortar with higher lime content than normally allowed (1:1½:6 proportions) at the request of the contractor. The completed test panels were of a high level of workmanship (see Fig. 6).

TABLE 4—*Oven dry to saturation data.*

	Brick A		Brick B	
Moisture Conditions	Weight, g	Moisture, %	Weight, g	Moisture, %
Oven dry weight	1785.0	0	2118.8	0
10-min soak weight	1881.7	83	2220.1	84
4-h soak weight	1890.3	91	2229.0	92
Saturated weight	1901.3	100	2238.8	100

NOTE—Weights based on the average of three bricks submerged in water tank.

TABLE 5—*Air drying from saturation data.*

	Brick A		Brick B	
Moisture Conditions	Weight, g	Moisture, %	Weight, g	Moisture, %
Saturated weight	1917.1	100	2232.7	100
1-h dry weight	1908.7	93	2225.9	94
6-h dry weight	1876.8	67	2195.8	70
24-h dry weight	1851.8	46	2169.8	48
Oven dry weight	1796.5	0	2111.6	0

NOTE: Weights based on the average of four bricks at 58°F (14.44°C) mean temperature.

The tooling of the large mortar bed joint was accomplished in this phase using a metal cylinder with an approximate diameter of 60 mm (2½ in.), slightly smaller than that used in Phase I. This smaller diameter tool shape appeared to allow better compaction of the large mortar volume in the cope area of the brick specimen. A regular mason's jointer was used for tooling the 10-mm (3/8-in.) joints of the panels fabricated with regular stretcher bricks.

Based on the saturation and drying curves developed for Brick A and B, the moisture levels of the bricks were monitored by weighing to allow measurement of specific levels of moisture at the time the masonry test panels were constructed (Tables 5 and 6). Consideration for moisture levels of brick was based upon water leakage values of Phase I tests, the journeyman mason's "feel" as to proper timing for laying up and tooling of the mortar joints.

ASTM C 67 tests for Brick B are presented in Table 7. After a period of 28 days, ASTM E 514 water permeance and flexural bond strength tests, ASTM C 1072, were conducted on the masonry test panels. These results are given by Table 6 and Fig. 8.

Review of the ASTM E 514 water leakage and bond strength test data indicated the Brick A and B special cope feature did not have any significant influence when compared to stretcher brick performance. Therefore, it was decided that the coped feature of the brick was acceptable. The water leakage rates were satisfactory for Brick B as they were within the program criteria of 1.89 L/h (½ gal/h) for a typical good wall [*1*]. Brick A was very close to the program criteria with a leakage rate of 2.28 L/h (0.60 gal/h).

Phase III—Brick A Testing and Evaluation, Modified Surface

Based upon our previous testing results and observations, the manufacturer of Brick A resubmitted a brick with a modified bottom surface exhibiting a closed, finished surface for additional testing. The brick production process was altered to allow an enhanced strike-off mecha-

TABLE 6—*Phase II test results of masonry wall panels, Bricks A and B.*

Test Panels	Brick Prewetting	Brick Moisture Content, %	Modified E 514 Rate, L/h (gal/h)	Average Flexural Tension, kPa (psi)	Comments
Panel 1 (Brick A—Regular)	Saturated and dried	60	2.28 (0.60)	234 (34)	Powdery residue both surfaces, poor bond
Panel 2 (Brick A—Cope)	Saturated and dried	58	2.28 (0.60)	772 (112)	
Panel 3 (Brick B—Regular)	Saturated and damp	85	1.90 (0.50)	924 (134)	Wall slightly wet at lay up
Panel 4 (Brick B—Cope)	Saturated and dried	72	1.51 (0.40)	876 (127)	Low leakage rate
Panel 5 (Brick A—Cope)	Sprayed by mason	67	5.24 (1.40)	469 (68)	High leakage rate, poor bond

NOTES:

1. Type N mortar: Panels 1 through 4 (1:1:5 proportions), Panel 5 (1:1½:6 proportions).
2. Compressive strength of mortar cubes, ASTM C 270: Panels 1 through 4: 6812 kPa (988 psi), 9860 kPa (1430 psi), 10 115 kPa (1467 psi), and 10 329 kPa (1498 psi); Panel 5: 4799 kPa (696 psi).
3. Flexural tension determined by bond wrench tests, ASTM C 1072, on brick prisms (removed from test panel) [*4*]. Tooled bed joint in tension.
4. Moisture content based on 24-h cold soak capacity.

TABLE 7—*ASTM C 67 Tests, Brick B.*

Compressive strength	Average five bricks = 32.2 MPa (4678 psi) Range 24.8 to 38.7 MPa (3602 to 5617 psi)
Initial rate of absorption	Lower surface 33.3 g/min/30 in.^{2}
24-h cold soak	Average = 5.68%
5-h boil	Average = 10.37%
Coefficient of saturation	Cold soak/5-h boil = 0.55 ratio
Freeze/thaw test	To be executed later if brick is selected
Comments	Fine, powdery residue on upper surfaces of brick may affect bond strengths. Lower surface exhibited a closed, smooth surface.

nism after the clay was compacted in the molds. This represented an improvement in the finished surface texture and was expected to significantly improve both water leakage rate and bond strength properties. A further series of small masonry test panels were constructed with Bricks A and B, including the special cope feature, to allow a further understanding of the prewetting sensitivity characteristics of the bricks and to evaluate the modified surface of Brick A. The masonry test panels were constructed 81 cm (32 in.) wide by 46 cm (18 in.) high, one brick wythe in running bond and Type N mortar (1:1:5 proportions) ingredients identical to previous tests. The moisture levels of all bricks for the specimens were closely controlled during fabrication (weighing of oven dry bricks and adding measured amounts of water) to establish known moisture contents. The ASTM E 514 water permeance values were determined by utiliz-

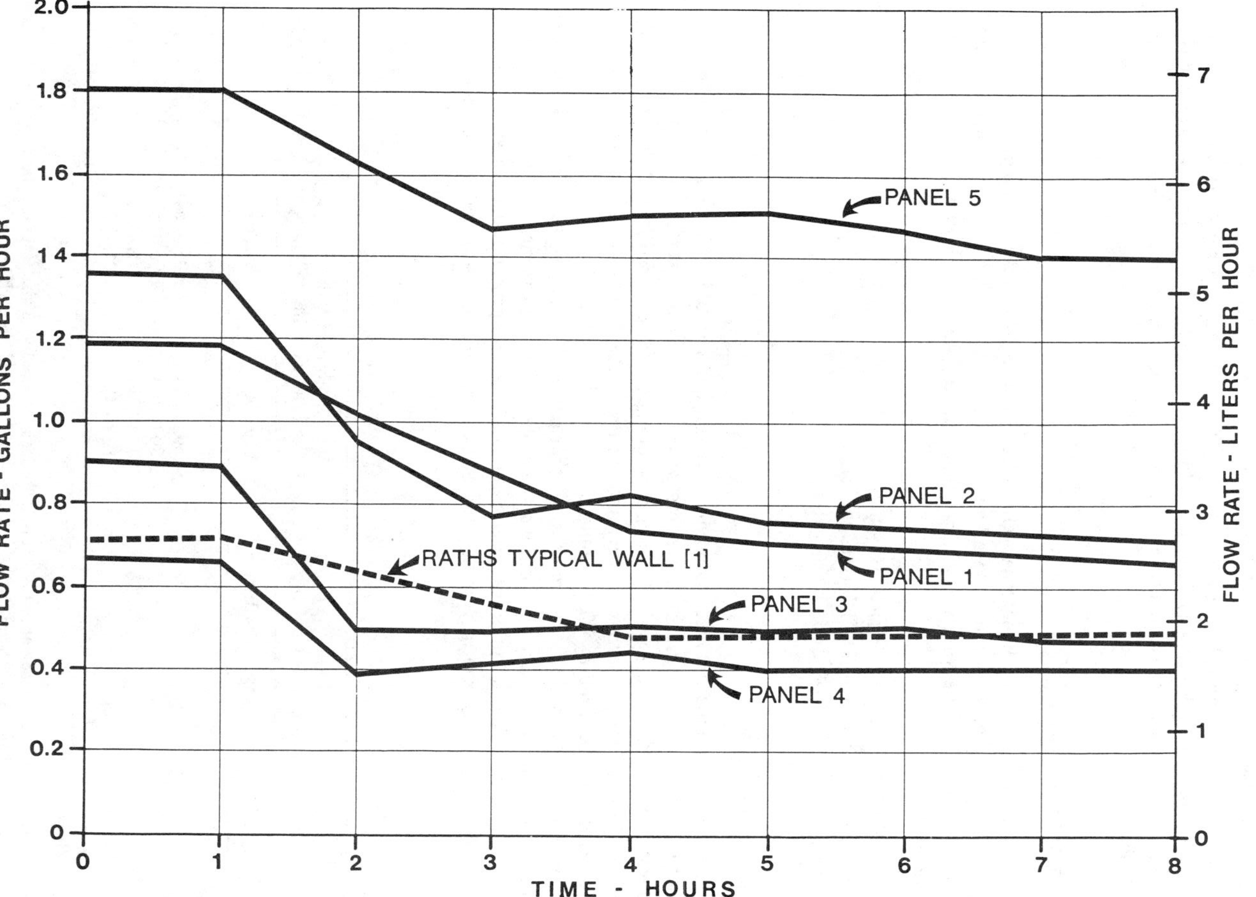

FIG. 8—*Phase II test results, modified ASTM E 514 water permeance test.*

ing a special small scale test chamber 30.0 cm (12 in.) high by 49.5 cm (19$^1/_2$ in.) wide. The flow rate and pressure differential were maintained to be equivalent to the standard size ASTM E 514 chamber. The water leakage test rates were then multiplied by the ratio of the standard size test chamber divided by the small-scale test chamber to allow direct comparison of all ASTM E 514 test values. The results of these tests are shown in Table 8.

Based on the Phase II and III test results for ASTM E 514 water permeance values, acceptable flexural bond strengths, freeze/thaw durability results, and the designed appearance of the bricks in the finished wall, the coped Brick A with the modified bottom surface was selected for construction.

ASTM E 514 Field Testing

The completed in-place brick masonry wall was tested during the construction period to monitor the workmanship of the completed wall and conformance to the established project criteria of 1.89 L/h ($^1/_2$ gal/h). The results from three field adapted ASTM E 514 tests established the following values:

1. Test Location No. 1—1.89 L/h ($^1/_2$ gal/h).
2. Test Location No. 2—1.18 L/h ($^5/_{16}$ gal/h).
3. Test Location No. 3—2.36 L/h ($^5/_8$ gal/h).

Testing Conclusions

The preconstruction masonry testing program established several important factors: compatibility of materials, sensitivity of brick wetting procedures to service performance of the completed wall, effects of 16-mm ($^5/_8$-in.) radius cope along the top exposed face of the brick, and the influence of surface properties (density and texture of the bottom brick surface) for sand-molded brick manufactured by the wet mud process.

The program established that the most important parameter, the degree of brick wetting at the time of construction, related to both water permeability performance and bond strength. The high initial rate of absorption required controlled wetting procedures in order to ensure watertight joints between the mortar and the brick units. Technical literature and industry

TABLE 8—*Test results field adapted E 514 water permeance.*

	Moisture Content, %		Equivalent Values, E 514 Leakage Rate, L/h (gal/h)	
Panel No.	Brick A	Brick B	Brick A	Brick B
1	0	10	6.80 (1.80)	5.20 (1.43)
2	21	25	4.50 (1.12)	2.05 (0.54)
3	44	44	2.30 (0.61)	1.96 (0.52)
4	60	56	2.10 (0.55)	N/A
5	70	82	1.63 (0.43)	1.21 (0.32)

NOTES:

1. E 514 water permeance tests run at 14 days with reduced size test chamber, 30.5 cm (12 in.) by 49.5 cm (19$^1/_2$ in.). One test executed on each test panel.
2. Tooling of bed joints with 6.35 cm (2$^1/_2$ in.) diameter cylinder.
3. Moisture contents based on 24-h cold soak capacity and based on average weight of five bricks.
4. Reduced size E 514 test chamber 30.0 cm (12 in.) high by 49.5 cm (19$^1/_2$ in.) wide.

practices may not provide adequate guidance for the designer to specify in detail the brick prewetting requirements for bricks with high initial rates of absorption. Therefore, it becomes critical to conduct preconstruction testing with ASTM E 514 as well as flexural bond strength tests to establish the criteria for the specific brick and mortar selected. The preconstruction testing program described above resulted in the selection of a brick and mortar proportions yielding a water permeability rating of approximately 1.89 L/h (1/2 gal/h) for brick test panels, thus meeting the program criteria. This was accomplished after refinement of brick wetting procedures and modification of the texture of the lower surface of Brick A.

The influence of the special cope feature on the exposed top edge of the brick created a large mortar volume at the exterior bed joint. The results of preconstruction testing for water permeability and bond strength indicated this feature was neither significant nor detrimental to performance. However, examination of demolished masonry test panels revealed that proper tooling was important to obtain full compaction of mortar in the area of the cope (see Fig. 2). The radius and width of the jointer were varied in order to establish better mortar compaction techniques for the large exposed joint. The tooling technique was unique for this bed joint shape and resulted in establishment of tooling methods with a specially designed jointer.

The fine powdery residue that adheres to the upper surfaces (constructed position) of the sand-molded brick was related to the molding process in manufacturing. Testing indicated its presence reduced bond strengths by up to 15% but did not affect permeance performance.

The preconstruction testing program yielded positive results by quantifying the critical items of performance and identifying required construction methods and procedures. Also, the designer, contractor, mason subcontractor, and owner jointly participated directly in the brick masonry test program, which resulted in understanding of the project requirements among all the parties. Procedures developed during the masonry testing program led to the development of additional field quality control measures regarding brick prewetting and mortar tooling. These items critical to the performance of the completed masonry wall were identified well in advance of construction to allow the contractor and masonry subcontractor to plan and develop wetting procedures for construction.

The preconstruction and construction testing program identified some important factors which may apply to other masonry construction:

1. Initiate preconstruction testing program sufficiently in advance of construction to allow evaluation of brick and mortar and time to effect any material changes without delaying the project.
2. Importance of a closed finish surface on sand-molded brick as it relates to water permeance.
3. Determination of compatible mortar proportions.
4. Degree of brick prewetting and identification of initial rate of absorption level where this is no longer required.
5. Development of field procedures to achieve the required prewetting.
6. The testing program results can be translated into actual construction methods which produce in-place walls having the same performance as tested walls.

The benefits of a proper preconstruction testing program should promote understanding of project requirements, decrease construction problems, increase the owner's assurance of obtaining quality constructed masonry, and improve the service performance of the completed brick masonry wall.

References

[1] Raths, C. H., "Brick Masonry Wall Nonperformance Causes," *Masonry: Research, Application, and Problems, ASTM STP 871*, J. C. Grogan and J. T. Conway, Eds., American Society for Testing and Materials, Philadelphia, 1985, pp. 182-201.

[*2*] Monk, C. B., Jr., "Adaptations and Additions to ASTM Test Method E 514 (Water Permeance of Masonry) for Field Conditions," *Masonry: Materials, Properties, and Performance, ASTM STP 778,* J. G. Borchelt, Ed., American Society for Testing and Materials, Philadelphia, 1982, pp. 237–244.
[*3*] ASTM Test Method for Water Permeance of Masonry (E 514-74), American Society for Testing and Materials, Philadelphia, 1974.
[*4*] ASTM Method for Measurement of Masonry Flexural Bond Strength (C 1072-86), American Society for Testing and Materials, Philadelphia, 1986.
[*5*] "Recommended Practice for Engineered Brick Masonry," Brick Institute of America, McLean VA, 1978.
[*6*] ASTM Method of Sampling and Testing Brick and Structural Clay Tile (ASTM C 67-83), American Society for Testing and Materials, Philadelphia, 1983.
[*7*] ASTM Specification for Facing Brick (Solid Masonry Units Made from Clay or Shale) (ASTM C 216-84), American Society for Testing and Materials, Philadelphia, 1984.
[*8*] ASTM Specification for Mortar for Unit Masonry (ASTM C 270-84), American Society for Testing and Materials, Philadelphia, 1984.

Richard M. Gensert[1] and William C. Bretnall[1]

Problems and Solutions to Masonry Buildings

REFERENCE: Gensert, R. M. and Bretnall, W. C., **"Problems and Solutions to Masonry Buildings,"** *Masonry: Materials, Design, Construction, and Maintenance, ASTM STP 992,* R. A. Harris, Ed., American Society for Testing and Materials, Philadelphia, 1988, pp. 57–95.

ABSTRACT: The authors have been actively investigating masonry buildings for over 25 years and have documented problems associated with differential movement within masonry systems and between systems and their supporting structures. While they have found many buildings and components which perform well, it has been their experience that masonry buildings with problems have often been brought to the attention of the public without full and in-depth understanding of the source of the trouble.

One cannot analyze a building's problems by a quick visual inspection and a first reading of the drawings and specifications. It is necessary to document all signs of distress as an overlay upon the structural frame. Materials must be sampled and tested for quality and compatibility. Walls must be probed with fiber optic instruments to determine the quality of assemblages and their attachments to the structure. Architectural details, such as drainage, flashing, movement joints, bonding, etc., are important to the well-being of masonry walls and must be examined. Structural support of masonry and its interaction with the supporting structural component and with the structural frame as a whole is also critical in terms of differential movements. The authors have documented this experience by means of photography computer graphics instrumentation, and this paper will share with you the knowledge gained by this work effort.

KEY WORDS: masonry, prefabricated panels, rigid connections, deflection, differential movement, interaction, movement joints

Before we look at specific problems, we must first understand some basic principles of materials and the action of structural frames.

Masonry curtain walls, whether laid in place or prefabricated, must be free to move independently of their supporting structures. If a masonry wall is restrained by the supporting frame from expanding or contracting, excessive stresses will be introduced into the wall. Just as importantly, a structural frame which is more limber than the stiff masonry will induce excessive stresses into the masonry by means of deflection. The different properties of masonry and structural frames can be itemized to draw some comparisons. Table 1 compares the weaker properties of masonry to the comparable properties of structural steel and reinforced concrete.

When a structural frame deflects laterally from wind loads or seismic loads, the deflection is often called a drift and the drift comparison between various structural frames is called a drift ratio. Thus, a braced frame in steel or a shear panel system in concrete or masonry would be very stiff and would have a drift ratio of 1. By comparison, a steel frame with shear connectors and no moment-resisting joints would be very flexible with a drift ratio of 10. Table 2 gives reasonable values for the drift of different structural systems.

The shortening of the height of a structural frame may also induce restraints and resulting stresses in masonry walls. This shortening of the frame may be due to axial loads in the columns

[1]Principals, Gensert Bretnall Associates, Cleveland, OH 44114.

TABLE 1—*Properties of masonry, structural steel, and reinforced concrete masonry unit strength and mortar strength.*

Bricks 27 600 to 110 000 kPa (4000 to 16 000 psi) compression
2800 to 11 000 kPa (400 to 1600 psi) tension

Mortar 5200 to 20 700 kPa (750 to 3000 psi) compression 3000
70 to 2100 kPa (10 to 300 psi) tensile bond

Stresses (Ultimate)	Brick Masonry	Steel	Reinforced Concrete
Compression	14 000 to 32 000 kPa (2000 to 4600 psi)	414 000 to 690 000 kPa (60 000 to 100 000 psi)	41 400 to 70 000 kPa (6000 to 10 000 psi)
Tension	70 to 2100 kPa (10 to 300 psi)	414 000 to 690 000 kPa (60 000 to 100 000 psi)	103 400 to 206 800 kPa (15 000 to 30 000 psi)
Flexure	14 000 to 32 000 kPa (2000 to 4600 psi)	248 000 to 414 000 kPa (36 000 to 60 000 psi)	41 400 to 70 000 kPa (6000 to 10 000 psi)
Shear	690 to 1400 kPa (100 to 200 psi)	138 000 to 207 000 kPa (20 000 to 30 000 psi)	690 to 2100 kPa (100 to 300 psi)
Torsion	1400 to 2100 kPa (200 to 300 psi)	207 000 to 345 000 kPa (30 000 to 50 000 psi)	1380 to 3450 kPa (200 to 500 psi)
Ductility	Brittle	Ductile	Semibrittle
Coefficients			
Thermal	5.4 to 6.3 $\times 10^{-6}$ mm/mm/°C (3.0 to 3.5 $\times 10^{-6}$ in./in./°F)	11.7 $\times 10^{-6}$ mm/mm/°C (6.5 $\times 10^{-6}$ in./in./°F)	9.0 to 10.0 $\times 10^{-6}$ mm/mm/°C (5.0 to 5.5 $\times 10^{-6}$ in./in./°F)
Moisture	0.00018 to 0.00025	...	...

TABLE 2—*Comparison of typical lateral deflections of structural frames.*

Structural Frame	Drift Ratio
Braced frame steel	1
Rigid frame steel	3
Semirigid frame steel	6
Shear connections only steel	10
Rigid frame concrete	2
Shear panel metal	2
Shear panel concrete and masonry	1

and is instantaneous for elastic behavior in steel or concrete or may be time related for concrete columns, walls, or posttensioned slabs as a result of material creep. For steel, the ratio may be 1, but for concrete it may be 3. Volume changes in masonry can work against deflections of the structure or individual components, inducing cumulative stresses which exceed the strength of the masonry.

Figure 1-a illustrates the deflection of an unrestrained steel frame under lateral load without regard to the exterior curtain wall system. Figure 1-b illustrates the same frame under the same lateral load but this time with the exterior curtain wall added. The frame deflects without restraint by the curtain wall because the movement joints within the curtain wall are effective. The frame deflects the same amount as it did without regard to the curtain wall. Although J-1 is less than J-2, as shown in Fig. 1-c, the reduction and extension of joints between panels is within the workable tolerances of the joint materials. The deflection of the frame applies no stress to the curtain wall.

Figure 2-a illustrates the same frame, without regard to the exterior curtain wall, deflecting the same amount under the same lateral loads. The frame is illustrated again in Fig. 2-b but this time with a curtain wall which has inoperable, insufficient, or even no movement joints (Fig. 2-c). The curtain wall now becomes a significant factor in resisting lateral loads, and the frame deflects less than it would without the curtain wall. If this wall is masonry, an examination of Table 1 will clearly show that it does not take much relative force to induce high stresses into the masonry.

Before we look at a classic case history of this type of problem, we should be aware of the tensile failure mode of masonry when it is under a compressive stress. Figure 3 shows a brick pier under compression to the point of failure with vertical cracks that result from lateral tension stresses resulting from Poisson's ratio. The brick pier splits.

What we learn from this basic background is that if short masonry walls or hollow piers are supported by a structural frame that deflects, and if horizontal movement joints do not exist, the masonry is subject to compression stresses and splits vertically at the point of highest stress. Points of highest stress are typically at corners and returns, as this is where the masonry is stiffest.

Differential Movement Between Masonry and Structural Frames

Case History I

Several case histories illustrate what can happen when no allowance is made for these differential movements. Figure 4 shows a 23-story steel moment resistant frame building. The exterior curtain wall is partly comprised of piers 1.22 m (4 ft.) deep by 0.61 m (2 ft.) wide, built with 102 mm (4 in.) of brick without any backup. The brick is supported at every other floor, and a 6-

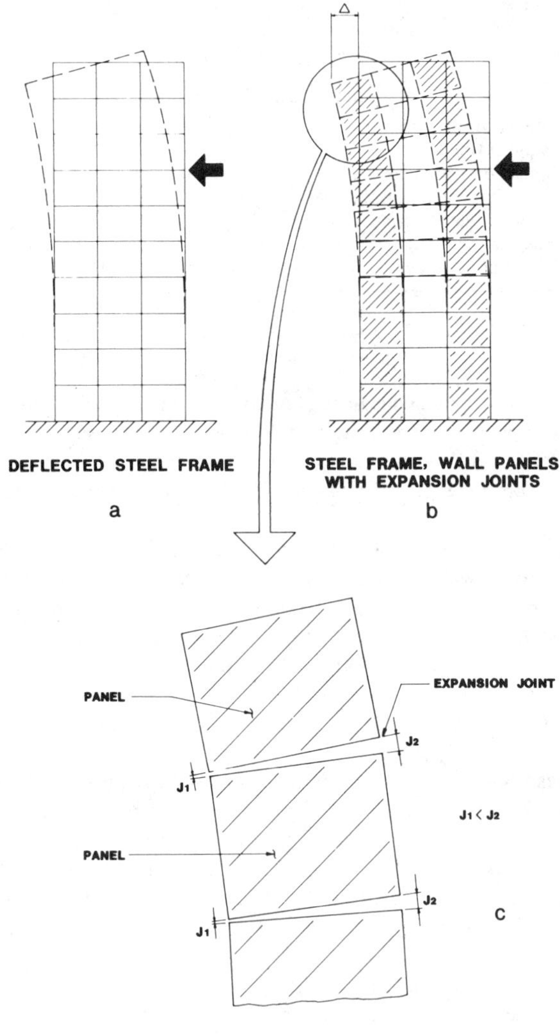

FIG. 1—*Action of curtain wall with expansion joints.*

mm (1/4-in.)-thick movement joint was provided under the supporting angle at every fourth floor. Within eight years of completion, cracks were observed near the corners of these piers (Figs. 5,6).

It is not hard to understand why these cracks occurred. Movement joints were provided only at every other supporting shelf angle. Where movement joints were not provided, the normal and expected volume increase in the brick, due to moisture absorption and thermal expansion, was restrained by the supporting shelf angle and by the shelf angle above. Where the movement joint was provided, these volume changes significantly reduced the thickness of the joint; the thickness had already been reduced during construction by deflection of the shelf angle support

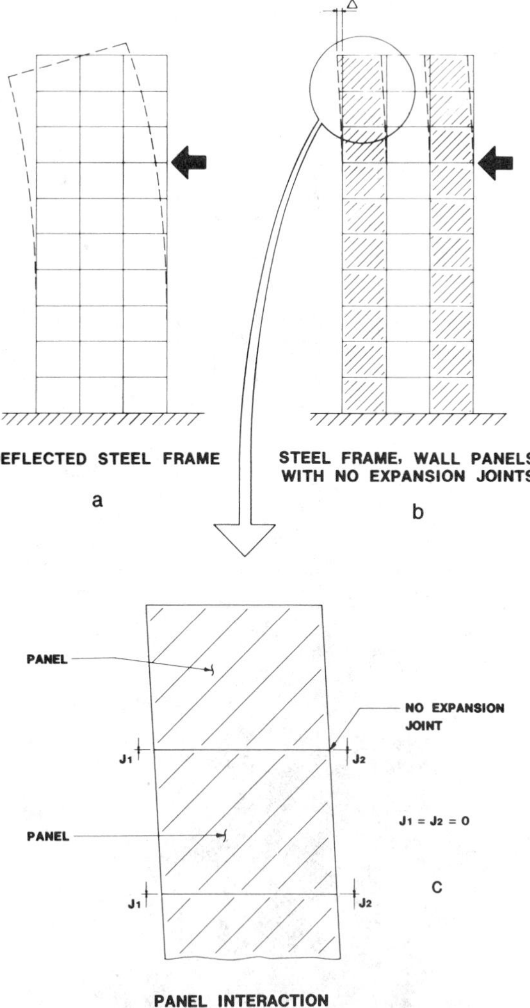

FIG. 2—*Action of curtain wall without expansion joints.*

system. The masonry was restrained and compressed by the jaws of a giant vise, the jaws of the vise being each succeeding set of shelf angle assemblies. Building drift due to wind caused additional compressive stresses in the masonry, sufficient to cause the masonry to crack at the stiffest place: the corners of the piers (Figs. 7,8).

Figure 9 shows lack of vertical expansion. Figure 10 shows compression stress contours as a result of a finite-element analysis. Interestingly, ten years after the building was built all of the cracked masonry was replaced, but the system of movement joints in the reconstruction followed the original design. Today, seven years after this masonry was replaced, it is cracking in the same location and in the same general pattern as the original masonry (Figs. 11-14). If the problem had been understood and a new movement joint system properly designed, there would be no cracks in the masonry today.

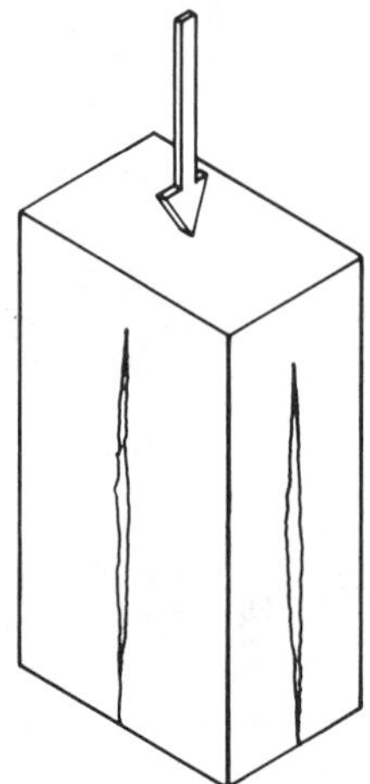

FIG. 3—*Tensile splitting due to compressive load.*

FIG. 4—*Case History I: Overall view of building.*

Case History II

Prefabricated masonry panels can have just as much trouble with overdeflecting frames as laid-in-place masonry. As another example, Fig. 15 shows a portion of a four-story unbraced steel frame building with precast concrete floors. For reasons that baffle the authors, the building has lateral tie beams only at the roof; a futile attempt was made to provide a moment connection between the precast concrete slabs and the steel columns.

Figure 16 shows a cross section through the structure. This 15.24-m (50-ft)-high structure

FIG. 5—*Case History I: Compression cracks (original masonry).*

FIG. 6—*Case History I: Compression cracks (original masonry).*

FIG. 7—*Case History I: Compression cracks (original masonry).*

FIG. 8—*Case History I: Spalling of brick (original masonry).*

FIG. 9—*Case History I: No vertical expansion.*

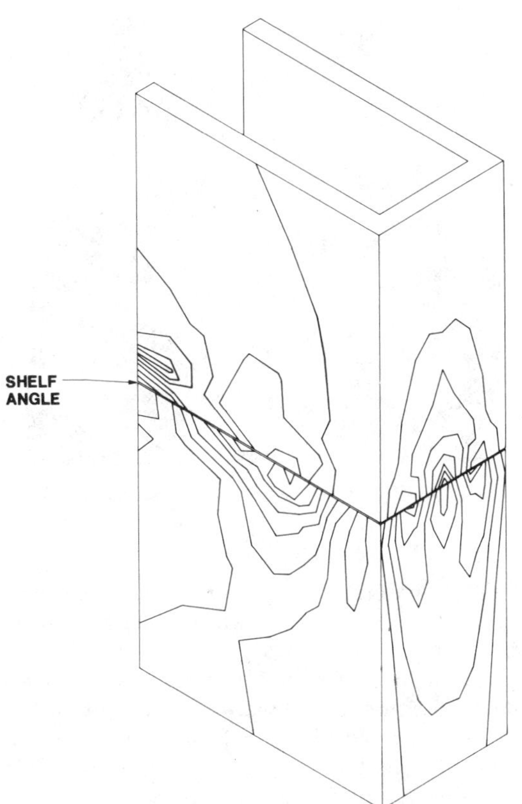

FIG. 10—*Case History I: Compression stress contours.*

FIG. 11—*Case History I: Compression cracks (new masonry).*

FIG. 12—*Case History I: Compression cracks (new masonry).*

would theoretically deflect 300 mm (11.5 in.), a Δ/H ratio of 0.019, under a code specified wind load (Fig. 17). A properly designed steel frame for this building would deflect only 38 mm (1.5 in.), or a Δ/H ratio of 0.003 (Figs. 18,19).

Any four-story building with this amount of drift would be uninhabitable, yet this particular building has been in use for over ten years. What keeps the drift within tolerable limits? The prefabricated 122-mm (4-in.)-thick masonry panel curtain wall, being much stiffer than the building frame, resists all of the wind loads (Fig. 20). However, they are not designed to do so and have become distressed as a result. The panels are cracked at the corners of the window and louver openings, exactly in the pattern one would expect from racking panels. Window frames were also trying to resist these racking forces, and shear cracks occurred in the panels where

FIG. 13—*Case History I: Compression cracks (new masonry).*

FIG. 14—*Case History I: Compression cracks (new masonry).*

window frame connections to the masonry occurred. All in all, the building was on the verge of collapse. Figures 21-23 show the distressed condition.

Figures 24 and 25 indicate the shear displacement between adjacent masonry panels as reflected by shear displacements in panel-to-panel caulking.

One other aspect of this building, unrelated to differential movement, needs to be mentioned. Figure 26 details the belt course of masonry. Original design drawings called for a thicker brick at the belt course so that the back of the panel would be flush. The panel as-built uses a brick the same size as other bricks in the panel, effectively destroying the structural integrity of the panel for wind loads and allowing a great potential for water infiltration.

B

FIG. 15—(A) *Case History II: Overall view of building;* (B) *Case History II: Typical wall panel.*

Case History III

At the other extreme, consider a concrete building which experiences no significant drift from lateral loads and which is clad with prefabricated masonry panels with all connections rigidly welded to the supporting structure. This building, Fig. 27, consists of one-way posttensioned lightweight concrete slabs supported on lightweight concrete bearing walls. The bricks in the prefabricated panels were not compatible with the mortar, resulting in very low bond strength, and rather than change either the brick or the mortar, an attempt was made to provide the necessary strength by threading reinforcing rods through the brick cores and then attaching these bars at three points to a structural steel T, the assumption being that the brick and the steel T would act as a composite section (Fig. 28). However, the principles of composite action were poorly understood, and horizontal cracks soon appeared in the panels (Fig. 29).

Case History IV

In yet another building with no significant wind drift, the prefabricated brick curtain wall panels failed partly because prefabrication methods led to weak mortar with poor tensile bond. Not only was the bond of poor quality, but the process of grouting the joints and cores of the

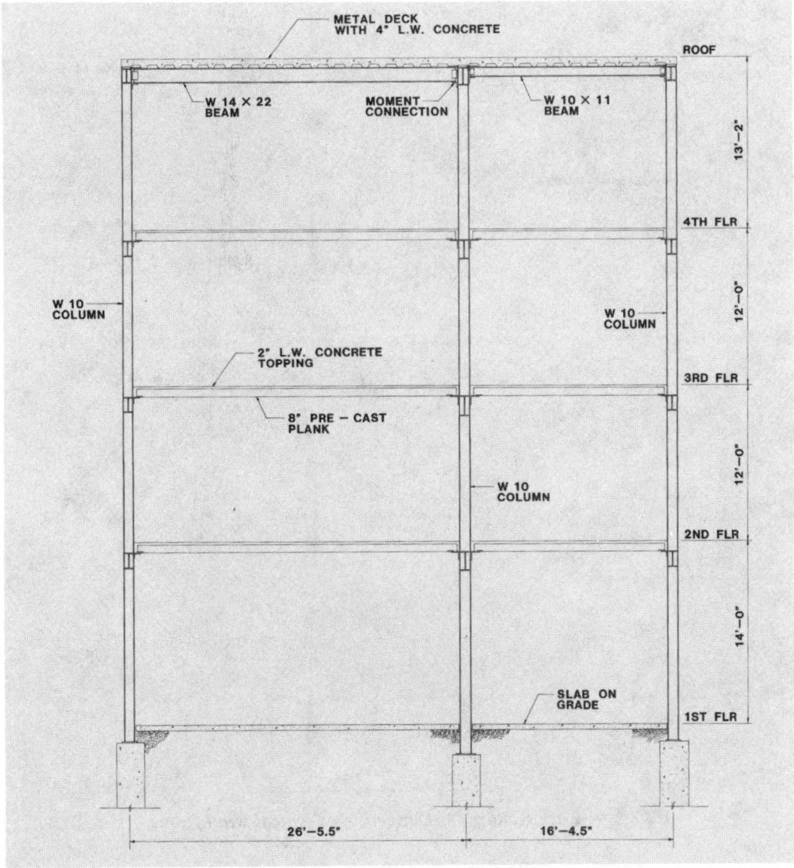

FIG. 16—*Case History II: Typical wind frame as designed and built.*

brick required the panels to be fabricated face down. As a result, the exterior mortar joints were not tooled and are not water resistant.

As with the previous example, all connections of these panels to the structural steel frame of this building were welded and are rigid, and shims between adjacent panels were not removed. And also as with the previous example, the tension field induced in the panels as they cooled with respect to the supporting structure caused cracks at the reinforcing bars, where the cross section of the masonry was reduced. Due to the weakness of the mortar and the relatively close spacing of the reinforcing bars, the cracks also travelled from bar to bar, causing the panels to split at midthickness. Figures 34 through 43 show the distressed condition of these panels.

More interestingly, vertical cracks soon developed at the location of the outer reinforcing rods. During the first winter after construction, the panels cooled while the supporting structure remained at a relatively uniform elevated temperature. This created a tension field between the rigid connections, and the panels cracked at the reinforcing rods where the cross section was reduced (Fig. 30). Several years later, these same panels, which are supported by the posttensioned slab, cracked at the center reinforcing rod. This crack, which telegraphed through from the back of the panel, resulted from high compression forces, induced both by creep shortening of the posttensioned lightweight concrete slab and thermal and moisture expansion of the brick.

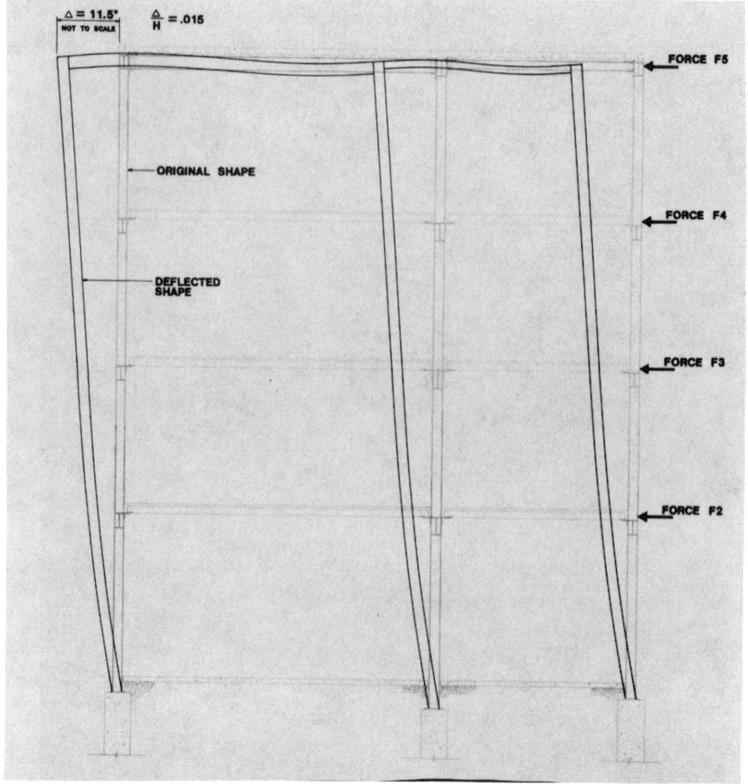

FIG. 17—*Case History II: Deflection of typical wind frame.*

These forces, applied to the panel eccentrically through its rigid connectors, caused the panel to fold like an accordion (Fig. 31). The progression of these cracks—horizontal and vertical—was documented in a series of three detailed surveys made over a five-year period.

One must always ask the question: Can these stresses be induced into a panel before it is destroyed at the connections? After all, the stresses must be higher at the connections where the transfer of force occurs between panel and structural frame within a small area. A computer analysis was made using the finite-element method, and although the stresses were higher at the connectors, they were confined to a relatively local area where the panel was reinforced. Figure 32 indicates the tension force contours by analysis and Fig. 33 shows the reinforcing that was present.

Case History V

Tall parapet panels sometimes require extra supports in the form of steel bracing between the parapet panel and the roof deck. One building in particular (Fig. 44) had parapets with diagonal bracing and short parapets without bracing (Figs. 45, 46). Bracing for the tall parapets restrained the panels from bowing under a temperature rise on the outside face of the panels (Fig. 47). The restraint occurred at the roof deck and cracked the panels due to a buildup in negative bending (Fig. 48). Short parapet panels, however, had two lines of support rather than three lines and so were not restrained and did not crack (Fig. 49).

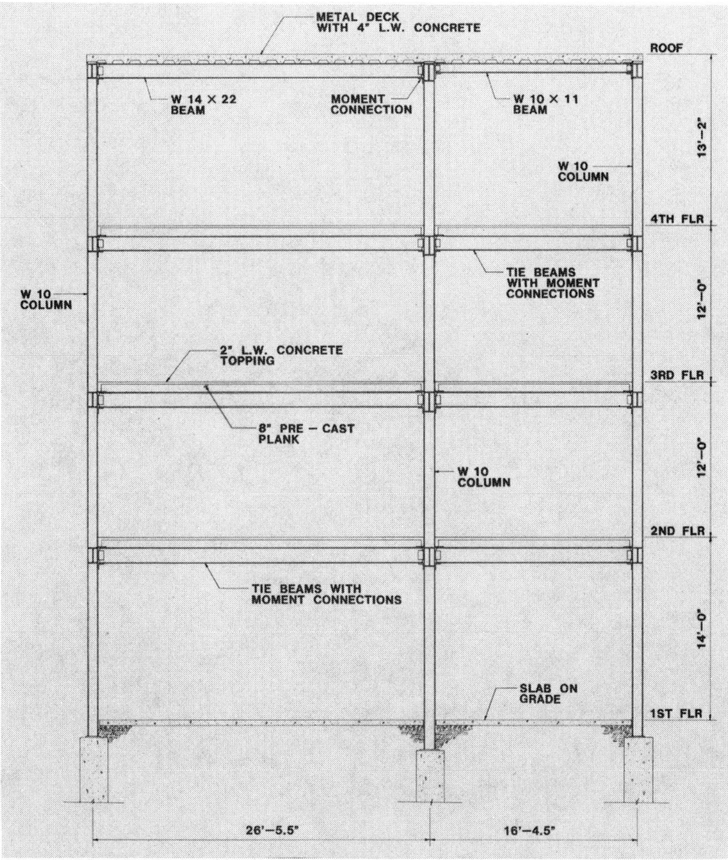

FIG. 18—*Case History II: Typical wind frame as should have been designed.*

Case History VI

Moisture and thermal expansion properties vary with the brick and with respect to the bricks principle axis. These factors can present problems when different bricks are used in the same wall and when they are not laid in the same configuration. One building which presented this problem had a panelized curtain wall consisting of a bottom soldier course of jumbo brick topped by standard brick laid in running bond. The bricks in running bond expanded at a rate 4 1/2 times that of the soldier course bricks across their width. This differential in expansion cracked the mortar joints of the soldier coursing as shown in Fig. 50.

Conclusions

These few case histories, selected from the hundreds of buildings the authors have documented, show that masonry becomes distressed when it is restrained by its supporting structure or when the supporting structure imposes loads on the masonry which the masonry is not designed to accommodate. Different properties or materials incorporated into a single system can also lead to distress of the system. The clear lesson to be learned is that designers must closely

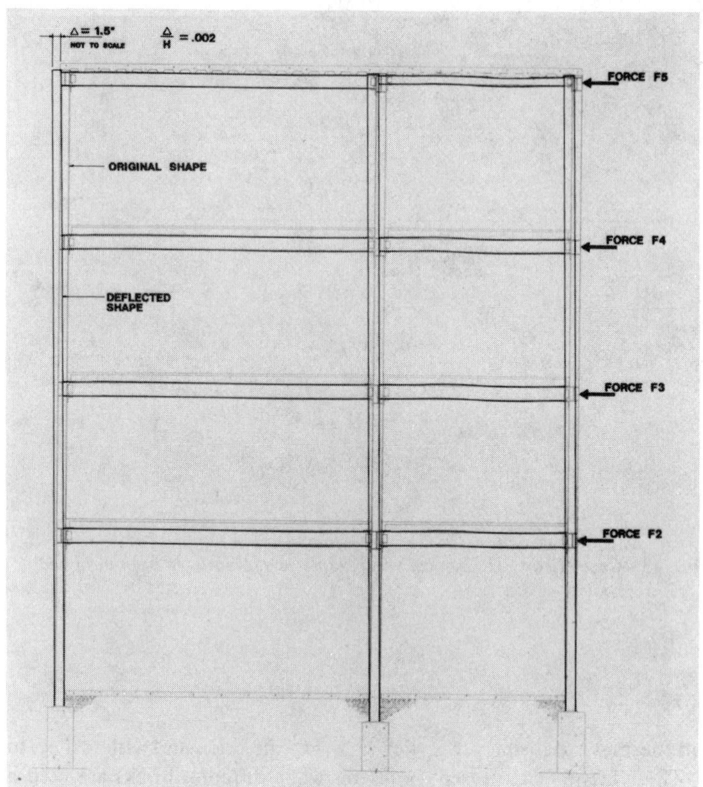

FIG. 19—*Case History II: Deflection of typical wind frame as should have been designed.*

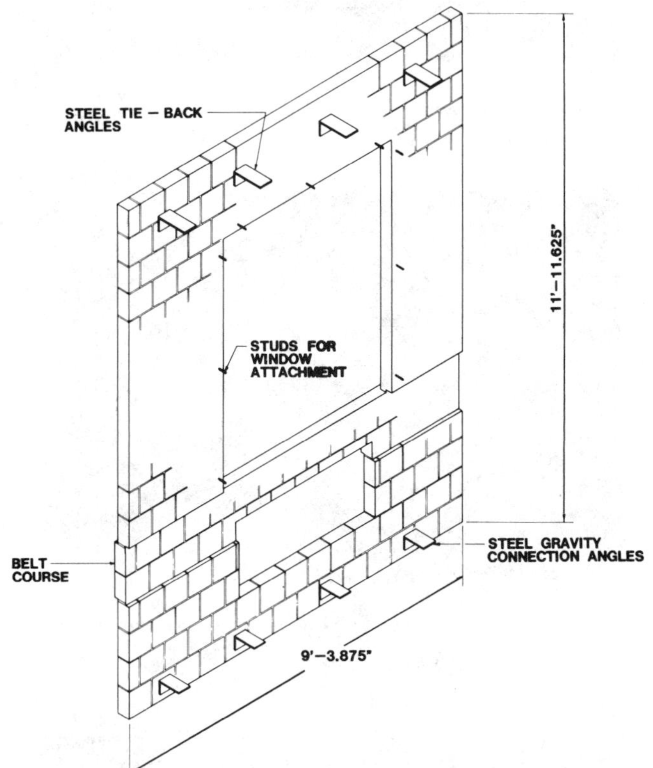

FIG. 20—*Case History II: Single window panel, back view.*

FIG. 21—*Case History II: Crack at unreinforced opening.*

FIG. 22—*Case History II: Crack at unreinforced opening.*

FIG. 23—*Case History II: Cracks at window frame connection.*

FIG. 24—*Case History II: Shear displacement cracks in caulking.*

FIG. 25—*Case History II: Shear displacement cracks in caulking.*

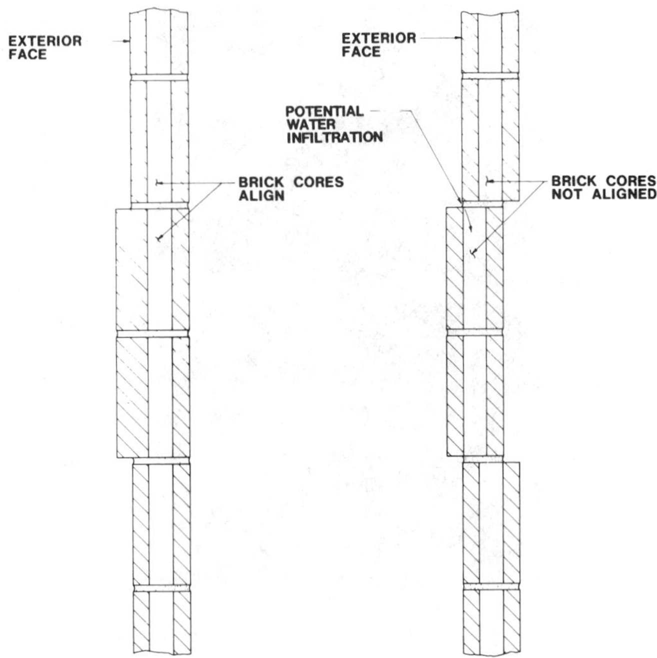

FIG. 26—*Case History II: Belt course sections.*

FIG. 27—*Case History III: Overall view of building.*

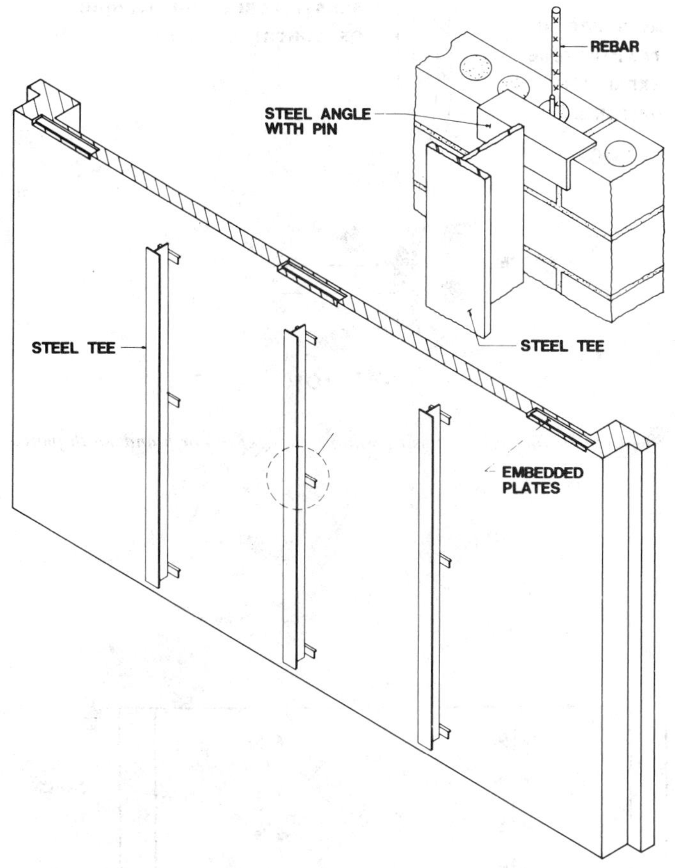

FIG. 28—*Case History III: Back view of north and south panels.*

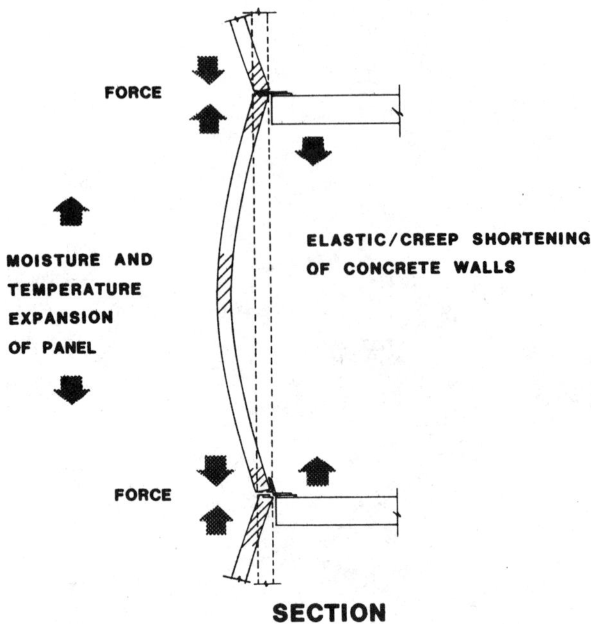

FIG. 29—*Case History III: Opposing volume changes in north and south panels.*

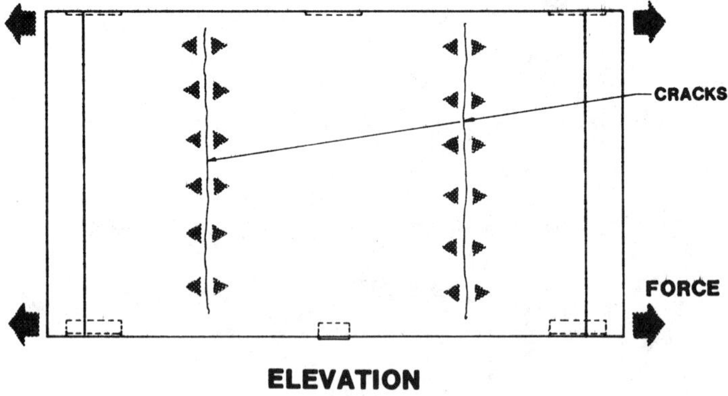

FIG. 30—*Case History III: Tension due to temperature changes in north and south panels.*

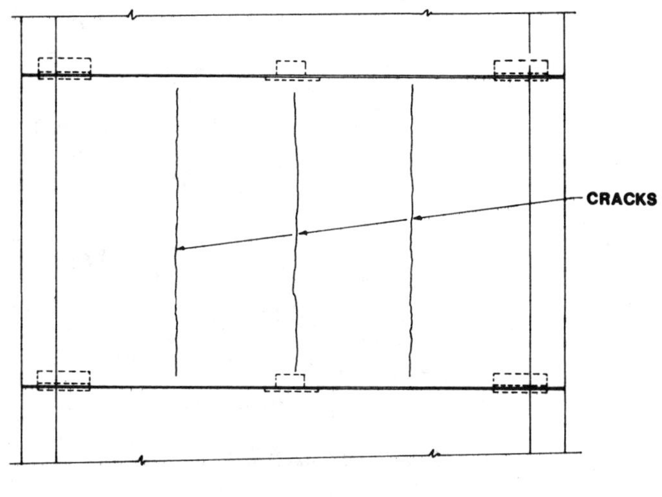

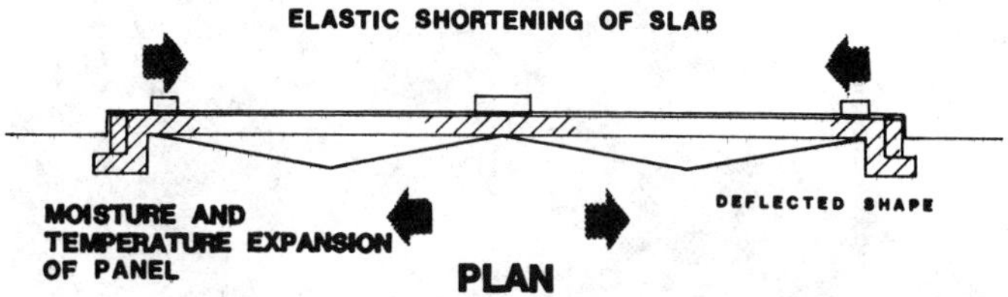

FIG. 31—*Case History III: Induced lateral force in north and south panels.*

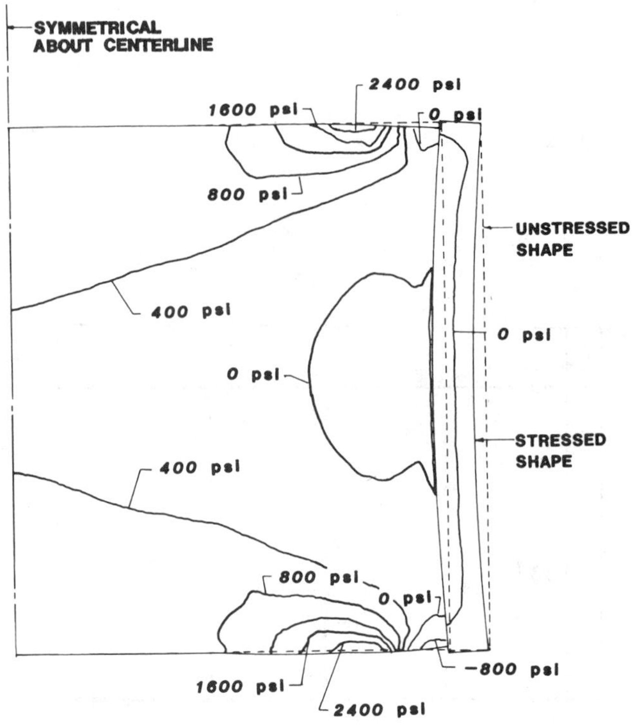

FIG. 32—*Case History III: Stress map of north and south panels due to lateral tension forces.*

FIG. 33—*Case History III: Vertical reinforcing at connectors.*

FIG. 34—*Case History IV: Building wall.*

FIG. 35—*Case History IV: Rusted X-bracing.*

FIG. 36—*Case History IV: Cracked and spalled brick.*

FIG. 37—*Case History IV: Cracked and spalled brick.*

FIG. 38—*Case History IV: Welded connections.*

FIG. 39—*Case History IV: Cracked panel.*

FIG. 40—*Case History IV: Cracked panel.*

FIG. 41—*Case History IV: Split panel.*

FIG. 42—*Case History IV: Cracked mortar.*

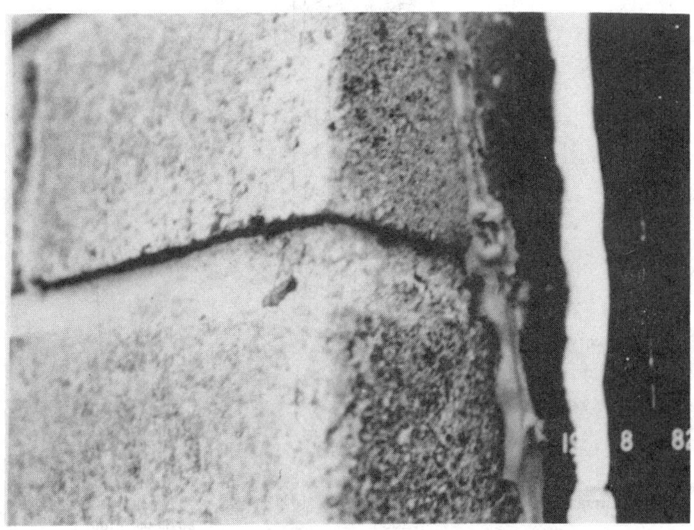

FIG. 43—*Case History IV: Eroded mortar.*

FIG. 44—*Case History V: Overall view of building.*

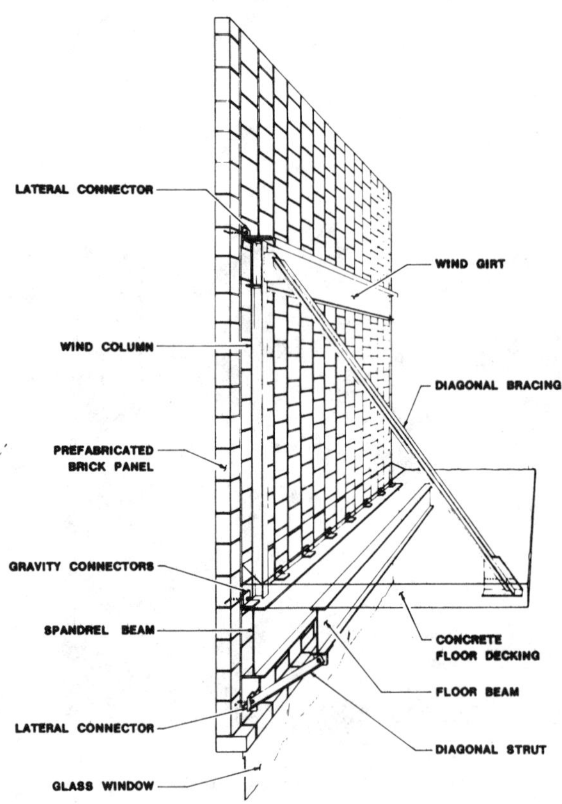

FIG. 45—*Case History V: Typical deep parapet panel.*

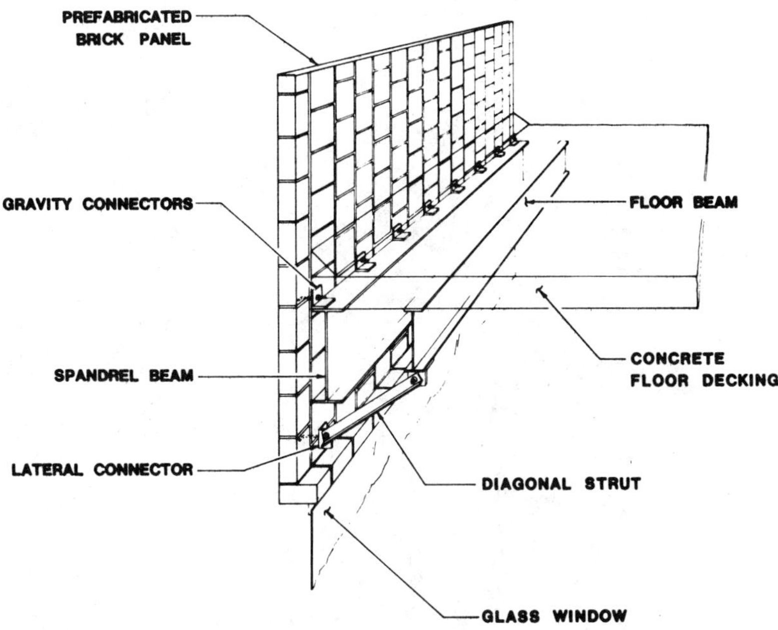

FIG. 46—*Case History V: Typical shallow parapet panel.*

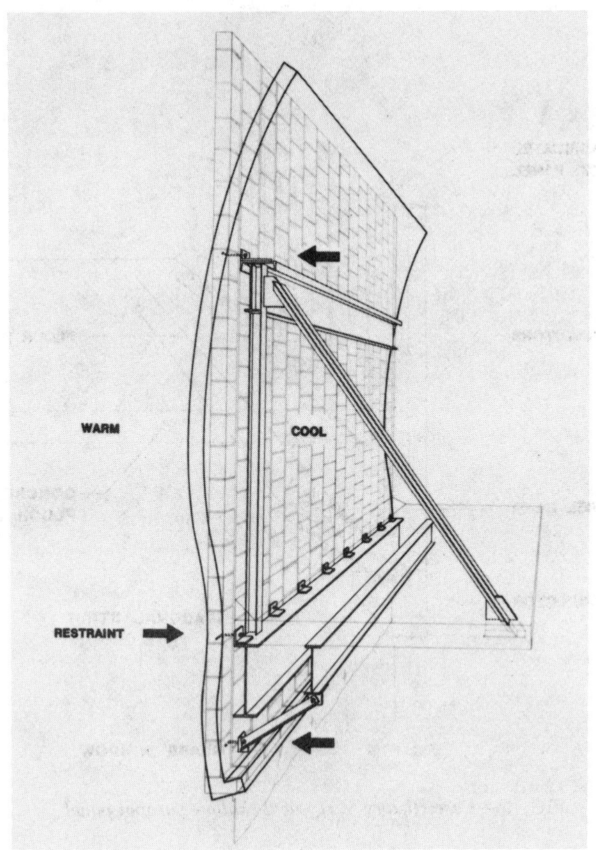

FIG. 47—*Case History V: Deep parapet panel: restrained bowing.*

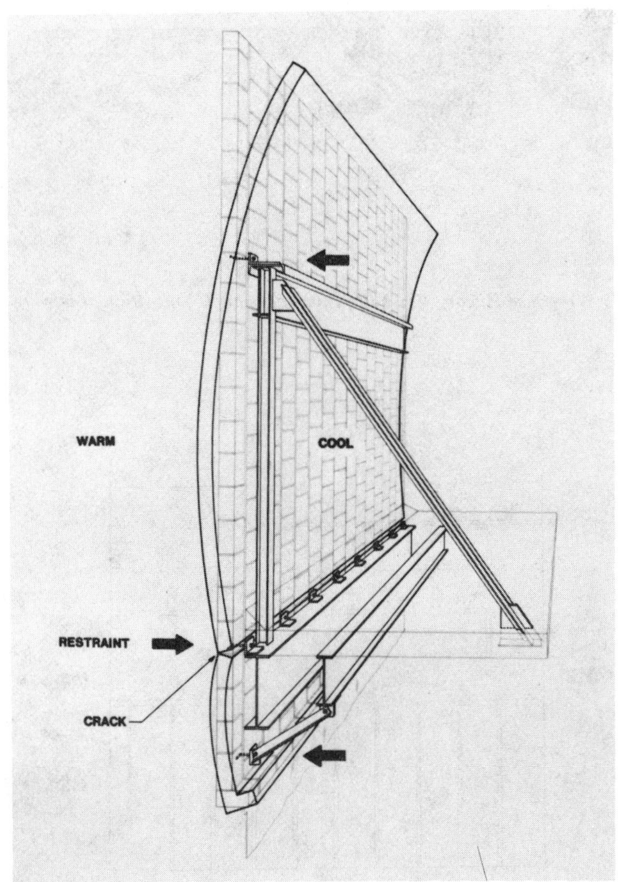

FIG. 48—*Case History V: Deep parapet panel: restrained causes cracking.*

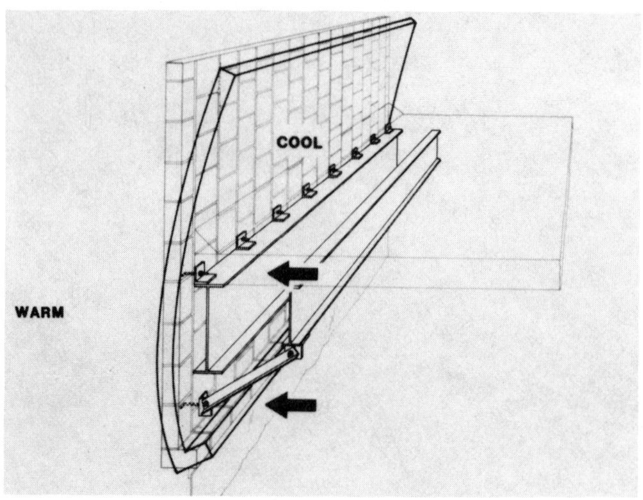

FIG. 49—*Case History V: Shallow parapet panel: unrestrained bowing.*

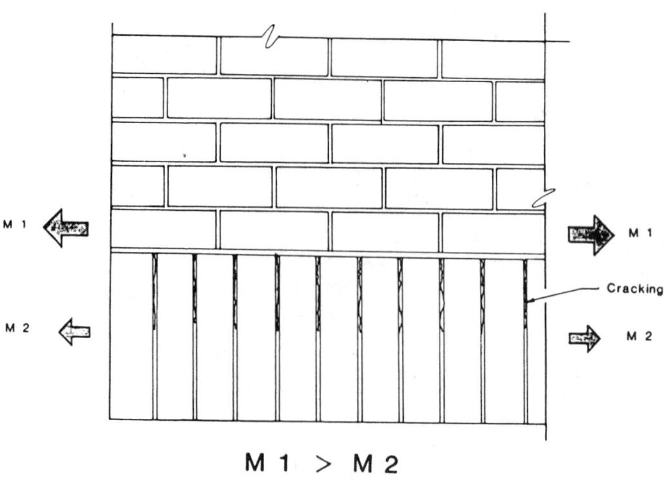

FIG. 50—*Case History VI: Differential moisture expansion.*

and carefully examine the interaction of the masonry and its supporting structure, as well as the interaction of the masonry materials with each other, and to design the overall system to interact freely and without restraint.

DISCUSSION

J. G. Stockbridge[1] *(written discussion)*—I personally had the opportunity to investigate the cause of the masonry distress in three of the six buildings discussed in this paper, Case Histories I, III, and V. In two cases, I worked for the building owner and on the third I worked for the developer. In all three cases, Gensert and Bretnall worked for the company that provided the mortar which was used in the buildings.

The masonry distress in all three buildings was caused by corrosion-induced cracking rather than the causes cited in the Gensert/Bretnall paper. The mortar used was breaking down and releasing tremendous amounts of chloride. Laboratory tests on samples removed from the buildings discovered chloride levels 10 to 40 times (not percent) higher than the amount known to cause corrosion of embedded steel. Galvanized coatings were being completely eaten away, the base metal of embedded steel elements were severely corroding, and the buildup of the rust product on the steel was causing cracking and spalling of the masonry.

Case History I: a bank—The paper fails to mention that the reason the horizontal joints were unable to accommodate volume movements was because they were filled with rust product from the severely corroding horizontal leg of shelf angles. Also, when inspection openings were cut into the piers, it could be seen that severe rust buildup on the face of the vertical legs of the angles were prying the piers apart and also causing cracking. The corrosion buildup on the leading edges of the horizontal legs of the shelf angles at the alternate levels without movement joints was shoving whole sections of mortar laterally out of the joints.

Case History III: a hotel—Vertical cracking developed at the vertical reinforcing bars because they were severely corroding. When inspection openings were cut into the panels, it could clearly be seen that the rust buildup on bars was causing the cracking. The cracks were radiating out from the corroding steel. The cracks were not externally induced.

In addition to the cracking at the vertical bars, corrosion-induced cracking and spalling was occurring at corroding pencil rods at free edges of panels, vertical cracking was occurring at corroding lifting lugs, and crescent-shaped spalling was occurring at corroding connections.

Cracks took longer to develop at the center bar than at the outer bars because rust-induced outward bowing of the panel was restrained at the center bar.

Case History V: a mall—The cracking at the roof connections of the high parapets was not caused by restraint of panel bowing. It was caused by pressure from rust buildup on the severely corroding anchors at the roof line.

Corrosion-induced cracking was also occurring at anchors in spandrel panels and at lifting lugs. There was even cracking occurring at anchors which had never been used and were attached to nothing. The reason the cracking was more severe on the high parapet panels than on the low parapet panels was because the severely corroding connections in the high parapet panels were closely spaced at 16 in. on center while in the low parapet panels they were spread out at 40 in. on center.

W. G. Hime[2] *(written discussion)*—It is distressing that Messrs. Gensert and Bretnall present an analysis of curtain walls of six buildings without noting that: the masonry of five of the six was made with mortar additive of the company that employed them; that severe corrosion occurred to metals in contact with or embedded in the mortar; that engineers, chemists, and metallurgists have attributed most of the distress experienced by those buildings to corrosion

[1]Wiss, Janney, Elstner Associates, Inc., Northbrook, IL 60062.
[2]Erlin, Hime Associates Division, Northbrook, IL 60062.

caused by the production of chloride ion through degradation of the saran latex component of this mortar; and that they (Gensert and Bretnall) were experts employed by this company to investigate those buildings. (The author of this discussion was employed by plaintiffs in lawsuits filed against this company in relationship to the distress.)

Gensert and Bretnall's first example is particularly of note because it went to trial, and a jury held against the company that employed them for over 25 million dollars including punitive damages. According to public records, this company appealed the verdict and later settled for about 19 million dollars. The records also indicate that the other buildings the authors describe, but do not name, and which contained the same mortar, were the subject of suits that were settled by this company for amounts ranging from a few million dollars to over ten million dollars.

Considering that five of their examples contained the masonry in question, none of these authors addressed the significance of such statistics.

C. H. Raths[3] *(written discussion)*—Messrs. Gensert and Bretnall in the verbal presentation of their paper elected to discuss masonry building problems using a bank and a hotel, both located in Cleveland, a manor in the Philadelphia area, and an apartment building, also near Philadelphia, as examples. The bank project was constructed using in-situ brick masonry and the mortar in question. Prefabricated brick panels were employed on the other three projects in which both the hotel and the manor had panels fabricated with this mortar. The apartment house used a proprietary grout in the panel instead of mortar.

The presentation of the paper regarding the three panel buildings indicates that a main cause of the observed cracking and distress is differential structural responses between the panels and the building frame. These structural responses, according to Gensert and Bretnall, result from *rigid* connections attaching the panels to the structural frame. Because the connections are rigid and restrain movements, they induce cracking stresses into the panels when the panels are acted upon by temperature, creep behavior of concrete building frames, moisture expansion of brick, building lateral shear forces, weak planes within panels, etc. Also, it is offered that panel cracking directly over reinforcement results from a stress riser effect of the rebar within the brick cores.

No mention is made by the authors of the extremely excessive amounts of the Cl^- within the mortar in question nor the extensive corrosion suffered by the reinforcement and other embedded metal (with and without protective coatings).

This writer, as well as other qualified engineers, has been involved in investigating, analyzing, and testing these four projects. Relative to the panelized projects, visual evidence of distress and deformation or failure about connections was not typically present. And, considering the magnitude of forces which would be attracted to rigid connections, shearing and bearing stresses should occur which would cause complete failure of the connection itself and the adjacent brick. Yet, neither distress nor structural failure of the connections has occurred other than that resulting from corrosion caused by Cl^-.

The conclusions about panel cracking and distress generated by rigid connections rest upon analyses which do not recognize load-deformation characteristics of the connections but instead consider connections absolutely rigid. Structural testing by the writer has provided data on the load-deformation spring characteristics of these brick panel connections. Analyses using connection springs, which provide significantly reduced restraint to panel movements, has indicated that the magnitude of restraint forces induced into panels are small and will not produce cracking distress. The analyses coincide with the observed behavior.

Gensert and Bretnall suggest that the cracking in the panels of the apartment house resulted from a stress riser effect in the brick caused by vertical panel reinforcement. In-depth petrographic and related testing determined that the in-plane cracking of the panel was a direct

[3]Raths, Raths & Johnson, Inc., Willowbrook, IL 60521.

consequence of proprietary grout freeze/thaw deteriorations. Structural behavior of the panel was not a factor in the cracking. Reinforcements within the cracked panels were not corroded, and no chlorides were present in the grout.

Restrained differential movements between in-situ or panelized brick masonry and the building frame can and do lead to a certain amount of distress as correctly characterized by Gensert and Bretnall. But, once cracking occurs about connections, the restraint is basically relieved and further cracking does not develop. And, for mortars and grouts not containing Cl^-, corrosion of reinforcements at these relief crack locations does not occur. The cracking described in the paper cannot happen without some other factor being involved. As discussed by this writer, Cl^--caused reinforcement corrosion and freeze/thaw deterioration of grout are the other factors leading to the paper's reported cracking and distress. Thus, the conclusions reached and presented by Gensert and Bretnall relative to problem causes are in error and the result of incorrect analyses and/or improper site investigations.

G. P. Chacos[4] (written discussion)—As the structural-engineer-of-record of the building described in Case History III, I am obliged to submit additional information for those interested in understanding the facts of this case. In addition to providing the structural design of the building, I reviewed the fabricator's design and shop drawings of the brick panels and have closely studied the distress in these panels.

I take exception to many of the assumptions and conclusions presented by the authors in their discussion of Case History III. Their investigation was one of many done on behalf ofthe company that employed them to prepare for lawsuits centering around claimed corrosive tendencies of a mortar additive supplied by this company. It is surprising, if an unbiased investigation is the objective, that their discussions do not include some reference to possible detrimental effects of chemical interaction between the components of the panels.

There have been no problems with the structural performance of the brick panels. Load tests of full-sized panels verified the composite design of the stiffened, reinforced panels prior to their installation. Properly designed soft joints and flexible connections were used, and no instances of fully compressed horizontal joints or fully compressed vertical joints have been reported. The forces indicated on Figs. 29 and 30 cannot occur with the real connections.

On Fig. 31 the authors show the vertical cracks which occur at the locations of the three vertical rebar in each of the most common panels on the north and south faces of the building. The top floor panels, however, have four vertical rebar and four vertical cracks, totally inconsistent with the "accordian" pattern of failure described. It is my observation that the cracks are caused by corrosion of the rebar, and I dispute the authors' contention that corrosion of the rebar came after the cracks occurred.

I agree with the authors that it is necessary to examine all possible sources of distress when analyzing a problem, but the discussion of Case III does not present all of the facts and forms conclusions based on an inaccurate model.

Author's Closure

The number of discussion papers received indicates that there is keen interest in the subject of masonry curtain wall distress and that this topic should be a matter for lively debate. It should be pointed out that the discussers, except for one, are an organized group of professionals working collectively for the plaintiffs which they cultivated throughout the country.

In response, additional information is offered:

[4]Gregory P. Chacos, Inc., Cleveland, OH 44115.

Case History I

Horizontal movement joints were provided approximately every 15 m (50 ft) and were filled with 6.35-mm (1/4-in.) compressible material. Deflection of the supporting cantilever bracket, moisture expansion of the masonry, and a temperature rise of 28°C (50°F) will close the joint.

The vertical legs of the shelf angles do not have enough stiffness to pry the piers apart. For the prying force to be a factor, the piers on the front and side faces would have been displaced outwards. In fact, piers were cracked on the front face only.

It is interesting to note that cracking occurred at high stress concentrations even though rusting was not always present. Text books on masonry expansion and B.I.A. Tech Notes were totally disregarded.

The recent history of this building is particularly significant. All masonry near steel framing was replaced in 1979 using mortar containing no additives. All embedded metal was replaced with new metal protected with multiple coats of epoxy paint. The reconstruction of the facade followed the original design. Today, eight years later, the facade has become distressed in the same locations and in the same pattern as the original distress. The origin of the distress is clearly due to the structural action between the masonry and the supporting frame. The authors are not alone in this opinion. Charles Raths, who examined the building at about the same time as the authors, wrote in his summary report:

> The visual examinations made by RRJ resulted in identification of certain conditions which were of a repetitive nature. Among repetitive conditions noted for the . . . office building were the following:
>
> B2. Tapered vertical and horizontal brick cracks in the intermediate piers resulting from structural frame deformations.

Case History III

The owners of this building documented, very thoroughly and very carefully, the progression of the distress in three separate (but not consecutive) years. A detailed study of this documentation shows that there is a very clear pattern, consistent from panel to panel, to the progression of the distress. The total length of cracking at embedded metal is substantially less than the total length of cracking away from embedded metal. If corrosion of embedded metals caused the cracking, there would be no clear consistent pattern, and certainly no cracking where there was no metal.

The authors are aware of only one load test of a stiffened panel. This test was performed by laying a panel flat and having a group of men stand on it. The test was not conducted in a controlled, scientific manner.

The steel connections between the panels and the structure are rigid and show signs of distress due to structural interaction between the panel and the structure. There were many documented instances of fully compressed horizontal joints. These joints closed because of the expansion of the brick and the creep shortening of the concrete frame.

Case History V

It is significant that the tall parapet panels of this building cracked on only two sides of the building—those with the greatest thermal exposure. Yet support details and construction are identical on all four sides. Thermal differentials of as much as 22°C (40°F) were recorded across the thickness of a cracked panel on a sunny but relatively mild day, with the outside face being the warmest. Bowing of the panels did occur.

Cracking at the lifting bolts or lugs is due to extremely high prying forces generated during handling. These lugs were not only used to lift the panels but also to rotate them 90° from their as-built orientation to their in-place orientation. Entry of water and formation of ice during the winter propagated initial cracking.

Summary

The distress seen in all these masonry curtain walls has a common origin: structural interaction between the curtain wall and the supporting frame.

Of conclusive evidence: the patterns of distress in masonry using the mortar additive also occur in masonry which does not use the additive. Further, nonmasonry curtain wall systems investigated by the authors exhibited similar distress for the same reasons.

The lesson learned from these case histories is that curtain walls must be designed to allow for the movements that do and will occur between the curtain wall system and the supporting frame.

It is most surprising and distressing to the authors that many in the profession investigate only to the seemingly obvious and compelling conclusion for their clients, yet refrain from further research to seek out the underlying and motivating factor of a structural phenomenon, in this instance that a certain masonry additive causes masonry distress regardless of design application or construction technique. Yes! Some chemical additives can accelerate corrosion. But what is required to initiate corrosion? A basic fact of chemistry is that water and oxygen promote corrosion. If masonry cracks from structural distress, it is obvious that water enters the masonry and corrodes the steel. Conversely, if the masonry does not admit water to the steel, there is no corrosion, as has been documented.

It is our contention, corroborated by several decades of private practice experience and coupled with extensive university teaching and research, that the interaction of structural systems and the systems they support are so sensitive as to require isolation of the supported system.

Perhaps there is a greater lesson to be learned here. Let us benefit from the documented past. Let us understand the problems. And in so doing, reduce our forensic efforts and direct our energies toward creating structures that will stand the test of time.

Ian R. Chin,[1] *Norbert V. Krogstad,*[1] *and Clare B. Monk, Jr.*[1]

Influence of Tie Flexibility, Relative Length, and End-Boundary Condition on Brick Veneer-Metal Stud Flexural Bond Stress[2]

REFERENCE: Chin, I. R., Krogstad, N. V., and Monk, C. B., Jr.,**"Influence of Tie Flexibility, Relative Length, and End-Boundary Condition on Brick Veneer-Metal Stud Flexural Bond Stress,"** *Masonry: Materials, Design, Construction, and Maintenance, ASTM STP 992,* H. A. Harris, Ed., American Society for Testing and Materials, Philadelphia, 1988, pp. 96-117.

ABSTRACT: Based on insights from the Clemson test report on the wall system under discussion, the authors analytically examined the issue of the relative stiffness between the brick veneer and the metal stud.

Much controversy has surrounded the contention that the inherent stiffness of the brick veneer will result in critical flexural bond stresses in the veneer when laterally supported on metal studs. Using ordinary conventional relative stiffness methods of analysis, the brick veneer can attract 85% or more of the wind load, whereas the metal studs attract 15% or less. Even if the studs are designed to take full wind load at a span-deflection ratio of 360 to 600, ordinary analysis may still indicate that critical flexure bond stresses will occur in the veneer prior to full participation of the stud.

This paper shows that if the inherent tie flexibility, normal shorter metal stud length, and usually free top boundary wall condition are taken into account, critical flexural bond stresses do not exist in the brick veneer on typical brick veneer/metal stud walls. Recommendations on the utilization of inherent boundary conditions to minimize the development of critical flexural bond stresses in the veneer are presented.

KEY WORDS: ordinary conventional analysis, relative stiffness, computer analysis, brick veneer, metal studs, flexural bond stress, span-deflection ratio, metal tie flexibility, spring constant, fixity, soft joint, relative height

Brick veneer with metal studs (BV/MS) construction has been used in the United States since the mid 1960s. Early field experience was obtained in the walls of the Malaysia Pavilion buildings of the New York World's Fair (1963). Thus, there has been over twenty years of commercial application. Development of the concept was initially made by the metal stud producers through their trade association, the Metal Lath/Steel Framing Association (ML/SFA), and was based upon the successful use of brick veneer/wood stud walls in low-rise residential construction. By 1979 the Brick Institute of America (BIA) published Technical Note No. 28B, incorporating the brick industry's recommendation for BV/MS construction. The BIA criteria was at variance with the earlier practice of ML/SFA in at least two ways:

[1]Senior consultant, architect II, and senior consultant, respectively, Wiss, Janney, Elstner Associates, Inc., Northbrook, IL 60062.

[2]Metric conversion for units in this paper: in. = 2.54 cm; 1 ft = 30.48 cm; 1 psi = 6894.0 N/m^2; 1 psf = 47.88 N/m^2; 1 lb/in. = 175.1 N/m^2.

1. The use of adjustable 9-gauge wire ties instead of 16 gauge, 7/8-in.-wide corrugated ties to tie the brick veneer to the metal stud back-up wall.
2. Limiting span-deflection ratios of the metal studs under full wind load to 600 to 720 with 900 preferred instead of 240 to 360.

By 1980 ML/SFA had had over a decade of apparently successful BV/MS wall installations on thousands of buildings with their criteria. The BIA recognized that BV/MS walls were increasingly being used on high-rise construction, well over three stories, where high lateral forces are encountered. Accordingly, their criteria were directed at assuring greater wall stiffness to minimize the development of critical flexural bond stresses in the brick veneer.

To resolve these conflicting criteria, the two trade associations (BIA and ML/SFA) sponsored a research program at Clemson University to investigate the BV/MS system. A report was issued 23 April 1982 [*1*].

During the 1970s designers became aware that while the majority of BV/MS walls performed satisfactorily, some did not. Five papers plus a workshop were dedicated to this subject at the recent Third North American Masonry Conference (June 1985). This demonstrated an active professional interest in the subject. No paper offered evidence of any structural collapses; however, some papers reported on ordinary structural analysis, which showed that ultimate flexural bond stresses in the brick veneer on BV/MS walls were sometimes exceeded, which could presumably lead to excessive water penetration. Following the June conference, the BIA sponsored a round-table conference in September 1985 to resolve the growing concern over this matter. The results of this conference have not yet been made public.

The authors have been challenged in reviewing this issue to explain the apparent success of the majority of BV/MS installations. In their own and in their colleagues' field experiences, many more successes than failures of BV/MS walls have been observed. This paper shows why this is the case. A careful review of the Clemson tests provided some helpful insights.

The classical ordinary strength of materials analysis, based on the relative stiffness between the brick veneer and the steel stud, usually indicates that critical flexural bond stresses exist in the brick veneer. However, the Clemson tests show that if the difference in length between the brick veneer and the metal studs, the boundary conditions of the brick veneer and the metal studs, and the flexibility of the metal ties connecting the brick veneer to the studs are taken into account, critical stresses may be avoided. This paper examines these conditions.

In examining these conditions, a metal stud span-deflection ratio of 600 under full wind load, as recommended by the BIA, is principally utilized. A metal stud span-deflection ratio of 360, as earlier recommended by ML/SFA, is also utilized to explain the apparent success of typical BV/MS walls designed under this earlier criteria.

Material Properties

At the outset it is well to establish a reasonable range of brick masonry properties based on the best available test data. The brick industry, through its Structural Clay Products Research Foundation (SCPRF), published in October 1964 the results of a National Testing Program based on 31 varieties of regional brick. Much of this work can be found in "Progress Report No. 1 Small Scale Specimen Testing, October 1964" by SCPRF [*2*]. Figures 1 and 2 show the results of the test program, which used Type S mortar throughout. From this work the statistics in Table 1 have been compiled.

The recommended allowable flexure bond stress for masonry walls by the Building Code Requirements for Engineered Brick Masonry [*3*] is:

1. Type S and M mortar = 36 psi.
2. Type N = 28 psi.

FLEXURAL STRENGTH OF BRICK MASONRY SPECIMENS

CODE	FLEXURAL STRENGTH OF WALLETTES				BRICK PHYSICAL PROPERTIES											MORTAR PHYSICAL PROPERTIES							
						COMPRESSIVE STRENGTH			MODULUS OF RUPTURE			SUCTION			COMPRESSIVE STRENGTH				TENSILE STRENGTH			
	n	X̄ psi	s psi	v %	n	X̄ psi	s psi	v %	X̄ psi	s psi	v %	X̄ ***	s ***	v %	n	X̄ psi	s psi	v %	n	X̄ psi	s psi	v %
LL	5	133	19.8	14.9	5	6066	784	12.9	679	51.7	7.6	46.6	6.6	14.1	6	1439	59.1	4.1	6	181	14.8	8.2
LM	5	85	13.6	16.0	5	6306	955	15.1	686	58.7	8.6	24.1	5.1	21.0	6	1284	55.2	4.3	6	171	16.3	9.6
FL	5	49	15.2	31.2	5	6340	2420	38.2	601	315.0	52.4	*50.7	11.7	23.0	6	1346	44.0	3.3	6	194	15.7	8.1
KM	5	126	28.4	22.5	5	6868	278	4.0	968	71.0	7.3	16.9	1.7	10.0	6	1911	114.1	6.0	6	216	21.0	9.7
GM	5	151	5.7	3.8	5	7504	963	12.8	563	73.0	13.0	35.2	5.3	14.9	6	1322	62.7	4.8	6	185	18.3	9.9
JM	5	86	15.5	18.1	5	7806	400	5.1	1224	114.0	9.3	*34.4	1.4	4.0	6	1608	40.6	2.5	6	243	18.1	7.5
MM	5	164	25.0	15.3	5	8039	1256	15.6	1066	194.0	18.2	18.8	7.8	41.6	6	1348	68.0	5.0	6	214	31.5	14.7
DM	5	131	19.3	14.8	5	8717	1639	18.8	915	208.0	22.8	7.6	1.1	14.1	**	1348	68.0	5.0	**	214	31.5	14.7
FM	5	170	10.4	6.1	5	9161	1198	13.1	825	75.0	9.1	21.2	1.6	7.3	**	1322	62.7	4.8	**	185	18.3	9.9
AL	5	73	17.5	24.0	5	9267	2327	25.1	615	227.0	36.8	*30.8	8.8	28.5	5	1343	94.9	7.1	6	165	11.8	7.2
KH	5	140	26.1	18.6	5	9708	624	6.4	1022	93.9	9.2	8.8	1.7	19.3	**	1911	114.1	6.0	**	216	21.0	9.7
EM	5	158	28.6	18.1	5	9730	1209	12.4	623	65.3	10.5	14.8	3.6	24.6	6	1925	83.8	4.4	6	245	25.2	10.3
CL	5	142	18.8	13.3	5	9810	647	6.6	950	213.0	22.5	23.7	0.9	3.8	6	1338	119.3	8.9	6	209	20.3	9.7
MH	5	99	22.7	22.9	5	10711	547	5.1	1105	102.0	9.2	14.8	4.1	27.6	6	1276	68.8	5.4	6	215	16.8	7.8
CM	5	118	17.8	15.1	5	11502	2004	17.4	832	137.0	16.5	11.5	2.0	17.3	6	1244	27.4	2.2	5	199	13.9	7.0
AH	5	158	15.4	9.7	5	12015	531	4.4	876	61.6	7.0	17.3	0.5	3.1	6	1979	70.2	3.6	6	200	18.9	9.4
GH	5	130	14.4	11.1	5	12138	975	8.0	1299	117.3	9.0	5.4	1.5	28.0	6	1395	115.8	8.3	6	172	10.9	6.3
DH	5	107	12.6	11.7	5	12264	984	8.0	889	186.0	20.9	3.4	1.4	39.4	6	1680	39.3	2.3	6	245	21.3	8.7
LH	5	144	19.8	13.8	5	12336	1494	12.1	1014	114.0	11.3	8.7	1.4	16.0	**	1284	55.2	4.3	**	171	16.3	9.6
FH	5	87	10.5	12.1	5	12782	1031	8.1	1002	91.0	9.0	12.4	1.7	13.[illegible]	**	1338	119.3	8.9	**	209	20.3	9.7
JH	5	172	32.7	19.0	5	13052	945	7.2	1179	86.9	7.4	22.1	4.6	20.6	6	1680	39.3	2.3	6	245	21.3	8.7
AM	5	133	25.9	19.5	5	14388	254	1.8	1056	95.6	9.0	13.8	2.3	16.4	6	1979	70.2	3.6	6	200	18.9	9.4
BH	5	100	10.0	10.0	5	14760	876	5.9	1424	80.8	5.7	2.3	0.5	19.4	6	1145	73.7	6.4	6	135	5.1	3.8
HM	5	148	15.6	10.5	5	15300	1802	11.8	1468	61.0	4.2	2.2	0.3	15.5	6	1228	82.7	6.7	6	162	11.6	7.2
HH	5	101	7.9	7.6	5	16093	2801	17.4	1568	116.0	7.4	4.0	0.7	17.2	**	1228	82.7	6.7	**	162	11.6	7.2
EH	5	84	7.0	8.3	5	16440	1461	8.9	1407	207.0	14.7	*1.6	0.5	32.8	**	1925	83.8	4.4	**	245	25.2	10.3
CH	5	132	10.9	8.2	5	17838	474	2.7	1166	85.0	7.3	2.7	1.6	59.1	6	1169	42.5	3.6	6	179	11.2	6.2
NL	5	48	8.0	16.5	5	4489	546	12.2	680	95.4	14.0	172.0	13.1	7.6	6	1491	82.6	5.5	6	165	7.3	4.4
NH	5	103	14.5	14.1	5	6370	338	5.3	1191	184.0	15.4	45.0	5.0	11.1	6	1343	94.9	7.1	6	165	11.8	7.2
NM	5	39	14.3	37.0	5	3968	736	18.6	1037*	191.0	18.4	52.4	11.7	22.4	6	1491	82.6	5.5	6	165	7.3	4.4
SCR	5	125	12.5	10.0	5	11771	642	5.5	1214	75.4	6.2	10.6	1.5	14.3	6	1432	69.8	4.9	6	203	24.3	12.0

*n = 4

**Indicates that more than one set of wallettes was made from one batch of mortar.

***g/min/30 in.^{2}

6 - 5

FIG. 1—*Results from the National Testing Program published by the Structural Clay Products Research Foundation.*

TABLE 1—*Statistics based on SCPRF brick testing program.*

FLEXURAL BOND STRENGTH
Average = 123.18 psi
Standard deviation = 31.62 psi
Coefficient of variation = 25.67%
Number of samples = 140
MODULUS OF ELASTICITY, E
Average = 2.245×10^6 psi
Standard deviation = 0.882×10^6 psi
Coefficient of variation = 39.28%
Number of samples = 218

For the purposes of this paper, the authors assume that these "allowable" stresses are applicable to BV/MS walls. These values may be increased 33 1/3% when wind is acting to 48 and 37 psi, respectively. For Type S mortar, the 36 psi could be increased three-fold to 108 psi without the brickwork cracking. It is conceivable that some walls in service have experienced 108 psi without failure. Further, very high wind loadings are events that happen relatively infrequently in the life of a building. In the work to follow, the authors examine the conditions of BV/MS walls that results in calculated stresses that fall within the 48 to 108 psi range for Type S or M mortar (or 37 psi to 84 psi range for Type N mortar). Examination of the modulus of elasticity of brickwork between the values 1 to 3 million psi, which reflect average values, is incorporated in the analysis.

Ordinary Strength of Materials Analysis

The deflection criteria established by BIA and ML/SFA for BV/MS walls assumed that the metal studs take the total design wind load. Ordinary relative stiffness analysis indicates that the brick veneer frequently takes more than half the load, which commonly stresses the masonry to well over the "allowable" stresses, even in excess of ultimate stresses in some cases. Although the recommended span-deflection ratio of 600 for the metal studs by the BIA is intended to limit the deflection of the studs, inherently the veneer is very stiff and not able to tolerate span-deflection ratios of 600 or less without cracking. To examine these ideas, the following plot was made.

From ordinary flexural theory it may be shown that: $(L/\Delta) = (24E/5f) \div (L/d)$. This can be plotted on a log-log plot as a series of straight lines as follows:

$$\log (L/\Delta) = \log (24E/5f) - \log (L/d)$$

Figure 3 is a plot of this equation. The abscissae are L/d values or wall slenderness ratios; the ordinates are L/Δ values or span-deflection ratios. Two regions of brickwork modulus of elasticity are shown: E = 1 000 000 and 3 000 000 psi. Each region is bounded by two levels of flexural bond stress: f_b = 48 and 108 psi. The first represents the "allowable" for Type S mortar increased by 33 1/3% for wind or 36 psi × 1.33 = 48 psi. The second is an assumed high stress level where brickwork cracking is not yet expected to develop, 36 × 3 = 108 psi. Values within this range exceed the "allowable" stress for masonry walls but are below the ultimate cracking stress. Also plotted are the ranges of slenderness ratios recommended by BIA and ML/SFA for the design of the backup system. Because the backup system and the brick veneer are assumed to deflect together in an ordinary analysis, it is clear that the brick veneer have to take the

BRICK PRISM COMPRESSIVE STRENGTHS (SCR Laboratory Tests)

CODE	PRISMS							MORTAR COMPRESSIVE STRENGTH psi AIR CURED WITH PRISMS** $\bar{X}$	BRICK COMPRESSIVE STRENGTH psi $\bar{X}$
	n	COMPRESSIVE STRENGTH - f'_m				E_m*	F_m/f'_m		
		$\bar{X}$ psi	R psi	s psi	v %	psi x 10^{-6}			
LL	3	3805	661	340	8.9	1.82	478	1439	6066
LM	3	4553	555	283	6.2	1.74	382	1439	6306
FL	3	2658	126	64	2.4	1.33	500	1773	6340
KM	3	4230	811	452	10.7	1.49	352	1773	6868
GM	3	4243	521	269	6.3	2.73	643	1157	7504
JM	3	4790	1018	519	10.8	1.89	395	1319	7806
MM	3	4849	522	299	6.2	2.63	542	1665	8039
DM	3	4584	350	194	4.2	3.45	753	1748	8717
FM	3	6516	483	265	4.1	2.02	310	1440	9161
AL	3	4002	1039	572	14.3	1.41	352	1580	9267
KH	3	5880	498	251	4.3	3.33	566	1773	9708
FM	3	5891	582	306	5.2	1.74	295	1748	9730
CL	3	4583	510	255	5.6	1.83	399	1440	9810
MH	3	5851	99	52	0.89	2.70	461	1319	10711
CM	3	4689	371	197	4.2	3.66	781	1363	11502
AH	3	6186	298	154	2.5	2.70	436	1319	12015
GH	3	6439	594	326	5.1	3.57	554	1157	12138
DH	3	4413	314	164	3.7	2.70	612	1665	12264
LH	3	6105	99	265	4.4	2.38	390	1166	12336
FH	3	6087	194	112	1.8	3.03	498	1773	12782
JH	3	6474	459	230	3.6	3.03	468	1665	13052
AM	3	6349	296	164	2.6	2.44	384	1580	14388
BH	3	4867	629	345	7.1	3.08	441	1336	14760
HM	3	4803	74	43	0.89	3.33	693	935	15300
HH	3	4937	210	112	2.3	2.63	533	880	16093
EH	3	5595	946	498	8.9	3.70	661	1748	16440
CH	3	6808	495	261	3.8	4.17	613	1363	17838
NL	3	1964	55	28	1.4	0.67	341	1491	4489
NH	3	4292	585	293	6.8	2.11	492	1580	6370
NM	3	1314	305	153	11.7	0.69	525	1580	3968
SCR	5	6281	296	130	2.1	3.03	482	1287	11771

*Modulus of Elasticity, initial tangent. **See Table 3-3

FIG. 2—*Results from the National Testing Program published by the Structural Clay Productions Research Foundation.*

BRICK PRISM COMPRESSIVE STRENGTHS
(Commercial Laboratory Tests)

BRICK	PRISMS n	COMPRESSIVE STRENGTH - f'_m X̄ psi	R psi	s psi	v %	E^*_m psi x 10^{-6}	E_m/f'_m	MORTAR COMPRESSIVE STRENGTH psi AIR CURED WITH PRISMS** X̄	BRICK COMPRESSIVE STRENGTH psi X̄
LL	4	2265	1141	543	24.0	0.792	350	2082	6066
LM	5	3626	1644	617	17.0	0.952	263	2082	6306
FL	5	2391	889	399	16.7	0.645	270	2082	6340
KM	5	2437	767	313	12.8	0.833	341	1708	6868
GM	5	2631	1093	437	16.6	0.901	342	713	7504
JM	5	3800	620	316	8.3	2.085	549	2779	7806
MM	5	3825	995	357	9.3	1.490	390	713	8039
DM	5	3807	380	142	3.7	3.448	906	897	8717
FM	5	4443	492	212	4.8	1.520	342	2082	9161
AL	5	3729	1480	576	15.5	0.952	255	1380	9267
KH	5	3429	1007	416	12.1	1.350	394	1708	9708
EM	5	3914	495	251	6.4	1.099	281	2423	9730
CL	5	5182	1380	502	9.7	2.280	440	1588	9810
MH	5	3544	615	294	8.3	1.500	423	713	10711
CM	5	4952	820	294	5.9	3.060	618	1588	11502
AH	5	5066	485	220	4.4	1.750	345	1380	12015
GH	5	4719	1098	501	10.6	2.220	470	713	12138
DH	5	4207	575	226	5.4	3.571	849	897	12264
LH	3	5617	463	261	4.7	1.480	263	2082	12336
FH	5	5357	1055	507	9.5	1.730	323	2082	12782
JH	5	5358	1520	581	10.9	1.482	277	2779	13052
AM	5	6018	635	227	3.8	2.060	342	1380	14388
BH	5	5556	1680	678	12.2	2.450	441	2154	14760
HM	5	5650	382	169	3.0	2.439	432	2875	15300
HH	5	5722	534	198	3.5	2.632	460	2875	16093
EH	5	4408	1090	440	10.0	2.469	560	2423	16440
CH	5	5760	530	225	3.9	2.730	474	1588	17838

*Modulus of Elasticity, initial tangent

**Duplications of values in this column indicate that one set of 6 air cured 2 in. by 2 in. cubes was made from the mortar batch but that two or more sets of prisms were made from that batch.

FIG. 2—*Continued.*

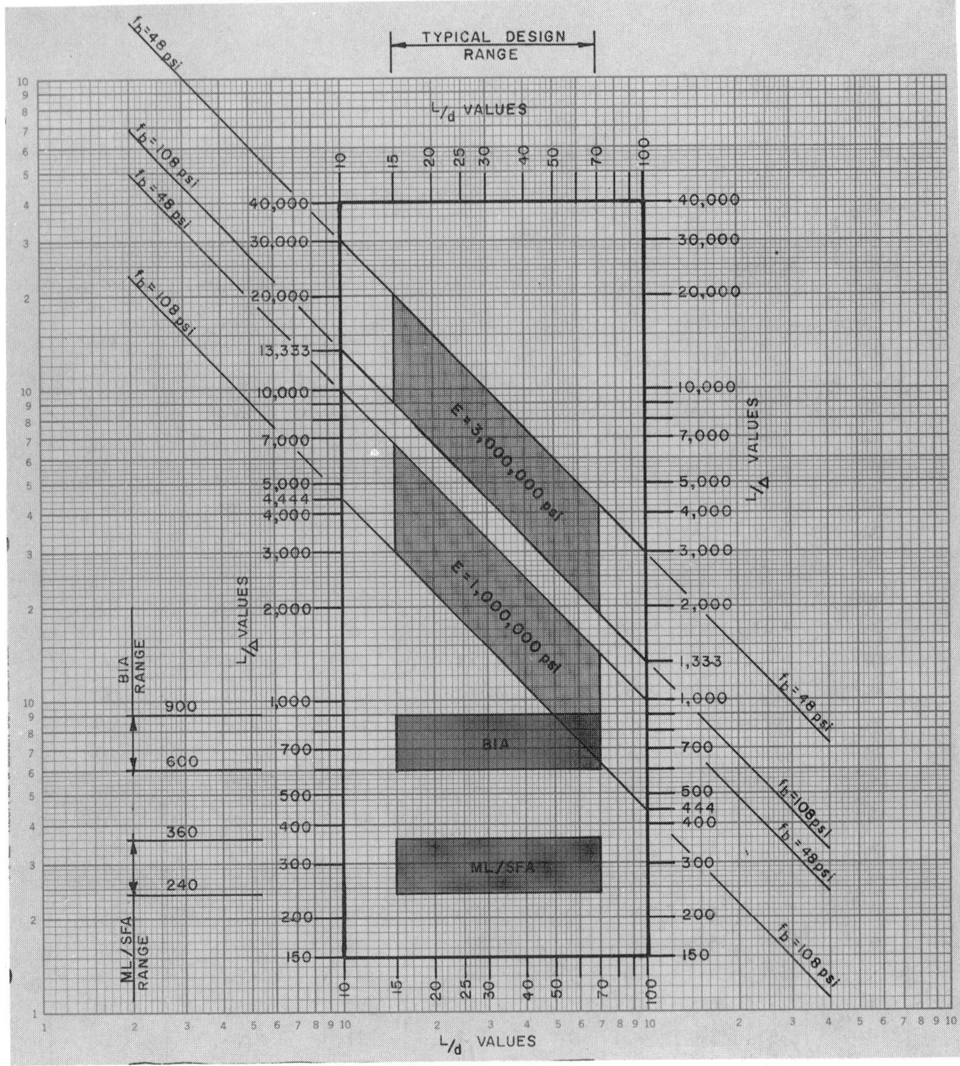

FIG. 3—*Parametric study showing height/deflection values as a function of slenderness ratio values for constant values of wall modulus and wall stress.*

majority of the load or crack. Brick masonry walls cannot normally tolerate, without cracking, span-deflection ratios below the $L/\Delta = 900$ limit.

To examine this issue with greater precision, graphs were developed which express the flexural bond stress of the brick veneer, (f, psi) based on the following equation

$$f = [(q\,L^2)/17.521] \div [1 + (zq\,L^3)/(25.405E)]$$

where

q = wind load, psf,
L = wall height, ft,

E = brick modulus, psi, and
z = span-deflection ratio of metal studs.

The equation is derived from ordinary strength of materials analysis, assuming that the total wind load is shared between the brick veneer and the metal studs according to their relative stiffnesses. For the purposes of simplifying the equation, the reduction of tensile stress due to the weight of the masonry has not been included. Figure 4 is a plot of the stresses in the brick veneer for the case of $z = 600$ assuming the stud is hinged top and bottom. For this case, the maximum stress in the brick veneer ranges from about 120 psi, when $E = 1\ 000\ 000$ psi, to about 260 psi, when $E = 3\ 000\ 000$ psi, showing that the flexural bond stress is sensitive to values of E.

The effect of metal stud fixity on stresses in the brick veneer with $z = 600$ was also determined by the ordinary analysis, as shown in Figs. 5 and 6. When the bottom of the metal studs is fixed and the top of the studs is hinged, the maximum stress in the brick veneer ranges from about 60 psi when $E = 1\ 000\ 000$ psi to about 140 psi when $E = 3\ 000\ 000$ psi. When the top and bottom of the studs are fixed, the maximum stress in the veneer ranges from about 40 psi when $E = 1\ 000\ 000$ psi to about 80 psi when $E = 3\ 000\ 000$ psi. The stresses in the brick veneer are therefore also very sensitive to stud fixity. Generally, the stresses in the veneer are greater than the assumed 48 psi "allowable" limit, but many are less than 108 psi when stud fixity exists.

In actual construction practice it is probable that neither a hinged-hinged or fixed-fixed stud end boundary condition exists. The bottom of the metal studs is normally square ended and fits directly against the web of the runner. The top of the metal studs, on the other hand, is generally held clear of the web of the top runner to allow for vertical building movement.

It is interesting to note that the span-deflection ratio of metal studs hinged at top and bottom changes from 600 to about 1445 when the bottom of the studs are fixed. This benefit requires no increase in the metal stud cross sectional area; however, some design effort is needed to assure bottom fixity.

While taking advantage of stud fixity can reduce flexural bond stresses, other benefits are possible by a careful review of the Clemson report as discussed in the next section.

Computer Analysis Including Tie Flexibility

The ordinary analysis assumes that the brick veneer and the backup system move together and share the same span. In actual walls, however, the backup and veneer have complex boundary conditions which allow the walls to move separately. This section examines the benefits of the following:

1. The typical soft joint installed at the top of the wall under the shelf angle to accommodate vertical movements.
2. The difference in height, between the usually taller brick veneer and the shorter metal stud.
3. The utilization of the inherent spring property of the metal ties used to tie the brick veneer to the metal studs.

The top soft joint, the taller brick veneer, and the spring action of the metal ties all contribute to reduce wall stresses as will now be shown.

A computer model was developed using the brick veneer, the metal studs, and the metal ties as a frame network. Figure 7 is a schematic layout of the computer model prepared for running on a STRESS program (Version 1 11-81). The ties were modeled as pin-ended springs. The range of spring constants: 200, 2000, and 20 000 lb/in. is based on metal tie tests contained in the Clemson report. The modulus of elasticity for the brick veneer was again bounded between

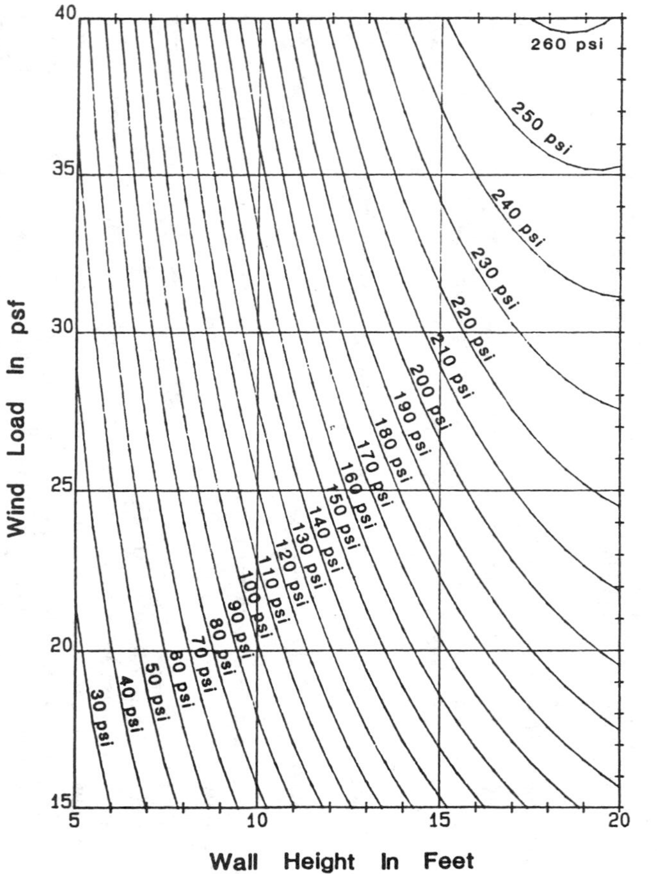

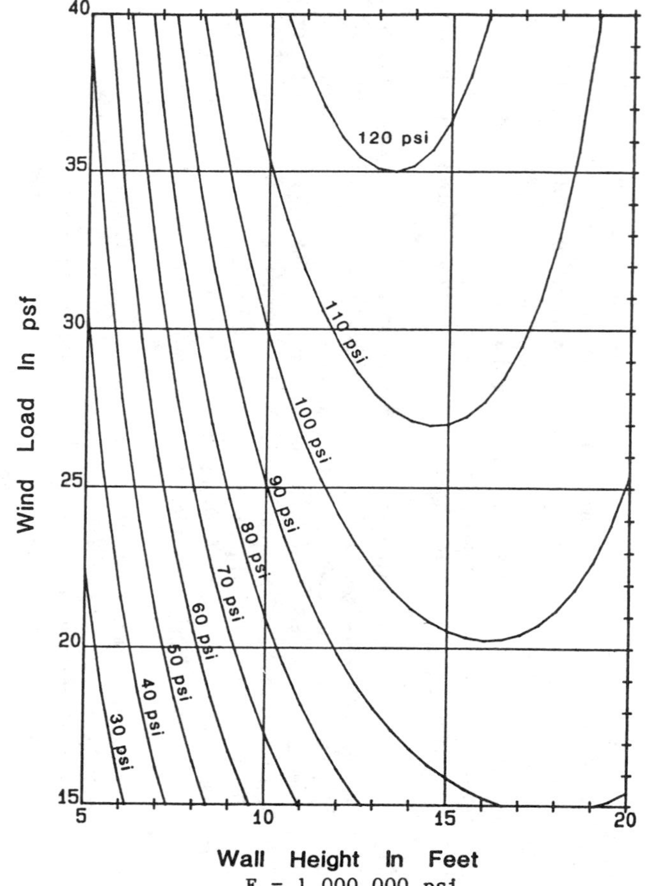

FIG. 4—*Plot of wall stress* (f) *as a function of wall height* (L) *and wind load* (q) *with studs hinged-hinged.*

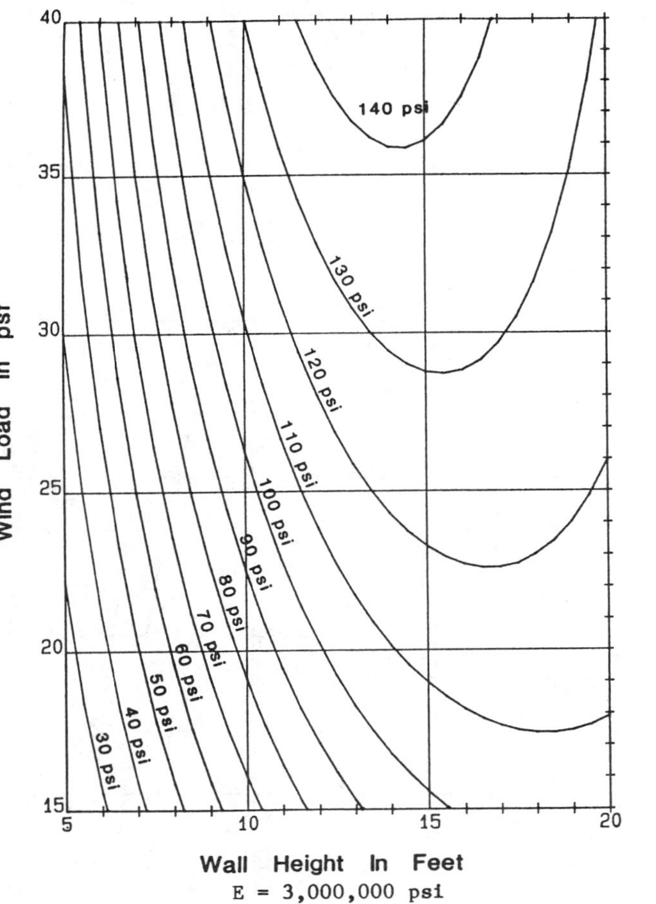

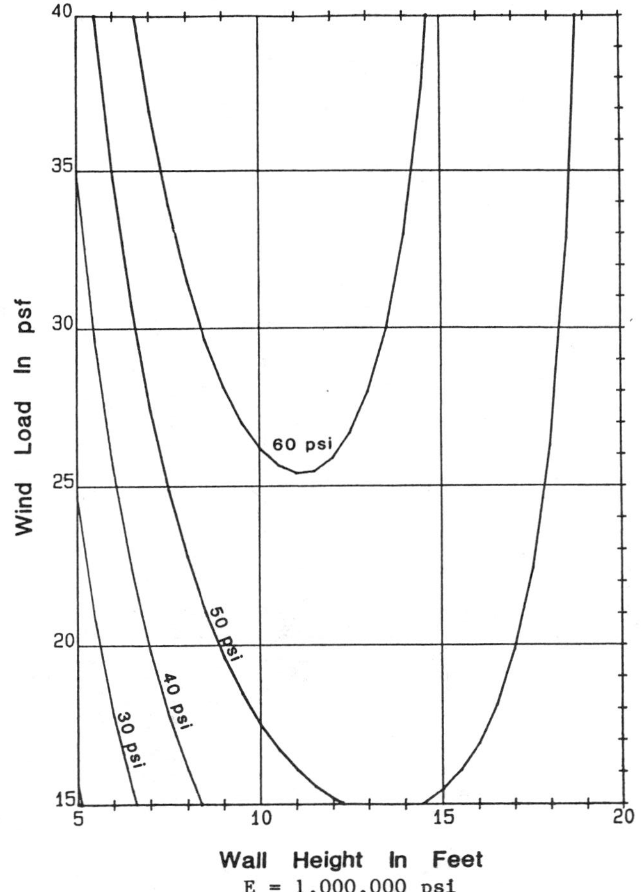

FIG. 5—*Plot of wall stress* (f) *as a function of wall height* (L) *and wind load* (q) *with studs fixed-hinged.*

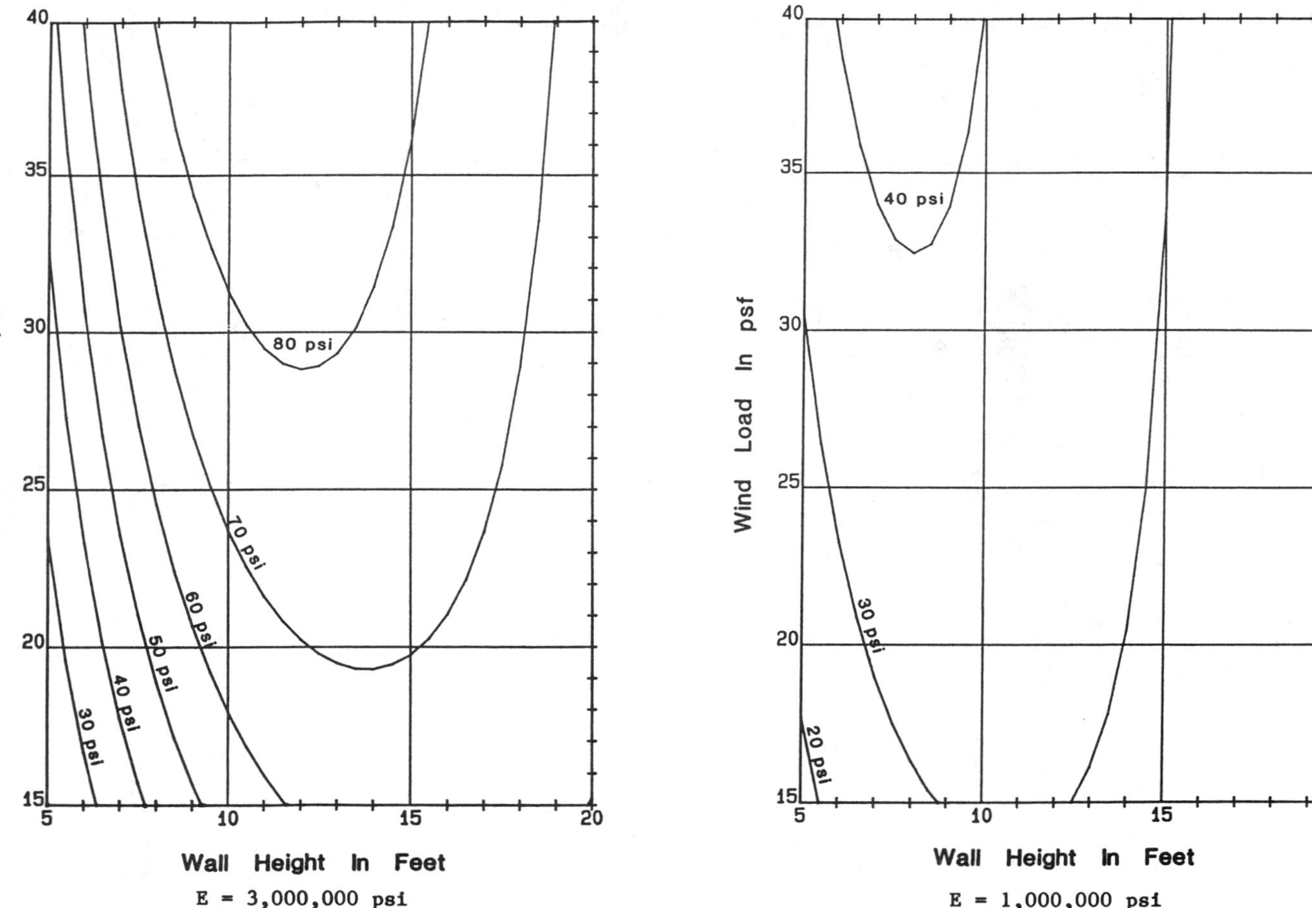

FIG. 6—*Plot of wall stress* (f) *as a function of wall height* (L) *and wind load* (q) *with studs fixed-fixed.*

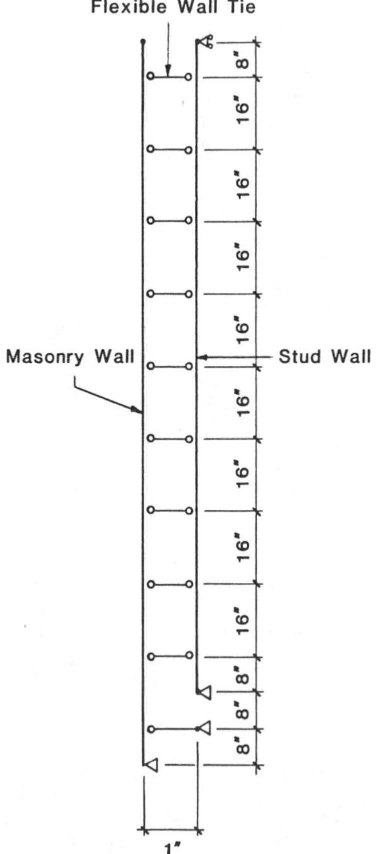

FIG. 7—*Diagram of computer model used.*

1 000 000 and 3 000 000 psi. Thus, for each wall height studied there are six combinations of material properties. Four stud heights were studied: 8 ft 0 in., 9 ft 4 in., 10 ft 8 in., and 12 ft 0 in. These were selected as being the range of most practical interest. Common stud heights in the majority of BV/MS walls range from 8 to 10 ft. The height of the brick veneer was assumed to be 16 in. taller than the studs to reflect actual installations, yielding veneer heights of 9 ft 4 in., 10 ft 8 in., 12 ft 0 in., and 13 ft 4 in. All cases studied assumed the base of the brick veneer to be hinged; the top was free. For the metal studs, the base was hinged or fixed and the top hinged. Six levels of wind loads were selected: 15, 20, 25, 30, 35, and 40 psf. In order to compare the results with the previous analysis, the reduction of tensile stress due to the weight of the wall was ignored.

Several parametric studies were undertaken to study the sensitivity of the flexural bond stress in the brick veneer. As the program modeled each mortar joint as a separate network joint, it was possible to scan all the joints and print only the maximum flexural bond stress. Figure 8 shows results with the metal studs designed with a span-deflection of 600 under full load. In addition to selecting the maximum flexural bond stress, the maximum tie load and the displacement at the top of the brick veneer are shown. Figure 9 shows the influence of fixing the base of the stud.

MAXIMUM STRESS IN MORTAR

WALL HT.	SPRING	E MASONRY	15PSF	20PSF	25PSF	30PSF	35PSF	40PSF
112.0	200.	1000000.	31.6	38.5	45.3	51.9	58.4	64.8
112.0	200.	3000000.	33.6	41.2	48.5	55.6	62.7	69.6
112.0	2000.	1000000.	46.7	56.3	64.4	71.5	78.0	83.7
112.0	2000.	3000000.	56.5	70.3	82.6	93.6	103.9	113.4
112.0	20000.	1000000.	44.5	56.3	67.0	76.6	85.3	93.3
112.0	20000.	3000000.	55.9	72.4	88.1	102.8	116.7	129.9
128.0	200.	1000000.	41.5	50.4	59.0	67.3	75.6	83.7
128.0	200.	3000000.	45.9	56.0	65.7	75.2	84.5	93.7
128.0	2000.	1000000.	56.6	67.5	76.6	84.4	91.2	97.2
128.0	2000.	3000000.	72.5	89.9	105.3	119.0	131.5	143.0
128.0	20000.	1000000.	53.5	66.8	78.5	88.8	98.0	106.1
128.0	20000.	3000000.	70.3	90.7	109.7	127.3	143.8	159.2
144.0	200.	1000000.	51.4	61.6	71.5	81.3	91.0	100.5
144.0	200.	3000000.	59.6	72.1	84.0	95.7	107.3	118.7
144.0	2000.	1000000.	65.7	77.1	86.3	93.8	100.2	105.7
144.0	2000.	3000000.	89.2	109.6	127.4	143.1	157.3	170.3
144.0	20000.	1000000.	61.5	75.5	87.5	97.8	106.9	114.8
144.0	20000.	3000000.	84.3	107.7	129.1	148.8	167.0	183.7
160.0	200.	1000000.	60.9	72.6	83.6	94.3	104.8	115.0
160.0	200.	3000000.	74.5	89.9	104.4	118.4	132.1	145.5
160.0	2000.	1000000.	73.5	84.9	94.0	101.2	107.3	112.6
160.0	2000.	3000000.	104.9	127.9	147.6	164.7	179.9	193.5
160.0	20000.	1000000.	67.9	81.9	93.4	103.0	111.3	118.3
160.0	20000.	3000000.	98.2	124.2	147.6	168.8	188.1	205.5

MAXIMUM LOAD IN TIE

WALL HT.	SPRING	E MASONRY	15PSF	20PSF	25PSF	30PSF	35PSF	40PSF
112.0	200.	1000000.	48.7	61.3	73.7	86.0	98.1	110.2
112.0	200.	3000000.	50.1	63.1	75.8	88.4	101.0	113.4
112.0	2000.	1000000.	74.8	92.1	108.0	122.7	136.6	149.8
112.0	2000.	3000000.	81.3	101.4	119.9	137.1	153.5	169.2
112.0	20000.	1000000.	92.8	116.9	139.2	160.1	179.8	198.4
112.0	20000.	3000000.	105.4	134.5	161.9	187.9	212.9	236.7
128.0	200.	1000000.	50.0	62.6	74.8	86.9	98.9	110.9
128.0	200.	3000000.	52.3	65.5	78.4	91.0	103.6	116.1
128.0	2000.	1000000.	75.7	92.5	107.7	121.8	135.0	147.6
128.0	2000.	3000000.	84.6	104.9	123.5	140.8	157.2	172.7
128.0	20000.	1000000.	106.1	129.4	155.7	179.8	202.1	222.8
128.0	20000.	3000000.	111.0	140.2	167.5	193.1	217.5	240.7
144.0	200.	1000000.	50.6	62.9	74.9	86.7	98.5	110.1
144.0	200.	3000000.	54.0	67.3	80.2	92.9	105.5	117.9
144.0	2000.	1000000.	75.2	91.2	105.7	119.0	131.5	143.4
144.0	2000.	3000000.	86.5	106.7	125.1	142.0	158.0	173.1
144.0	20000.	1000000.	123.5	155.3	183.4	208.5	231.4	252.3
144.0	20000.	3000000.	114.6	143.4	169.9	194.7	218.0	240.0
160.0	200.	1000000.	50.4	62.4	74.1	85.6	96.9	108.2
160.0	200.	3000000.	55.2	68.5	81.4	94.0	106.5	118.8
160.0	2000.	1000000.	73.8	88.9	102.4	115.0	126.8	138.1
160.0	2000.	3000000.	87.5	107.2	125.0	141.4	156.8	171.2
160.0	20000.	1000000.	145.5	179.2	208.3	233.7	256.5	277.0
160.0	20000.	3000000.	118.3	152.2	183.2	211.7	238.1	262.6

FIG. 8—*Wall results from studs designed for* L/*600.*

MAXIMUM TOP DISPLACEMENT

WALL HT.	SPRING	E MASONRY	15PSF	20PSF	25PSF	30PSF	35PSF	40PSF
112.0	200.	1000000.	0.294	0.362	0.429	0.495	0.560	0.625
112.0	200.	3000000.	0.303	0.374	0.444	0.512	0.579	0.647
112.0	2000.	1000000.	0.056	0.067	0.076	0.085	0.093	0.100
112.0	2000.	3000000.	0.061	0.073	0.085	0.095	0.105	0.114
112.0	20000.	1000000.	0.011	0.013	0.014	0.016	0.017	0.018
112.0	20000.	3000000.	0.012	0.014	0.016	0.018	0.020	0.022
128.0	200.	1000000.	0.295	0.362	0.428	0.492	0.555	0.618
128.0	200.	3000000.	0.310	0.381	0.451	0.519	0.586	0.652
128.0	2000.	1000000.	0.054	0.064	0.073	0.081	0.088	0.095
128.0	2000.	3000000.	0.060	0.072	0.083	0.094	0.103	0.112
128.0	20000.	1000000.	0.010	0.012	0.013	0.014	0.016	0.017
128.0	20000.	3000000.	0.011	0.013	0.015	0.017	0.019	0.020
144.0	200.	1000000.	0.293	0.358	0.421	0.483	0.545	0.605
144.0	200.	3000000.	0.314	0.386	0.455	0.522	0.589	0.655
144.0	2000.	1000000.	0.051	0.061	0.069	0.076	0.083	0.090
144.0	2000.	3000000.	0.059	0.071	0.082	0.091	0.100	0.109
144.0	20000.	1000000.	0.009	0.011	0.012	0.013	0.014	0.015
144.0	20000.	3000000.	0.011	0.013	0.015	0.016	0.018	0.019
160.0	200.	1000000.	0.288	0.351	0.411	0.471	0.529	0.587
160.0	200.	3000000.	0.317	0.388	0.456	0.523	0.588	0.653
160.0	2000.	1000000.	0.049	0.057	0.065	0.072	0.078	0.084
160.0	2000.	3000000.	0.058	0.069	0.080	0.089	0.097	0.105
160.0	20000.	1000000.	0.009	0.010	0.011	0.012	0.013	0.014
160.0	20000.	3000000.	0.010	0.012	0.014	0.015	0.017	0.018

FIG. 8—*Continued.*

Figure 10 compares the flexure bond stress in the brick veneer as determined by ordinary analysis and by computer analysis. As shown in this figure, the flexural bond stresses in the brick veneer, as determined by the computer analysis, are lower than the stresses as determined by the ordinary analysis. For a metal stud wall with a span-deflection ratio of 600 and a metal tie spring constant of 2000, the stresses analyzed are as shown in Table 2.

Based upon Table 2 information, it can be seen that the stresses in the brick veneer as determined by the computer analysis are significantly less than the stresses determined by the ordinary analysis and that fixing the base of the studs is very effective in reducing flexural bond stress in the brick veneer.

A study of the computer results shows that as the tie stiffness is reduced, flexural bond stress in the brick veneer is reduced. A 1:100 change in tie stiffness changes the stress roughly 1:2. For low tie stiffness (200 lb/in.) the maximum top displacement becomes excessive; for high tie stiffness (20 000 lbs/in.), the maximum load in ties becomes excessive.

When the top of the brick veneer is pinned, the beneficial effect of the spring ties is negated and will result in the brick veneer receiving more load than if the ties had been completely rigid.

A study of the computer results also shows that as the modulus of elasticity of the brickwork is reduced, flexural bond stress in the brick veneer is reduced. A 1:3 change in the modulus of elasticity changes the stress less than 1:2. For a 1 000 000 psi modulus of elasticity, the maximum flexural bond stress is under 100 psi. For a 3 000 000 psi modulus of elasticity the value is under 200 psi. The latter value falls to under 100 psi when the stud base is fixed.

Many BV/MS walls of commercial interest are not over 10 ft tall. The computer analysis reveals that these walls can withstand a wind load up to 40 psf without cracking when designed

WALL RESULTS FOR STUDS DESIGNED FOR L/600 ; BOTTOM OF STUDS FIXED

MAXIMUM STRESS IN MORTAR

WALL HT.	SPRING	E MASONRY	15PSF	20PSF	25PSF	30PSF	35PSF	40PSF
112.0	200.	1000000.	22.6	28.9	35.3	41.8	48.3	54.8
112.0	200.	3000000.	24.3	31.2	38.0	44.9	51.9	58.8
112.0	2000.	1000000.	26.5	31.0	34.7	38.1	41.2	44.1
112.0	2000.	3000000.	35.4	42.5	49.0	55.1	61.0	66.7
112.0	20000.	1000000.	30.6	37.7	43.7	48.8	53.3	57.1
112.0	20000.	3000000.	39.4	49.9	59.3	67.8	75.5	82.6
128.0	200.	1000000.	29.0	36.9	44.8	52.7	60.6	68.6
128.0	200.	3000000.	32.6	41.6	50.5	59.4	68.3	77.3
128.0	2000.	1000000.	31.8	36.2	40.0	43.5	46.6	49.5
128.0	2000.	3000000.	45.7	54.5	62.2	69.2	75.8	81.9
128.0	20000.	1000000.	36.6	44.4	50.9	56.4	61.2	65.3
128.0	20000.	3000000.	49.0	61.4	72.5	82.4	91.4	99.5
144.0	200.	1000000.	34.4	43.7	52.9	62.2	71.5	80.7
144.0	200.	3000000.	41.0	52.1	63.2	74.1	85.1	96.1
144.0	2000.	1000000.	36.5	41.2	44.9	48.0	50.7	53.2
144.0	2000.	3000000.	55.1	65.0	73.3	80.6	87.2	93.2
144.0	20000.	1000000.	42.0	50.1	56.7	62.1	66.6	70.5
144.0	20000.	3000000.	58.9	73.3	85.9	97.0	106.8	115.6
160.0	200.	1000000.	39.4	49.3	59.1	69.2	79.4	89.6
160.0	200.	3000000.	50.5	63.3	76.1	88.8	101.5	114.1
160.0	2000.	1000000.	40.6	45.2	48.6	51.4	53.6	55.6
160.0	2000.	3000000.	63.7	74.0	82.7	90.2	97.1	103.3
160.0	20000.	1000000.	46.2	54.1	60.2	65.1	69.2	72.6
160.0	20000.	3000000.	68.1	83.8	97.5	109.3	119.6	128.8

MAXIMUM LOAD IN TIE

WALL HT.	SPRING	E MASONRY	15PSF	20PSF	25PSF	30PSF	35PSF	40PSF
112.0	200.	1000000.	41.8	53.8	65.8	77.8	89.8	101.7
112.0	200.	3000000.	42.9	55.3	67.7	80.0	92.3	104.6
112.0	2000.	1000000.	59.2	72.4	84.7	96.3	107.7	118.7
112.0	2000.	3000000.	66.1	81.2	95.3	108.7	121.6	134.1
112.0	20000.	1000000.	83.9	104.2	122.4	139.1	154.6	169.0
112.0	20000.	3000000.	89.4	113.2	135.4	156.3	176.1	194.8
128.0	200.	1000000.	42.3	54.2	66.0	77.8	89.5	101.3
128.0	200.	3000000.	44.3	56.7	69.0	81.3	93.6	105.9
128.0	2000.	1000000.	60.2	73.0	84.9	96.2	107.1	117.7
128.0	2000.	3000000.	69.1	84.3	98.4	111.7	124.4	136.7
128.0	20000.	1000000.	92.5	112.4	129.9	145.6	160.1	173.6
128.0	20000.	3000000.	94.1	118.0	140.2	160.9	180.4	198.9
144.0	200.	1000000.	42.4	54.0	65.5	76.9	88.3	99.7
144.0	200.	3000000.	45.3	57.6	69.9	82.1	94.3	106.5
144.0	2000.	1000000.	60.1	72.6	84.1	95.0	105.5	115.7
144.0	2000.	3000000.	70.7	85.8	99.6	112.6	125.0	136.9
144.0	20000.	1000000.	98.0	116.6	132.8	147.3	160.5	172.8
144.0	20000.	3000000.	96.9	120.5	142.2	162.3	181.1	198.9
160.0	200.	1000000.	42.0	53.2	64.3	75.4	86.4	97.4
160.0	200.	3000000.	45.9	58.1	70.3	82.3	94.4	106.4
160.0	2000.	1000000.	59.4	71.4	82.6	93.2	103.4	113.4
160.0	2000.	3000000.	71.3	86.1	99.6	112.2	124.2	135.7
160.0	20000.	1000000.	100.7	117.9	132.6	145.7	157.8	169.1
160.0	20000.	3000000.	104.5	126.3	145.4	162.5	179.2	196.0

FIG. 9—*Wall results for studs designed for* L/1445—*bottom of studs fixed.*

MAXIMUM TOP DISPLACEMENT

WALL HT.	SPRING	E MASONRY	15PSF	20PSF	25PSF	30PSF	35PSF	40PSF
112.0	200.	1000000.	0.244	0.308	0.372	0.436	0.500	0.564
112.0	200.	3000000.	0.251	0.318	0.385	0.451	0.517	0.583
112.0	2000.	1000000.	0.044	0.052	0.059	0.065	0.071	0.077
112.0	2000.	3000000.	0.049	0.058	0.066	0.074	0.081	0.088
112.0	20000.	1000000.	0.009	0.011	0.012	0.013	0.014	0.015
112.0	20000.	3000000.	0.010	0.012	0.014	0.015	0.017	0.018
128.0	200.	1000000.	0.243	0.305	0.367	0.429	0.491	0.553
128.0	200.	3000000.	0.255	0.321	0.387	0.452	0.518	0.583
128.0	2000.	1000000.	0.043	0.050	0.057	0.063	0.068	0.074
128.0	2000.	3000000.	0.049	0.058	0.066	0.073	0.080	0.087
128.0	20000.	1000000.	0.008	0.010	0.011	0.012	0.013	0.014
128.0	20000.	3000000.	0.010	0.011	0.013	0.014	0.016	0.017
144.0	200.	1000000.	0.239	0.300	0.360	0.419	0.478	0.538
144.0	200.	3000000.	0.258	0.323	0.388	0.452	0.516	0.580
144.0	2000.	1000000.	0.041	0.048	0.054	0.060	0.066	0.071
144.0	2000.	3000000.	0.048	0.057	0.065	0.072	0.079	0.085
144.0	20000.	1000000.	0.008	0.009	0.010	0.011	0.012	0.013
144.0	20000.	3000000.	0.009	0.011	0.012	0.013	0.015	0.016
160.0	200.	1000000.	0.234	0.292	0.349	0.406	0.463	0.519
160.0	200.	3000000.	0.258	0.323	0.386	0.449	0.512	0.574
160.0	2000.	1000000.	0.039	0.046	0.052	0.057	0.063	0.068
160.0	2000.	3000000.	0.047	0.056	0.063	0.070	0.076	0.082
160.0	20000.	1000000.	0.007	0.008	0.009	0.010	0.011	0.011
160.0	20000.	3000000.	0.009	0.010	0.012	0.013	0.014	0.015

FIG. 9—*Continued.*

TABLE 2—*Stresses for a metal stud wall with a span-deflection ratio of 600 and a metal tie spring constant of 2000.*

	Ordinary Analysis, psi	Computer Analysis, psi
Studs hinged top and bottom	57.9 to 232.4	46.7 to 193.5
Studs hinged at top and fixed at bottom	44 to 145.1	26.5 to 103.3

with a metal stud span-deflection ratio of 600. At 8 ft tall, the flexure bond stresses in the brick veneer approach the "allowable" stresses (37 to 48 psi) when the wind is not over 25 psf.

As shown in Fig. 11, the stress in the brick veneer on earlier typical 8 to 10-ft-tall BV/MS walls designed with a span-deflection ratio of 360 and with corrugated wall ties with an assumed spring constant that approaches 500 lb/in. is less than the stress required to crack the wall. In contrast, by the ordinary analysis the flexure bond stress in some of these walls exceeds cracking stress levels. This computer analysis demonstrates why the vast majority of the typical BV/MS walls earlier constructed with span to deflection ratios of 360 and 600 have not cracked and are apparently successful.

The computer analysis has shown that above 10 to 12 ft the designer has a choice of either increasing the span-deflection ratio or fixing the base of the metal studs. This would be a matter of comparative economics.

COMPARISON BETWEEN ORDINARY AND COMPUTER ANALYSES
FOR STUDS HINGED ON TOP AND BOTTOM

MAXIMUM STRESS IN MORTAR

STUD HT.	SPRING	E MASONRY	15PSF	20PSF	25PSF	30PSF	35PSF	40PSF
112.0		1000000.	57.9	71.8	84.0	94.6	104.1	112.5
112.0	2000.	1000000.	46.7	56.3	64.4	71.5	78.0	83.7
112.0		3000000.	68.0	88.2	107.1	125.1	142.2	158.3
112.0	2000.	3000000.	56.5	70.3	82.6	93.6	103.9	113.4
128.0		1000000.	68.1	82.6	94.6	104.7	113.5	121.0
128.0	2000.	1000000.	56.6	67.5	76.6	84.4	91.2	97.2
128.0		3000000.	85.2	109.0	131.0	151.4	170.3	187.9
128.0	2000.	3000000.	72.5	89.9	105.3	119.0	131.5	143.0
144.0		1000000.	76.5	90.5	101.7	110.8	118.5	124.9
144.0	2000.	1000000.	65.7	77.1	86.3	93.8	100.2	105.7
144.0		3000000.	102.4	129.2	153.3	175.1	194.9	212.9
144.0	2000.	3000000.	89.2	109.6	127.4	143.1	157.3	170.3
160.0		3000000.	82.7	95.7	105.7	113.6	120.0	125.3
160.0	2000.	1000000.	73.5	84.9	94.0	101.2	107.3	112.6
160.0		3000000.	118.9	147.8	173.0	195.1	214.8	232.4
160.0	2000.	3000000.	104.9	127.9	147.6	164.7	179.9	193.5

COMPARISON BETWEEN ORDINARY AND COMPUTER ANALYSES
FOR STUDS FIXED ON BOTTOM

MAXIMUM STRESS IN MORTAR

STUD HT.	SPRING	E MASONRY	15PSF	20PSF	25PSF	30PSF	35PSF	40PSF
112.0		1000000.	44.0	51.7	57.6	62.5	66.4	69.8
112.0	2000.	1000000.	26.5	31.0	34.7	38.1	41.2	44.1
112.0		3000000.	60.6	76.0	89.7	102.0	113.0	123.0
112.0	2000.	3000000.	35.4	42.5	49.0	55.1	61.0	66.7
128.0		1000000.	47.9	54.5	59.6	63.4	66.5	69.1
128.0	2000.	1000000.	31.8	36.2	40.0	43.5	46.6	49.5
128.0		3000000.	72.4	88.9	103.1	115.2	125.9	135.2
128.0	2000.	3000000.	45.7	54.5	62.2	69.2	75.8	81.9
144.0		1000000.	49.8	55.4	59.4	62.4	64.8	66.7
144.0	2000.	1000000.	36.5	41.2	44.9	48.0	50.7	53.2
144.0		3000000.	82.7	99.3	112.9	124.3	134.0	142.3
144.0	2000.	3000000.	55.1	65.0	73.3	80.6	87.2	93.2
160.0		1000000.	50.4	54.9	58.0	60.3	62.1	63.5
160.0	2000.	1000000.	40.6	45.2	48.6	51.4	53.6	55.6
160.0		3000000.	90.9	106.9	119.4	129.6	138.0	145.1
160.0	2000.	3000000.	63.7	74.0	82.7	90.2	97.1	103.3

FIG. 10—*Comparison between ordinary analysis and computer analysis.*

MAXIMUM STRESS IN MORTAR

WALL HT.	SPRING	E MASONRY	15PSF	20PSF	25PSF	30PSF	35PSF	40PSF
112.0	500.	1000000.	46.1	55.7	64.3	72.2	79.6	86.5
112.0	500.	3000000.	50.7	62.0	72.3	81.7	90.5	98.8
112.0	1000.	1000000.	51.1	62.4	72.4	81.2	89.3	96.8
112.0	1000.	3000000.	57.7	72.0	84.9	96.6	107.4	117.4
112.0	1500.	1000000.	52.9	65.2	75.9	85.3	94.0	101.8
112.0	1500.	3000000.	60.6	76.4	90.9	104.0	116.1	127.5
128.0	500.	1000000.	59.4	71.3	81.7	91.0	99.8	108.0
128.0	500.	3000000.	68.3	83.6	97.2	109.6	121.2	132.1
128.0	1000.	1000000.	64.0	77.8	89.6	99.9	109.2	117.5
128.0	1000.	3000000.	75.9	94.8	111.7	126.9	140.9	153.8
128.0	1500.	1000000.	65.6	80.2	92.7	103.7	113.4	122.1
128.0	1500.	3000000.	78.5	99.1	117.7	134.6	150.2	164.6
144.0	500.	1000000.	72.2	86.1	98.1	108.8	118.5	127.5
144.0	500.	3000000.	87.0	106.6	123.8	139.5	154.0	167.6
144.0	1000.	1000000.	76.8	92.0	104.9	115.9	125.7	134.6
144.0	1000.	3000000.	95.2	118.4	139.1	157.7	174.7	190.4
144.0	1500.	1000000.	78.2	94.3	107.8	119.2	129.1	138.0
144.0	1500.	3000000.	98.1	123.0	145.3	165.5	183.8	200.9
160.0	500.	1000000.	83.9	98.9	111.4	122.4	132.6	142.1
160.0	500.	3000000.	106.7	130.2	150.8	169.3	186.5	202.4
160.0	1000.	1000000.	88.1	104.5	117.8	129.0	138.8	147.4
160.0	1000.	3000000.	114.8	142.2	166.2	187.5	206.8	224.4
160.0	1500.	1000000.	89.6	106.7	120.7	132.4	142.4	151.1
160.0	1500.	3000000.	117.3	146.3	172.0	194.9	215.6	234.3

MAXIMUM LOAD IN TIE

WALL HT.	SPRING	E MASONRY	15PSF	20PSF	25PSF	30PSF	35PSF	40PSF
112.0	500.	1000000.	66.6	82.2	96.7	110.5	123.9	136.8
112.0	500.	3000000.	69.5	86.2	101.8	116.6	130.9	144.8
112.0	1000.	1000000.	76.0	94.0	110.6	126.1	140.8	154.9
112.0	1000.	3000000.	80.2	100.1	118.6	135.9	152.4	168.1
112.0	1500.	1000000.	81.2	100.9	118.9	135.6	151.5	166.5
112.0	1500.	3000000.	86.3	108.2	128.6	147.7	165.8	183.1
128.0	500.	1000000.	68.6	84.3	98.7	112.4	125.5	138.2
128.0	500.	3000000.	73.1	90.4	106.5	121.7	136.3	150.4
128.0	1000.	1000000.	78.0	96.0	112.3	127.4	141.8	155.4
128.0	1000.	3000000.	84.1	104.7	123.6	141.4	158.2	174.2
128.0	1500.	1000000.	83.3	102.8	120.4	136.7	152.0	166.5
128.0	1500.	3000000.	90.4	113.0	133.8	153.3	171.6	189.1
144.0	500.	1000000.	69.5	84.8	98.9	112.2	124.9	137.2
144.0	500.	3000000.	75.6	93.3	109.7	125.1	139.8	153.9
144.0	1000.	1000000.	78.8	96.2	112.0	126.5	140.2	153.3
144.0	1000.	3000000.	86.9	107.7	126.8	144.5	161.3	177.3
144.0	1500.	1000000.	84.1	102.9	119.9	135.4	150.0	163.8
144.0	1500.	3000000.	93.4	116.1	137.0	156.4	174.7	191.9
160.0	500.	1000000.	69.5	84.3	97.8	110.6	122.7	134.5
160.0	500.	3000000.	77.4	95.2	111.5	126.8	141.5	155.5
160.0	1000.	1000000.	78.5	95.2	110.2	124.1	137.1	149.5
160.0	1000.	3000000.	88.8	109.5	128.4	145.9	162.4	178.0
160.0	1500.	1000000.	83.7	101.6	117.8	132.5	146.3	159.3
160.0	1500.	3000000.	95.4	118.0	138.6	157.6	175.4	192.3

FIG. 11—*Wall results for studs designed* L/*360—bottom of studs pinned. Spring constants from 500 to 1500 lb/in.*

MAXIMUM TOP DISPLACEMENT

WALL HT.	SPRING	E MASONRY	15PSF	20PSF	25PSF	30PSF	35PSF	40PSF
112.0	500.	1000000.	0.181	0.217	0.251	0.282	0.312	0.340
112.0	500.	3000000.	0.189	0.228	0.265	0.298	0.331	0.362
112.0	1000.	1000000.	0.111	0.133	0.153	0.171	0.188	0.205
112.0	1000.	3000000.	0.117	0.142	0.164	0.185	0.204	0.223
112.0	1500.	1000000.	0.083	0.100	0.115	0.128	0.141	0.153
112.0	1500.	3000000.	0.088	0.107	0.124	0.139	0.154	0.168
128.0	500.	1000000.	0.180	0.215	0.248	0.278	0.307	0.334
128.0	500.	3000000.	0.191	0.231	0.268	0.302	0.335	0.366
128.0	1000.	1000000.	0.109	0.130	0.149	0.167	0.183	0.198
128.0	1000.	3000000.	0.117	0.142	0.164	0.185	0.204	0.223
128.0	1500.	1000000.	0.081	0.097	0.111	0.124	0.135	0.146
128.0	1500.	3000000.	0.087	0.106	0.123	0.138	0.153	0.166
144.0	500.	1000000.	0.177	0.211	0.242	0.271	0.298	0.324
144.0	500.	3000000.	0.192	0.233	0.269	0.304	0.336	0.367
144.0	1000.	1000000.	0.106	0.126	0.144	0.160	0.176	0.190
144.0	1000.	3000000.	0.116	0.141	0.163	0.184	0.203	0.221
144.0	1500.	1000000.	0.078	0.093	0.106	0.118	0.129	0.140
144.0	1500.	3000000.	0.086	0.105	0.121	0.137	0.151	0.164
160.0	500.	1000000.	0.172	0.205	0.235	0.262	0.288	0.312
160.0	500.	3000000.	0.192	0.232	0.269	0.302	0.334	0.364
160.0	1000.	1000000.	0.102	0.121	0.138	0.154	0.168	0.181
160.0	1000.	3000000.	0.116	0.140	0.161	0.181	0.200	0.217
160.0	1500.	1000000.	0.075	0.089	0.102	0.113	0.123	0.132
160.0	1500.	3000000.	0.085	0.103	0.119	0.134	0.148	0.160

FIG. 11—*Continued.*

Discussion and Recommendations

From the foregoing analysis it has been shown that if careful attention is paid to the stud fixity, to the tie stiffness and tie force, and to the wall movement at the top soft joint, it is possible to reduce flexural bond stresses in BV/MS construction to within tolerable limits.

In retrospect, favorable inherent boundary conditions in as-built BV/MS walls may have actually reduced flexural bond stresses in the brick veneer below the stress levels predicted by the ordinary analysis. As typically built, metal studs have some partial fixity to an unknown degree. It has been shown that the lower the tie spring constant, the lower the flexural bond stress and total tie force. Possibly the earlier practice of using corrugated wall ties may have had some beneficial effect. Since the mid 1950s the need for a soft joint under the shelf angle to accommodate vertical wall movement has been recognized. Until the Clemson tests, its beneficial influence on stresses in the brick veneer was not recognized. Given these three conditions, as-built BV/MS walls have behaved better than the ordinary analysis would suggest.

Many failures of BV/MS construction known to the authors are frequently due to poor workmanship practices. A 4-in. wall is most unforgiving to construction installation practices that increase water penetration. As all 4-in. brick wythe systems will leak to some degree even under the best of commercial workmanship practice, a proper flashing and weephole system is imperative.

While proper workmanship is necessary to the success of BV/MS construction, the authors would emphasize an equal need for better engineered designs. The following design recommendations are made to enhance structural performance.

WALL RESULTS FOR STUDS DESIGNED FOR L/600 ; BOTTOM FIXED

MAXIMUM STRESS IN MORTAR

WALL HT.	SPRING	E MASONRY	15PSF	20PSF	25PSF	30PSF	35PSF	40PSF
112.0	500.	1000000.	23.7	29.3	34.8	40.4	45.8	51.3
112.0	500.	3000000.	27.7	34.5	41.1	47.7	54.2	60.7
112.0	1000.	1000000.	24.6	29.5	34.2	38.9	43.4	47.9
112.0	1000.	3000000.	31.4	38.2	44.8	51.1	57.3	63.4
112.0	1500.	1000000.	25.7	30.1	33.9	37.9	41.8	45.6
112.0	1500.	3000000.	33.6	40.7	47.2	53.4	59.4	65.3
128.0	500.	1000000.	28.8	35.3	41.7	48.2	54.6	61.0
128.0	500.	3000000.	36.5	45.2	53.6	62.0	70.3	78.6
128.0	1000.	1000000.	29.9	35.0	39.7	44.0	48.3	52.8
128.0	1000.	3000000.	41.3	49.6	57.2	64.3	71.9	79.3
128.0	1500.	1000000.	30.8	35.5	39.7	43.4	47.0	50.4
128.0	1500.	3000000.	44.0	52.6	60.2	67.2	73.9	80.3
144.0	500.	1000000.	33.4	40.0	46.4	52.7	59.5	66.5
144.0	500.	3000000.	46.2	56.2	65.9	75.3	84.7	94.4
144.0	1000.	1000000.	34.0	38.8	42.9	47.1	51.5	55.8
144.0	1000.	3000000.	50.6	60.3	69.1	77.5	85.7	93.6
144.0	1500.	1000000.	35.3	40.0	43.8	47.0	50.0	52.7
144.0	1500.	3000000.	53.4	62.8	71.0	78.7	86.2	93.2
160.0	500.	1000000.	36.4	42.5	49.1	55.7	62.3	68.7
160.0	500.	3000000.	54.8	66.3	77.3	88.2	98.9	109.4
160.0	1000.	1000000.	37.0	41.5	45.5	49.2	52.7	56.1
160.0	1000.	3000000.	59.4	69.8	78.9	87.4	95.4	103.7
160.0	1500.	1000000.	39.0	43.3	46.7	49.4	51.8	53.9
160.0	1500.	3000000.	62.0	72.2	81.0	88.8	96.1	102.9

MAXIMUM LOAD IN TIE

WALL HT.	SPRING	E MASONRY	15PSF	20PSF	25PSF	30PSF	35PSF	40PSF
112.0	500.	1000000.	46.7	58.6	70.3	81.9	93.5	105.0
112.0	500.	3000000.	49.4	62.0	74.5	86.7	99.0	111.2
112.0	1000.	1000000.	52.3	64.5	76.2	87.6	98.9	110.0
112.0	1000.	3000000.	56.9	70.4	83.3	95.8	108.1	120.3
112.0	1500.	1000000.	56.3	68.9	80.9	92.4	103.6	114.6
112.0	1500.	3000000.	62.2	76.5	90.0	102.9	115.5	127.8
128.0	500.	1000000.	47.4	59.0	70.3	81.5	92.7	103.7
128.0	500.	3000000.	51.5	64.2	76.6	88.8	100.9	112.9
128.0	1000.	1000000.	53.1	65.0	76.3	87.2	98.0	108.6
128.0	1000.	3000000.	59.6	73.1	86.0	98.4	110.6	122.5
128.0	1500.	1000000.	57.2	69.5	81.0	92.1	102.8	113.4
128.0	1500.	3000000.	65.1	79.5	92.9	105.7	118.1	130.2
144.0	500.	1000000.	47.3	58.5	69.4	80.2	90.8	101.4
144.0	500.	3000000.	52.9	65.5	77.7	89.7	101.6	113.4
144.0	1000.	1000000.	53.1	64.5	75.4	85.9	96.2	106.3
144.0	1000.	3000000.	61.2	74.6	87.2	99.4	111.2	122.8
144.0	1500.	1000000.	57.1	69.0	80.2	90.8	101.2	111.3
144.0	1500.	3000000.	66.7	81.0	94.2	106.7	118.8	130.5
160.0	500.	1000000.	46.8	57.5	68.0	78.3	88.5	98.6
160.0	500.	3000000.	53.7	66.0	78.0	89.7	101.3	112.8
160.0	1000.	1000000.	52.5	63.5	74.0	84.1	94.1	103.9
160.0	1000.	3000000.	61.9	75.0	87.3	99.1	110.6	121.8
160.0	1500.	1000000.	56.5	68.0	78.8	89.1	99.2	109.0
160.0	1500.	3000000.	67.3	81.3	94.2	106.3	118.0	129.3

FIG. 12—*Wall results for studs designed for* L/1445*—bottom of studs fixed. Spring constants from 500 to 1500 lb/in.*

MAXIMUM TOP DISPLACEMENT

WALL HT.	SPRING	E MASONRY	15PSF	20PSF	25PSF	30PSF	35PSF	40PSF
112.0	500.	1000000.	0.117	0.143	0.168	0.193	0.217	0.241
112.0	500.	3000000.	0.125	0.152	0.179	0.206	0.232	0.258
112.0	1000.	1000000.	0.071	0.085	0.097	0.109	0.121	0.133
112.0	1000.	3000000.	0.078	0.093	0.107	0.121	0.134	0.147
112.0	1500.	1000000.	0.054	0.063	0.072	0.080	0.088	0.096
112.0	1500.	3000000.	0.059	0.070	0.081	0.090	0.100	0.109
128.0	500.	1000000.	0.116	0.141	0.164	0.188	0.211	0.234
128.0	500.	3000000.	0.127	0.154	0.181	0.207	0.233	0.258
128.0	1000.	1000000.	0.070	0.082	0.094	0.106	0.117	0.128
128.0	1000.	3000000.	0.078	0.093	0.107	0.121	0.134	0.147
128.0	1500.	1000000.	0.052	0.061	0.070	0.078	0.085	0.092
128.0	1500.	3000000.	0.059	0.070	0.080	0.090	0.099	0.108
144.0	500.	1000000.	0.113	0.137	0.159	0.182	0.203	0.225
144.0	500.	3000000.	0.128	0.155	0.181	0.206	0.231	0.256
144.0	1000.	1000000.	0.068	0.080	0.091	0.102	0.113	0.123
144.0	1000.	3000000.	0.078	0.093	0.107	0.120	0.132	0.144
144.0	1500.	1000000.	0.050	0.059	0.067	0.075	0.082	0.089
144.0	1500.	3000000.	0.059	0.070	0.079	0.088	0.097	0.105
160.0	500.	1000000.	0.110	0.132	0.154	0.175	0.196	0.216
160.0	500.	3000000.	0.128	0.154	0.179	0.203	0.227	0.251
160.0	1000.	1000000.	0.065	0.077	0.088	0.098	0.108	0.118
160.0	1000.	3000000.	0.077	0.092	0.105	0.117	0.129	0.141
160.0	1500.	1000000.	0.048	0.057	0.064	0.072	0.079	0.085
160.0	1500.	3000000.	0.058	0.068	0.078	0.086	0.095	0.103

FIG. 12—*Continued.*

1. *Stud fixity*—Fixing both ends of the stud while having dramatic influence on wall stress is probably not practical because of the need for the joint at the top of the stud to accommodate vertical building movements. However, fixing the base should be given serious consideration by the designer as an alternate to increasing the span-deflection ratio of the metal studs, especially for walls over 10 ft tall.

2. *Engineered ties*—Based on the studies made, ideal spring tie constants appear to be between 200 and 2000 lbs/in. Figure 12 has been prepared for the constant values of 500, 1000, and 1500 lb/in. As shown in the figure, for metal studs fixed at the bottom to yield an effective span-deflection ratio of 1445, the stress in the brick veneer in walls up to 13 ft, 4 in. is under the cracking stress. An ideal tie to embrace the conditions studied is one having a 1000 lb/in. spring constant with a capacity of 175 lb.

3. *Top joint displacement*—The maximum top joint displacement is limited by the deformational limit of the material used to seal the joint. On the assumption that for a 3/8-in. joint a 20% shear deformation is tolerable, a maximum top displacement is 0.25 in. Such a 1/4 in. displacement will be controlled by the spring constant selected as determined from Fig. 12.

Summary

This paper has shown that excessive flexural bond stresses associated with brick-veneer/metal-stud walls as determined by ordinary strength of materials analysis are significantly lowered by taking into account the effect of the wall top soft joint, the tie flexibility, the shorter stud length, and the stud fixity. The complexities of the actual wall boundary condition have necessitated the use of a frame network analysis (computer) for solution. The results indicate that for walls under 10 ft, flexural bond stress levels in the brick veneer can be not only below cracking limits but also can be within "allowable" values as traditional built with corrugated ties. For walls between 10 and 12 ft, the authors recommend engineering the tie stiffness or fixing the stud base or both as design conditions demand. For taller walls less than three stories, fixing both ends at the stud would permit designs to 20 ft.

References

[*1*] "Performance Evaluation of Brick Veneer with Steel Stud Backup," Clemson University, Department of Civil Engineering, Clemson, SC, 23 April 1982.

[*2*] "National Testing Program, Progress Report No. 1 Small Scale Specimen Testing," Structural Clay Products Research Foundation, Geneva, IL, October 1964.

[*3*] "Building Code Requirements for Engineered Brick Masonry," Structural Clay Products Institute, McLean, VA, August 1969.

Richard C. Arnold,[1] *L. John Dondanville,*[1] *Norbert V. Krogstad,*[1] *and Clare B. Monk, Jr.*[1]

Analysis and Test of a Torsional Sensitive C-Shaped Prefabricated Brick Spandrel Panel[2]

REFERENCE: Arnold, R. C., Dondanville, L. J., Krogstad, N. V., and Monk, C. B., Jr., **"Analysis and Test of a Torsional Sensitive C-Shaped Prefabricated Brick Spandrel Panel,"** *Masonry: Materials, Design, Construction, and Maintenance, ASTM STP 992,* H. A. Harris, Ed., American Society for Testing and Materials, Philadelphia, 1988, pp. 118–144.

ABSTRACT: Recognizing the inherent torsional flexibility of an open C-section, the authors analyzed in detail the dead and live load behavior of the section under review. The section consisted of a sloping sill, a vertical spandrel, and a horizontal soffit configuration fabricated as a continuous bent plate. The three planes when joined together form an open C-section. To explore the stress patterns in this shaped plate, the authors modeled the analysis for a hand solution. This was then compared to a more rigorous solution using a finite-element computer approach.

The hand approach determined the principal axes of inertia, the shear centers, and the principal states of stress by Mohr's circle. Comparison is shown with graphic plots of principal stresses from the finite-element analysis. The latter technique makes a detailed investigation possible, including plots of both principal tensile and shearing stresses. The modeling of the complex metal embedments at each reaction location was also possible by the finite-element solution.

Full-scale testing of the panel was made utilizing dead weight from brick loading platforms that exerted gravity and wind loading on the spandrel panel through a series of compound lever frames. Contrary to results of brittle torsional failure expected from the analytical work, the experimental behavior demonstrated a capacity to a load factor of five without cracking or collapse.

KEY WORDS: prefabricated brick panels, reinforced structural masonry, torsional sensitive, C-shaped cross section, full-scale testing, combined wind and gravity forces, unique compound lever loading system

Prefabricated brick spandrel panels, about 27 feet long, had been fabricated to replace sets of three shorter panels of equal total length for a facade replacement program. A single, longer panel had the advantages of fewer total panels, fewer attachment anchors, and fewer vertical joints. A design goal was to replace the facade from the exterior as fast as possible with minimum disturbance to the on-going activities of the owner's business. While spandrel panels of this length are not unique, it is believed that brick panels of this length have been rare. During the course of erection one panel sustained failure. The mode of failure suggested torsional twist near one end. When erected, the panels were lifted at their center of gravity as simple beams near their ends without the benefit of lateral wind connections at their third points when installed. Such lateral constraints would restrain any twisting moments.

The panel geometry is described in Fig. 1 together with the cross-sectional properties. Inher-

[1]Engineer II, senior engineer, engineer II, and senior consultant, respectively, Wiss, Janney, Elstner Associates, Inc., Northbrook, IL 60062.

[2]Metric conversion for units in this paper: in. = 2.54 cm; 1 ft = 30.48 cm; 1 psi = 6894.0 N/m^2; 1 pcf = 157.0 N/m^3; 1 lb = 0.4536 kg.

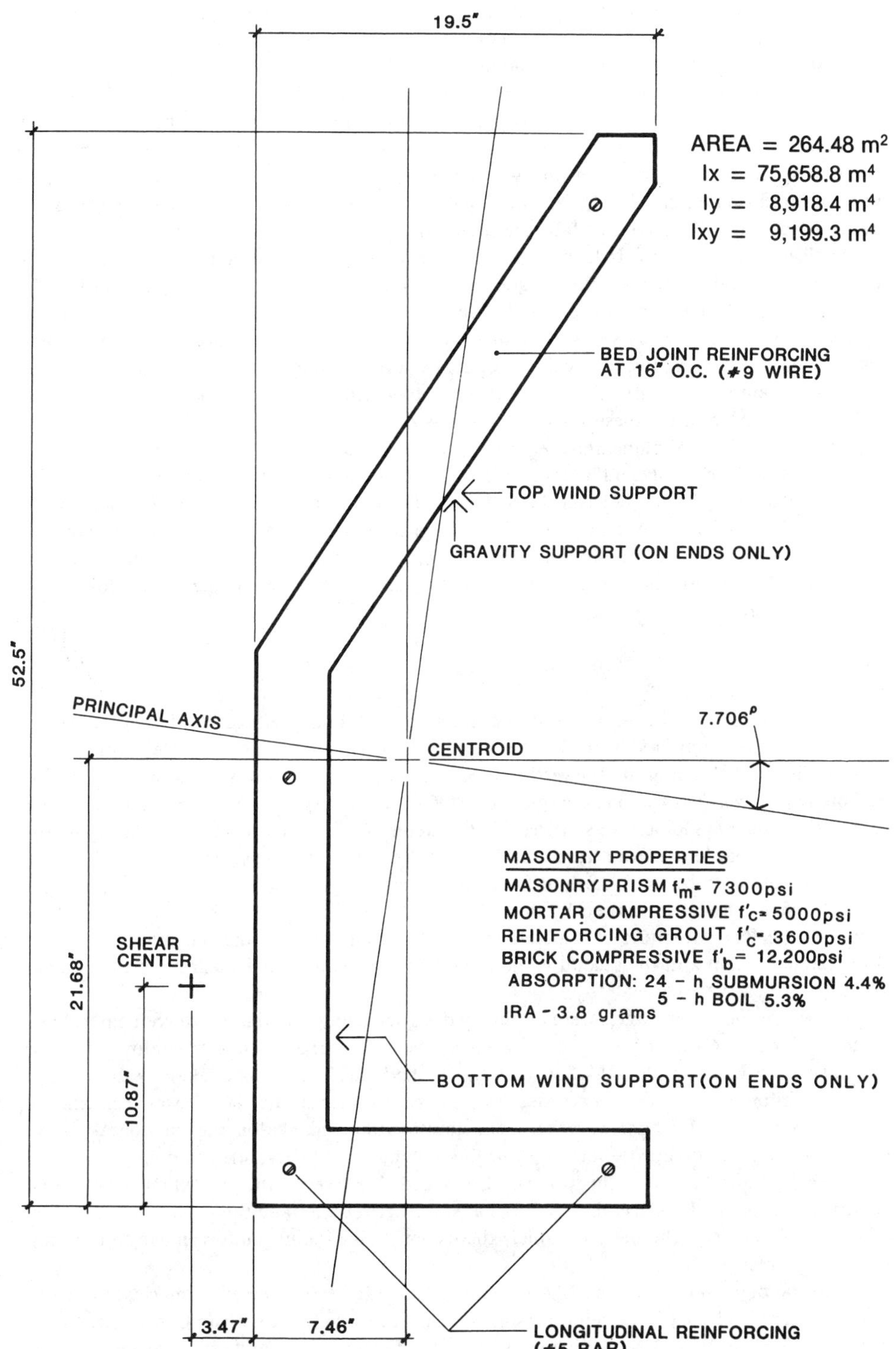

FIG. 1—*Spandrel cross section showing dimensions, steel, and structural properties.*

ently this configuration as an open C-section suggests torsional sensitivity, whereas its relatively low span-depth ratio suggests flexural strength. It is well recognized that prefabricated panels are frequently subjected to loading and boundary conditions more severe during handling when fabricated, transported, and erected than when finally installed. Such loading conditions are difficult to define rationally. The failed panel's condition was complicated by air voids that surrounded the longitudinal steel due to fabrication procedures. At the failed end a significant void was found in the outside, bottom steel rod. In retrospect, it is still unknown what precise mechanism caused failure. Nevertheless, the failure incident caused a full-scale investigation to assure adequacy of the panels as fabricated and installed.

A typical panel as fabricated, including representative steel grout void conditions, was selected for test. Generally the voids in question seldom exceeded a brick thickness in length, although the failed panel did have a void of several feet. Theoretically, the steel bond contact area that existed despite voids was capable of sustaining the flexural loading. Prior to the full-scale test described here, five full-scale tests were performed using lead weights to load the panel through its center of gravity. Generally, these tests withstood a vertical load factor of three without failure. However, these vertical load tests did not simulate to a sufficient degree the lateral loading torsional moment as required by code. It was decided to carry one panel to failure, loading both vertically and laterally by simulated gravity and wind forces. Not only was the lateral wind loading distributed over the surface of the panel but also the gravity loading as well. Such body loading is in sharp contrast to vertical gravity loading concentrated at the center of gravity using lead weights only in the prior testing. The latter, while relatively simpler to perform, failed to apply the proper ratio of torsional moment to flexural moment as the loading to failure was increased.

Code Review

The wind loading was based on the 1981 edition of the Building Official Conference of America (BOCA) code as required by the local authority. As will be shown, appropriate normal forces were applied to the sloping sill, the vertical fascia, and the horizontal soffit as interpreted by the authors and as verified by consultation with BOCA. With respect to torsion about the shear center, the torsional moment was increased by a factor of 1.81 when surface loadings were distributed as compared to concentrated test loads at the center of gravity. For an open C-section this difference could be significant, at least theoretically.

Two torsional issues are raised by the "Recommended Practice for Engineered Brick Masonry" of the Brick Institute of America (BIA). The first concerns the recommendation for shear reinforcement and the second is the calculation of the torsional shearing stresses themselves.

As shown in Fig. 1, the panels were originally designed using bed joint reinforcement as horizontal stirrups. BIA under Section 4.8.5.2 expressively omits the use of wire reinforcement for this purpose. Further, under BIA Section 4.8.5.1(5) *all* the shear is to be taken by the stirrups when unreinforced criteria are exceeded. For the case under study, the allowable maximum unreinforced shear of 50 psi was exceeded, necessitating reliance on the horizontal 5/8 in. rounds for possible stirrup steel action providing the maximum shear stress did not exceed 120 psi. It will be shown later that this value is also exceeded. However, the top rod failed the limits of stirrup spacing by BIA Section 4.8.5.5. Thus, BIA criteria limited the participation of steel inherently available in the design, namely the use of the bed joint reinforcement and the top steel bar for resisting shear.

It must be pointed out that the BIA recommended practice is directed primarily at flexural shear, not torsional shear. For helpful insight one must turn to the American Concrete Institute code: ACI 318, Section 11.6, dealing with combined flexural shear and torsional shear. First, ACI 11.6.1.1 limits the effective flange overhang to three times the flange thickness. Unlike

BIA, ACI 11.6.6 does allow the torsional resistance of the masonry material to be utilized (see ACI equation 11-22). However, calculation showed that, when the torsional resistance of the masonry was added to that of the steel, the combined strength is less than that of the applied factored torsional moment. Further, by ACI 11.6.8, stirrup spacing for torsion is very restrictive, not to exceed 12 in. Also, in the commentary on ACI 11.6.7.3, it is pointed out that "both longitudinal and closed transverse reinforcement are required to resist diagonal tension stresses due to torsion, and if one or the other types of reinforcement are not provided, the other will be relatively ineffective. The stirrups must be closed, since inclined cracking due to torsion may appear on all faces of a member." Thus the ACI criteria, which utilizes ultimate strength design, showed excessive shearing stresses and restricted steel participation to spacings less than 12 in. Further, the closed stirrup requirement would support the use of bed joint reinforcement but not centerline horizontal steel to avoid face cracking as cautioned by the commentary.

By both BIA and ACI (American Concrete Institute) design codes the torsional shear capacity was underdesigned. This occasioned the need for a more detailed study.

Strength of Materials Analysis

Initially, the design was reviewed by the ordinary strength of material analysis. The shear center was determined by using the basic definition as the locus at which loading must be applied to cause the torsional twist to vanish. Figure 2 shows the neighborhood of the plane of failure when the panel was lifted as a simple beam over a 24 ft, 2 in. span subjected to its own dead weight. An analysis was made at the failed section by ordinary strength of materials.

Several features of the analysis are pointed out. The cross section is assumed unreinforced. This is a very conservative assumption based on the uncertainties of the steel grouting voids. The calculation of the polar moment of inertia is based on the sum of the values for each thin rectangular element as known from basic mechanics. No reduction is made for overhanging flanges as required by ACI. The calculation of torsional shear (88 psi) and flexural tension (37 psi) is made by classic strength of materials formulae. Flexural shear less than 10 psi is ignored. As the calculated stresses are within 10% of the BIA allowables, it was believed that these considerations were not the cause of the field failure during erection lifting. Using Mohr's circle of stress, principal tensile stresses of 108 psi were determined. While this value approached the known modulus of rupture value (110 psi) for the bond of brick prisms, such a value occurs on a plane significantly inclined to the orientation of the bed and head joints. Given the interlocked arrangement of mortar and brick, critical inclined principal tensile failure was judged an unlikely event. It appeared that simple dead load lifting did not cause failure even if unreinforced. It should be noted, however, that this analysis did not take into account any lateral loading due to handling forces from fabrication, transportation, and erection. Such events are unknown in magnitude and position. Therefore a rational design investigation is not possible. However, based on this analysis, gravity alone is an unlikely cause of failure.

To use the ordinary strength of material theory for the as-built wind and gravity analysis had some difficulty. The original boundary conditions were so detailed as to cause simple beam action for any vertical loading but to cause continuous beam action over four supports for any horizontal loading. At the third point reactions, vertical slots allowed only vertical motion, but not horizontal. At the ends, both movements were prevented. Originally, top and bottom horizontal restraints were designed for all four reactions: two end reactions and two wind reactions. However, the location of the bottom inward projecting soffit with shallow clearance beneath the primary steel spandrel beam prevented field installation of the two bottom horizontal restraints at the wind reactions. Thus the brick spandrel acted as a bent plate supported horizontally along the top by four supports positioned at the third points and along the bottom by two supports at the ends along the bottom edge. Further, the design was complicated by the manner in

FIG. 2—*Closeup of failed panel at plane of failure.*

which the steel support mechanisms were embedded into the brickwork to avoid point concentration at all reactions.

To handle these as-built support conditions, a finite-element analysis was used.

Finite-Element Analysis

In an attempt to better understand the torsional behavior of the panel, a finite element method (FEM) analysis of a typical panel was performed. Both handling and "as-built" conditions were considered. Loadings consisted of gravity and wind forces. Analysis of the model was performed using a proprietary FEM processor based on SAP IV, the Structural Analysis Program for Static and Dynamic Response of Linear Systems developed at the University of California, Berkeley. The problem was executed on a Prime super minicomputer system.

Details of Model

A model of one half of a symmetric panel was constructed using a thin plate and shell quadrilateral element to represent the masonry. A total of 1200 elements were used to model the soffit, fascia, and sill components of the "C"-shaped panel. Three-dimensional beam elements were incorporated in the model to represent the steel plate and channel elements embedded in the masonry at support locations. Beam elements were also used to model portions of the supporting framework. Material properties used for the masonry plate elements include: E = 3 000 000 psi, Poisson's ratio = 0.25, and a unit weight of 132 pcf. For the steel members, a modulus of 29 000 000 psi, Poisson's ratio = 0.30, and a unit weight of 490 pcf were used.

Support conditions for the analysis included two distinct cases. An initial analysis with the panel supported at lifting points near each end of the panel was performed to determine handling stresses. A second analysis with the panel supported by beam elements representing members of the building's structural framework was performed to determine response after erection. In this "as-built" model, a support providing both vertical and horizontal reactions was provided near the end of the panel ("gravity support"). A second support providing only a horizontal reaction represents the panels "wind support" near the one-third point of the panel. In the actual construction, this connection uses a slotted hole to eliminate vertical load transfer.

Loading of the panel for both the handling and erected conditions included gravity loads. In addition, analysis of the "as-built" model included a uniformly distributed wind load as required by the BOCA Basic Building Code/1981, which governed the initial design of this building. After discussion with BOCA officials concerning interpretation of the requirements, the loadings shown in Fig. 3 were utilized.

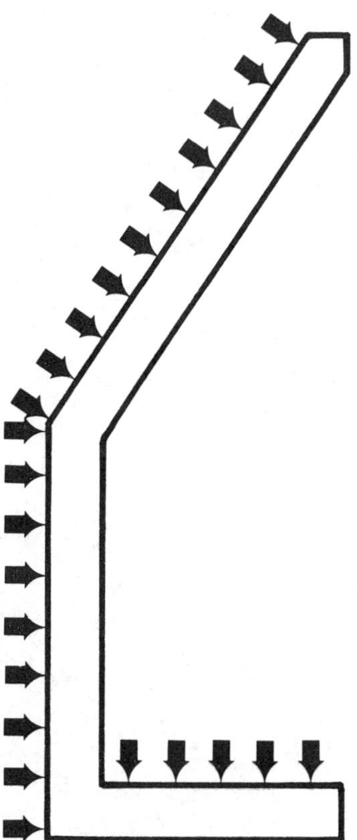

FIG. 3—*Orientation of positive wind loading for analysis and testing.*

Results

Results of the analysis described above were reviewed to determine the relative response of the structure to the dead load and dead load plus/minus wind load. Contour plots of the maximum principal stress on the faces of the brick panels are attached as Figs. 5A to 7B (stresses in these plots are in psi). Each group of plots shows either an inside or outside face of the panel. Stresses printed in large letters are the highest and lowest stresses on the plot and are located near the H (high) and L (low) printed on the plot. Stresses in the longitudinal direction of the panel for the dead load only case were also plotted as Figs. 4A and 4B to compare with hand calculations described earlier.

In all of these plots, high stresses in the immediate location of the support were considered as locations of critical stress, realizing that the magnitudes of stress directly at the point of support may be artificially high. No attempt was made to refine the model in this area or to determine the convergence of the solution. Stresses under consideration were primarily at midspan or a number of elements away from the support locations.

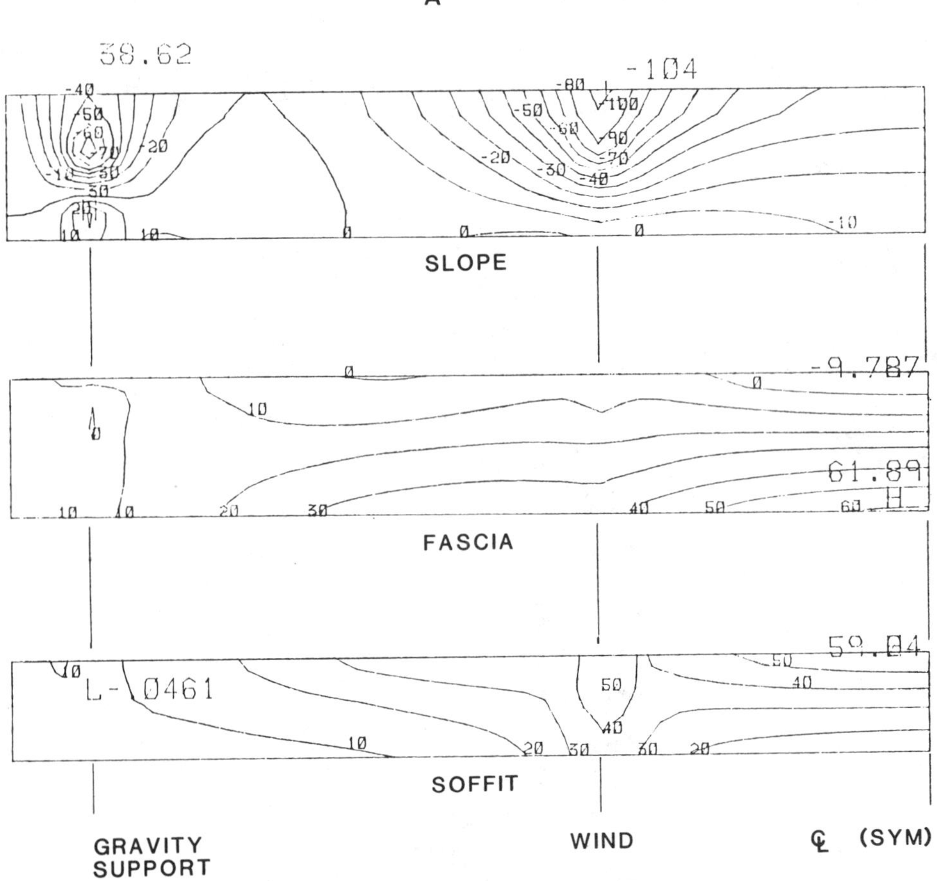

FIG. 4—(A) *Longitudinal stress distribution on inside face of panel due to dead load only.* (B) *Longitudinal stress distribution on outside face of panel due to dead load only.*

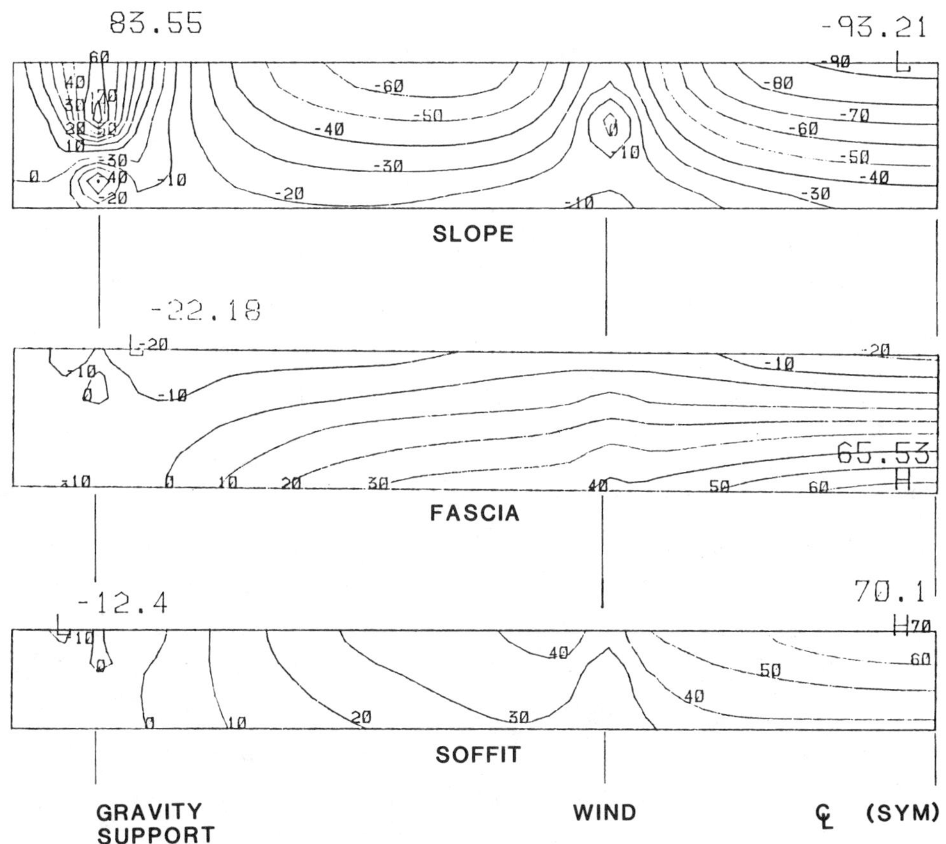

Discussion

Correlation of the model with hand calculations appears relatively good. Longitudinal stresses shown in Figs. 4A and 4B indicate tension at midspan near the soffit/fascia intersection of 60 (inside face) to 70 psi (outside face). The "A" figures represent stresses on the inside face of the panels whereas the "B" figures represent stresses on the outside face of the panels. At the top of the sloping panel near midspan, compression of approximately 93 psi is indicated on the outside face. These compare with hand calculations of approximately 73 psi at the extreme edges of the section when considering simple bending of the panel. Torsional effects can be noted in the reduction of tension at the free edge of the soffit relative to the fascia/soffit intersection and in the stress gradient from the inside to the outside face at the top of the sloping panel at midspan.

Principal stresses shown on Figs. 5A and 5B for dead load only should be compared to Figs. 6A and 6B for dead load plus wind load and compared to Figs. 7A and 7B for dead load minus wind load. The differences between the longitudinal and principal stresses are indicative of the effect of the torsion effect. Significant differences exist in stress levels between the inside face and the outside face: each must be examined to determine the location of maximum stress.

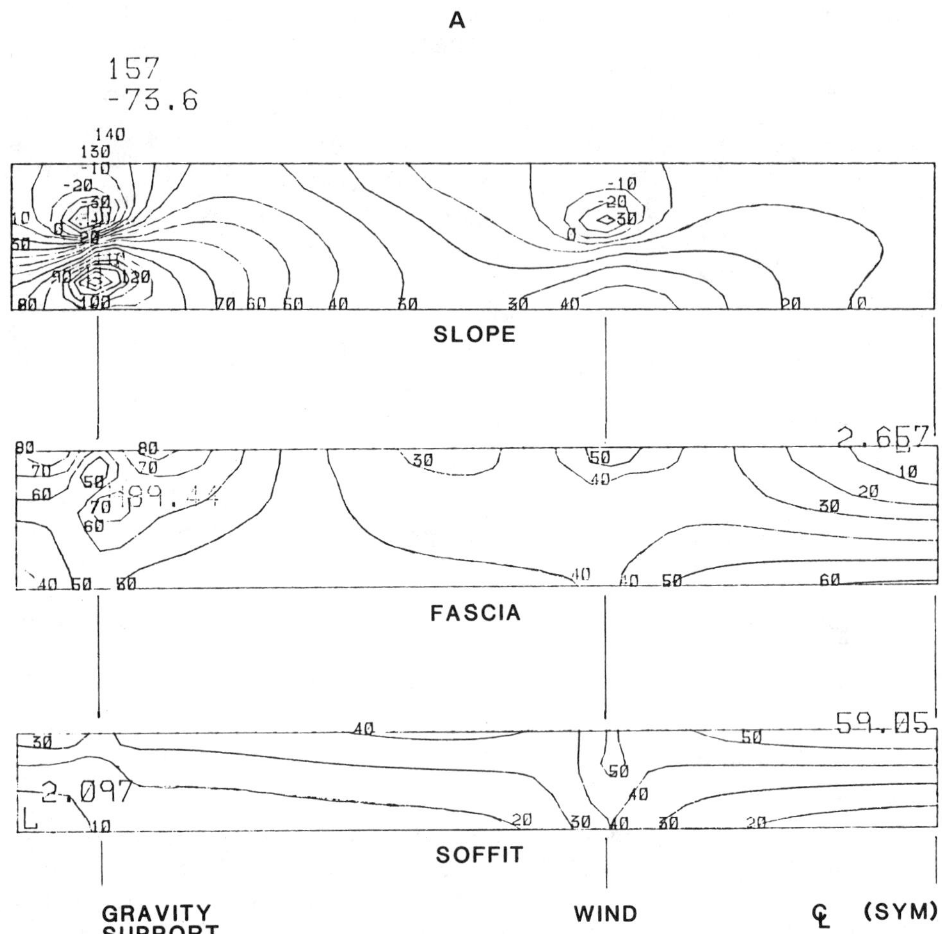

FIG. 5—(A) *Principal stress distribution on inside face of panel due to dead load only.* (B) *Principal stress distribution on outside face of panel due to dead load only.*

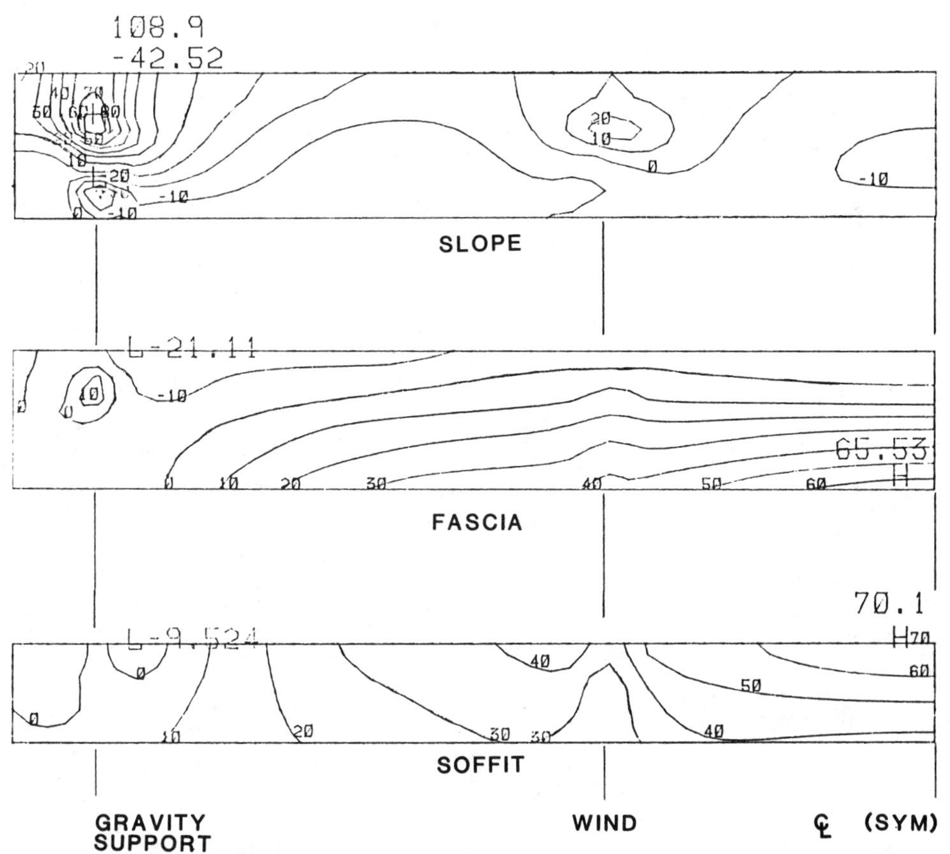
B
108.9
-42.52
SLOPE
L-21.11
65.53
H
FASCIA
70.1
H
L-9.524
SOFFIT
GRAVITY
SUPPORT
WIND
℄ (SYM)

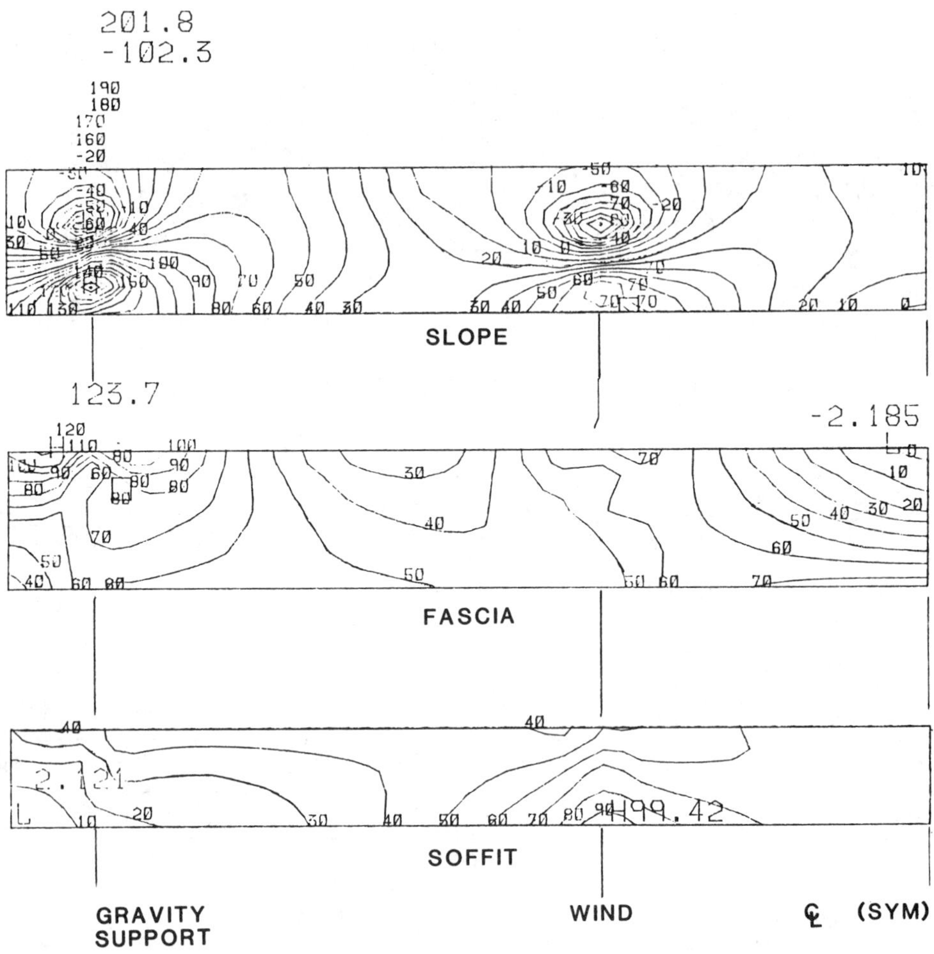

FIG. 6—*Principal stress distribution on inside face of panel due to dead load plus wind load.* (B) *Principal stress distribution on outside face of panel due to dead load plus wind load.*

B

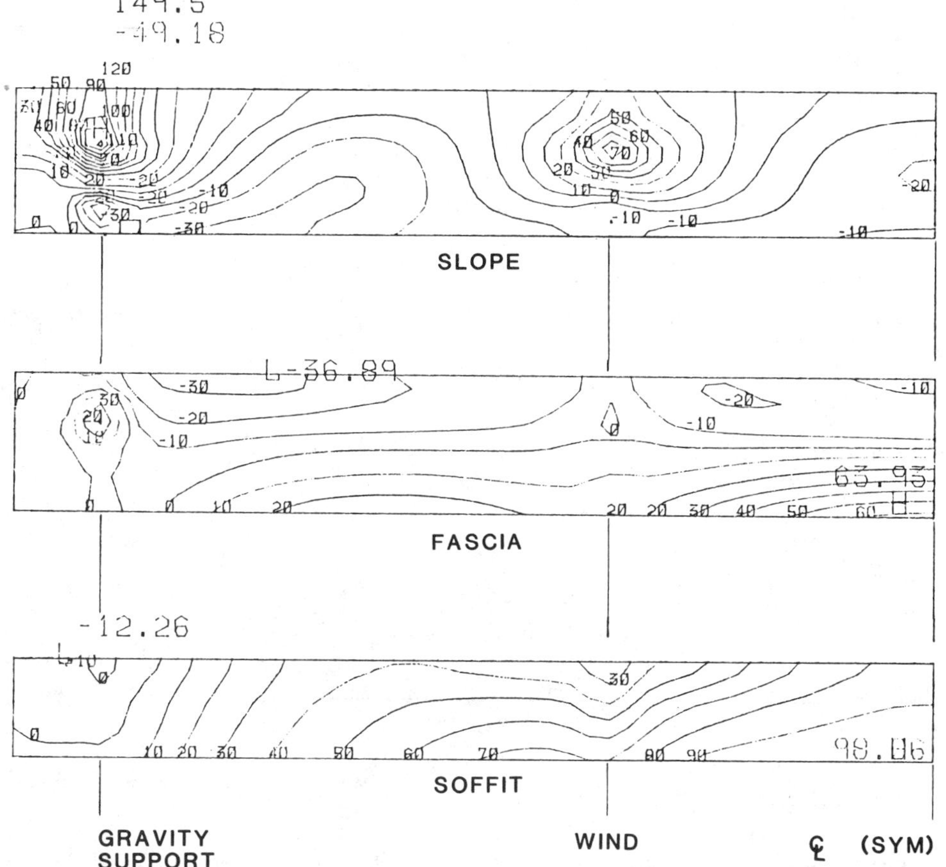
149.5
-49.18
SLOPE
-36.89
63.93
FASCIA
-12.26
98.06
SOFFIT
GRAVITY SUPPORT
WIND
℄ (SYM)

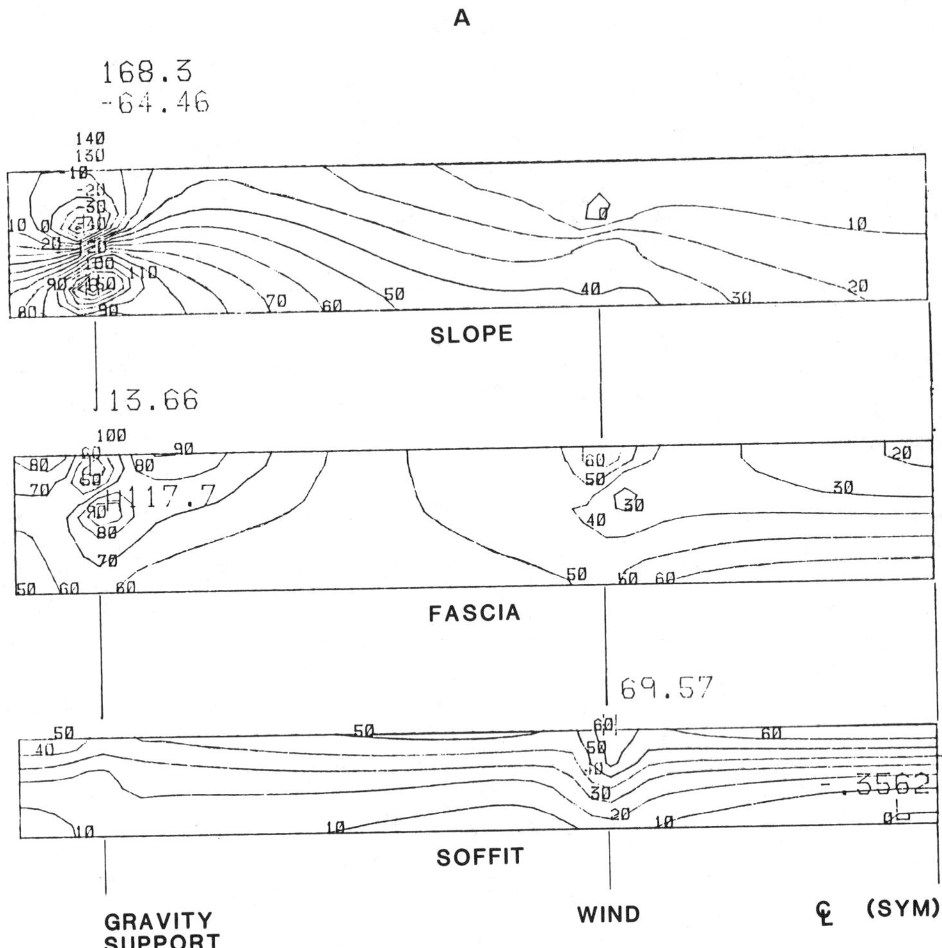

FIG. 7—*Principal stress distribution on inside face of panel due to dead load minus wind load.* (B) *Principal stress distribution on outside of panel due to dead load minus wind load.*

B

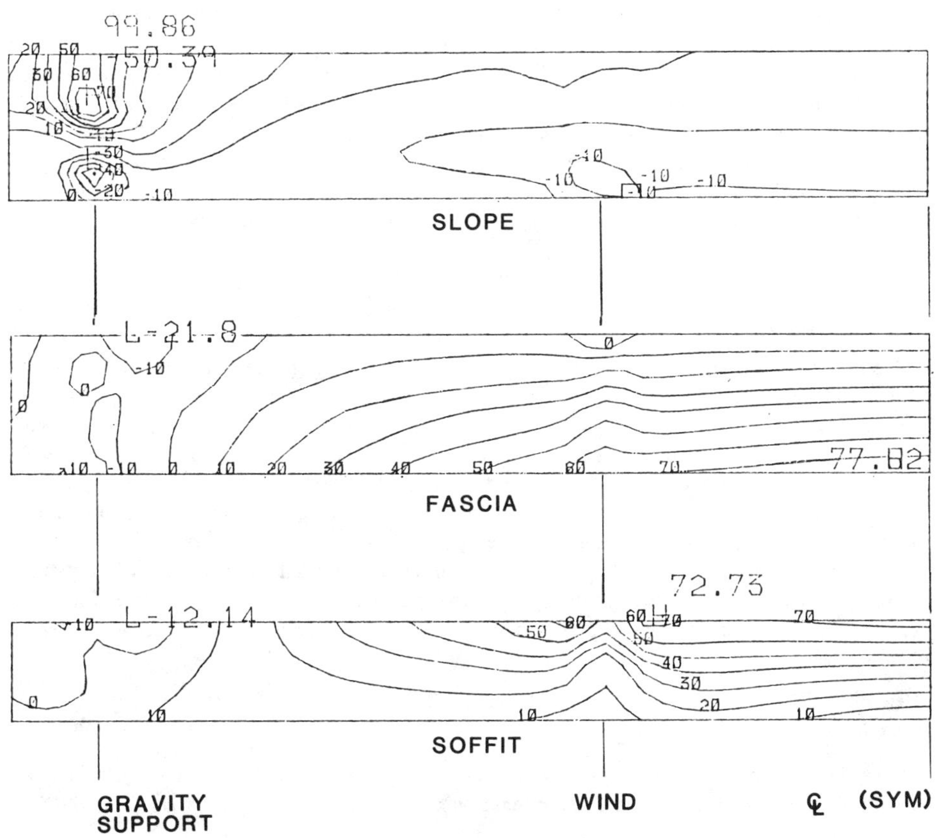
99.86
SLOPE
L-21.8
77.82
FASCIA
72.73
L-12.14
SOFFIT
GRAVITY SUPPORT
WIND
℄ (SYM)

Principal stresses are substantially increased when wind torsion effect is included. In the neighborhood of the gravity support, the maximum principal stresses exceed the BIA limit of 120 psi. It is this observation plus the concerns discussed in the code review discussion earlier that led to the decision to perform a full-scale test.

Full-Scale Test Program

Loading Mechanism

The loading was applied through vertical truss bents located at twelve stations along the total length of 26 ft, 10 in. Each bent applied load to a ten brick course region. The bents were braced in pairs to act as a coupled stable structure forming twelve frames, six inside and six outside the spandrel. See Figs. 8 and 9 for the loading plan layout and the loading bent elevations. The wind loading was uniformly applied through three sets of compound levers acting normal to the sill, fascia, and soffit surfaces. The gravity loading was similarly applied acting vertically to the three surfaces. The loadings were distributed through a surface grillage at all locations except the vertical gravity force on the fascia, which was necessarily an edge load along its top boundary. Grillage pads in contact with the brick surface were grouted with gypsum to minimize stress concentrations. The frame fabrication was made out of wood with the levers made out of wood and steel links connecting to wooden grillage. The wood loading platforms and the compound levers were connected by steel wires. Figures 10 through 14 are illustrative of the loading mechanism.

Loading Procedure

The loading mechanism was originally designed to be done in the field. Events cause the program to be done in a laboratory where better instrumentation was possible. The gravity loading was devised so that the weight of two 5-lb bricks on each of the 72 loading platforms constituted a loading increment on the spandrel panel surface. The lever system would necessarily follow the specimen as it deflected under load, avoiding any problems of the specimen creeping away from the loading mechanism. The system was designed to apply a load factor of six. At a load factor of three, the levers reached their allowable travel; the test was stopped and the travel clearances increased. The test was continued until a load factor of five was achieved, at which lever clearances were again exceeded. At the load factor of three and five, the loads were held for 24 h to measure creep and then were fully unloaded to measure recovery after an additional 24 h.

Instrumentation

Twenty-eight movement readings were taken on the test specimen itself: two vertical and two horizontal at each of the following seven locations:

1. Two at the end reaction supports.
2. Two at the wind reaction supports.
3. Two between the reaction and wind supports.
4. One at the panel center.

Six further additional movement readings were taken at the end reactions and wind supports as follows:

1. Two vertical at the reaction supports.
2. Two horizontal at the reaction supports.
3. Two horizontal at the wind supports.

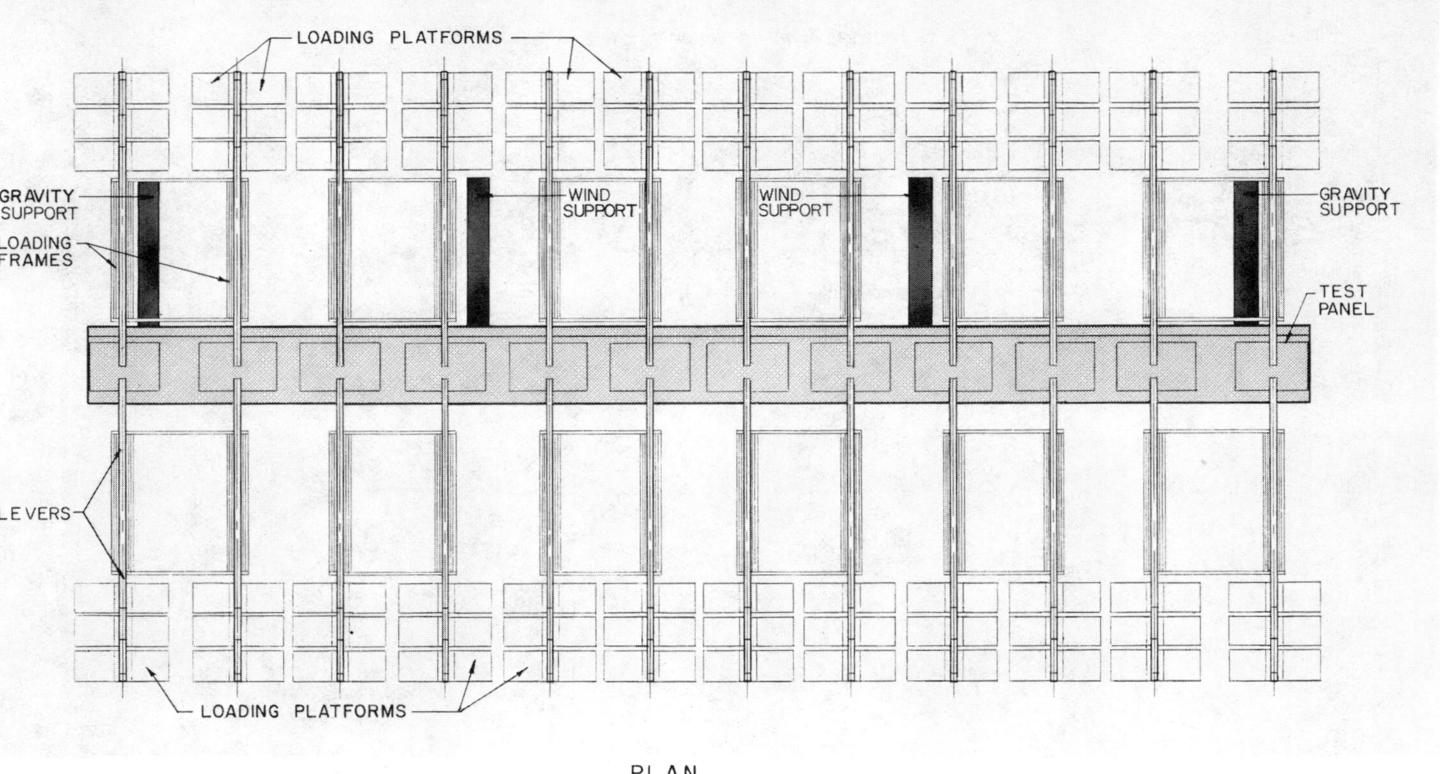

FIG. 8—*Loading plan layout for the test program.*

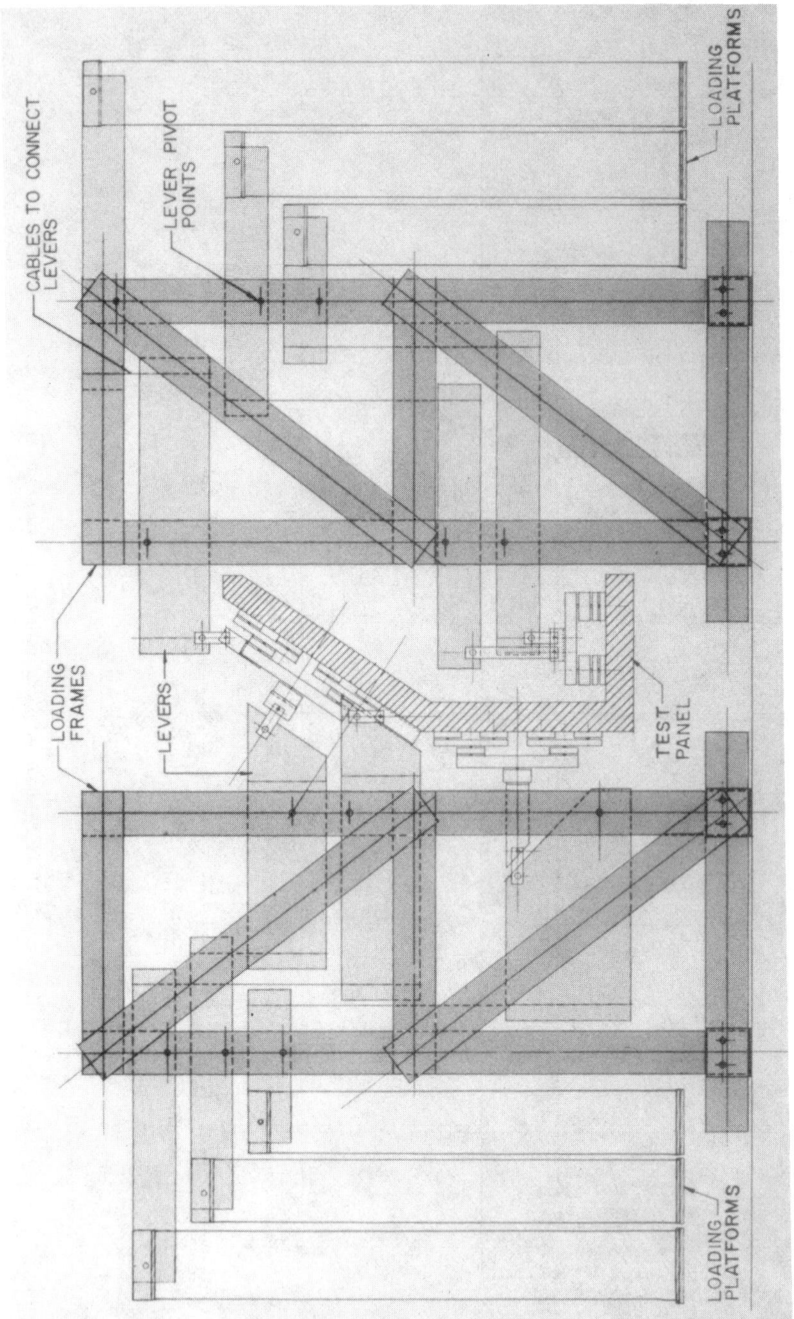

FIG. 9—*Elevation of loading bents.*

FIG. 10—*Overall view of panel with loading bents in position.*

These 34 movements were monitored by LVDTs (linear variable differential transducers) for automatic electronic collection and data processing. From these readings midspan deflection and rotations could be determined relative to the restraints at the supports. It is to be noted again that only a top horizontal restraint existed at the wind support.

In addition to these movement monitors, the loads at three bents were checked by electrical strain gauges. Special steel connecting links were machined to mount strain gauges on the links at the six delivered load locations: the wind and gravity loads on the sill, fascia, and soffit surfaces. This required 36 strain-gauge measuring channels. These gauge reading loads were compared to the primary loads delivered from the brick gravity platforms. During the test an automatic computation was made to compare the measured load to the expected load from the brick gravity platforms. This was read out as an actual load factor. The planned values ranged from 1.0 to 5.0 in 0.25 increments during the second test. Thus 20 readings were taken at each of 70 channels (34 movements plus 36 strain values) for a total of 1400. For the first test, which was stopped at a load factor of 3.0, 840 readings were taken. Without automatic data acquisition, such small incremental values would not be practical to do.

Test Results and Analysis

Figure 15 shows the center midspan vertical deflection as a function of the load factor and compares results from Test No. 1, Test No. 2, and the computer model. Figure 16 shows similar data for the relative midspan torsional rotation. Also shown are the 24-h creep deflection or rotations under sustained load and the 24-h deflection or rotational recovery upon load removal. Figures 17 and 18 show the individual vertical deflection readings under the soffit for each test. It is apparent that the outside edge deflected less than the inside, indicating substantial torsional twist.

FIG. 11—*View of typical end reaction frame.*

FIG. 12—*Loading system as seen from the left end.*

FIG. 13—*Closeup of grillage and loading fixtures on sloping sill.*

FIG. 14—*Closeup of bricks being applied to the loading platforms.*

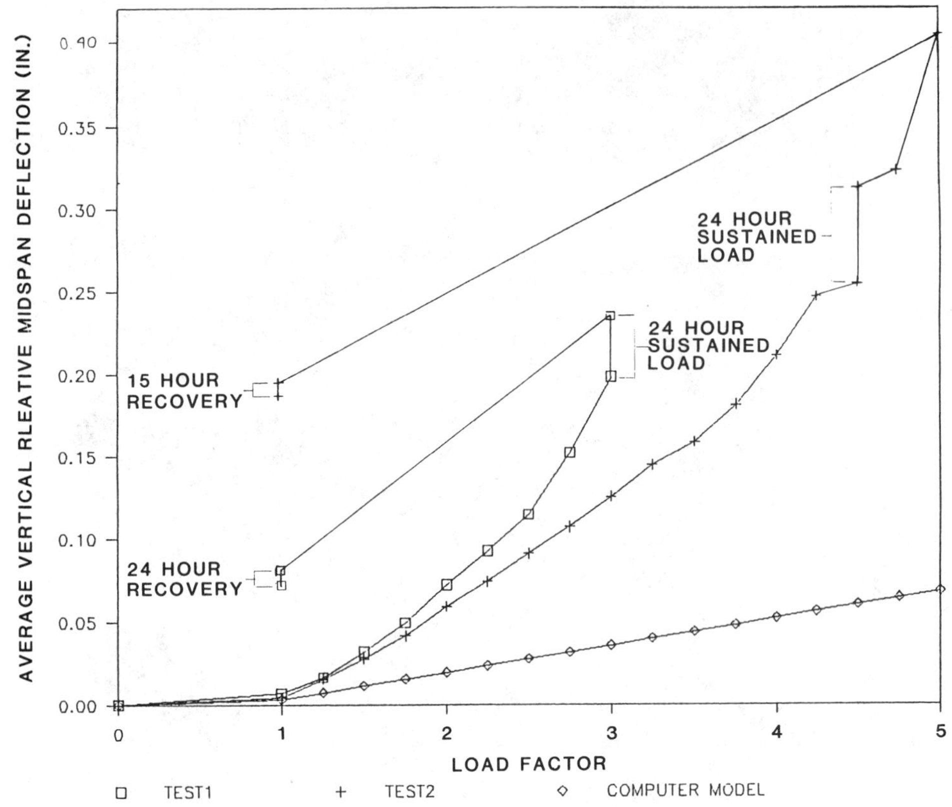

FIG. 15—*Plot of panel midspan deflection versus load factors.*

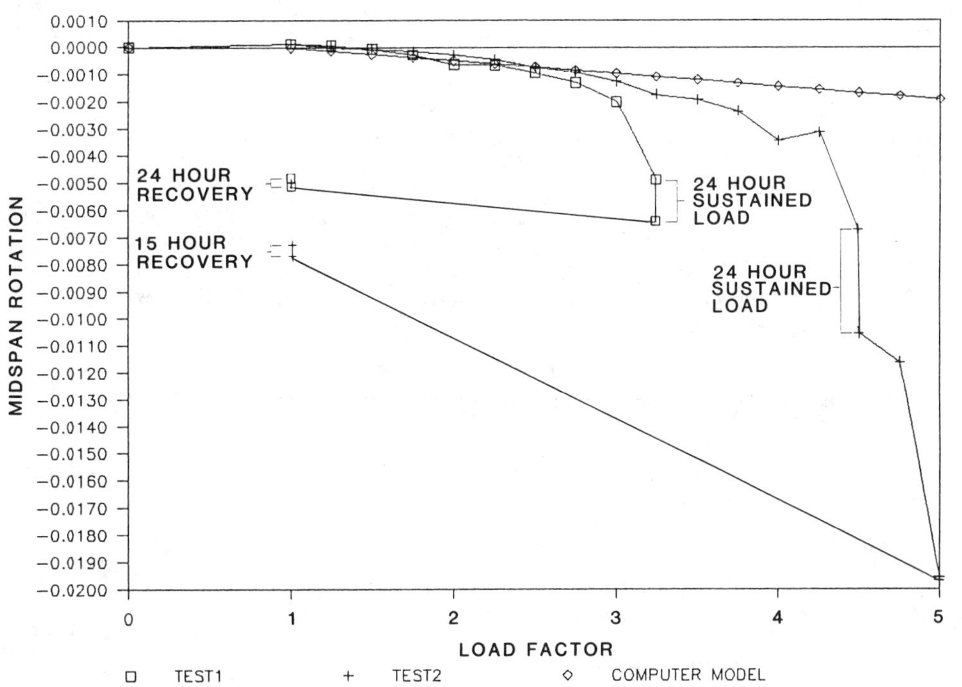

FIG. 16—*Plot of panel midspan rotation versus load factors.*

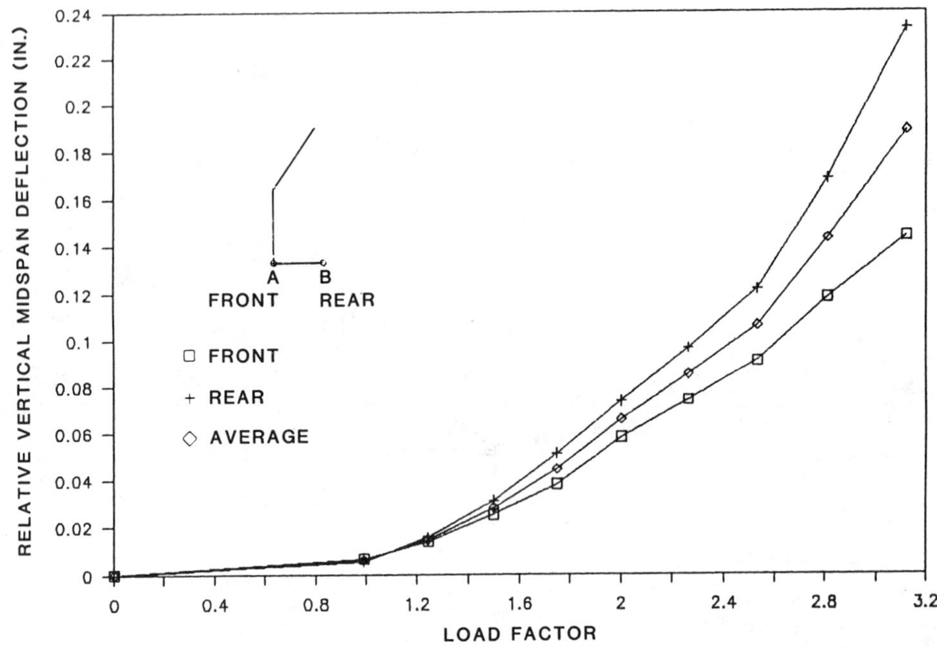

FIG. 17—*Plot of Test 1 vertical deflection versus load factors: front, rear, average.*

RELATIVE VERTICAL MIDSPAN DEFLECTION (IN.)

0.40
0.35
0.30
0.25
0.20
0.15
0.10
0.05
0.00

A
B
FRONT
REAR

FRONT
REAR
AVERAGE

24 HOUR
SUSTAINED
LOAD

0
1
2
3
4
5
LOAD FACTOR

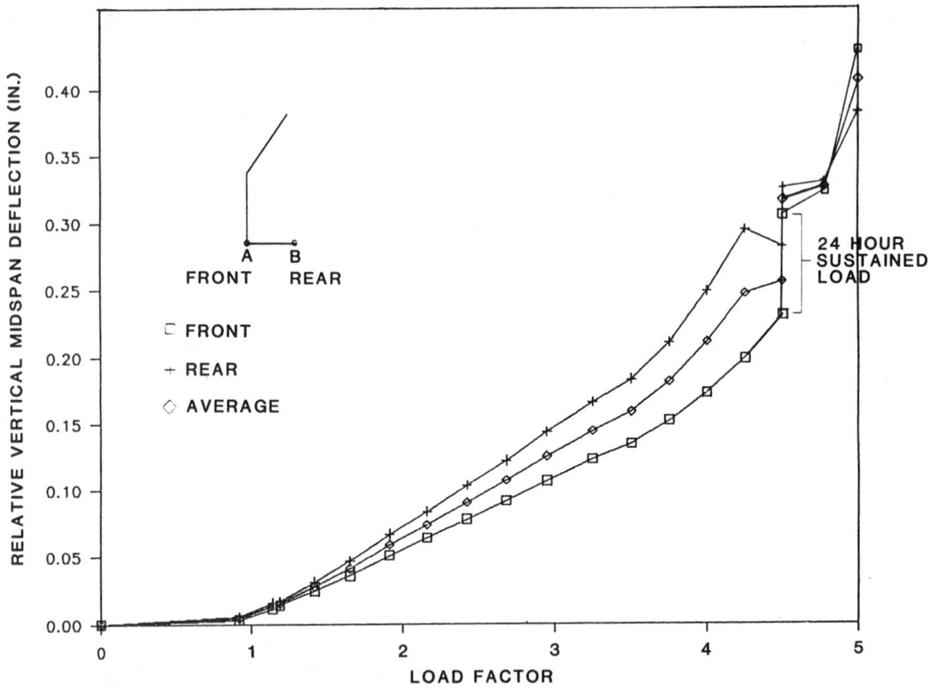

FIG. 18—*Plot of Test 2 vertical deflection versus load factors: front, rear, average.*

TABLE 1—*Load Test 1: Table of measured load and percentage of theoretical load by location on loading frame.*

Load Factor	Frame Number	Location on Loading Frame					
		A	B	C	D	E	F
1	1	*	136 (117%)	*	138 (110%)	74 (101%)	*
	2	*	112 (97%)	*	116 (92%)	82 (112%)	*
	3	*	114 (98%)	*	111 (88%)	71 (97%)	*
2	1	208 (102%)	288 (124%)	223 (99%)	223 (88%)	144 (99%)	118 (98%)
	2	230 (113%)	218 (94%)	234 (104%)	249 (99%)	149 (102%)	115 (96%)
	3	194 (95%)	225 (97%)	214 (95%)	208 (83%)	130 (89%)	105 (88%)
3	1	442 (108%)	408 (117%)	452 (100%)	466 (123%)	209 (97%)	255 (106%)
	2	456 (112%)	395 (114%)	491 (109%)	414 (110%)	260 (120%)	256 (107%)
	3	410 (100%)	358 (103%)	441 (98%)	300 (79%)	188 (87%)	239 (100%)

*No load applied at load Factor 1.

TABLE 2—*Load Test 2: Table of measured load and percentage of theoretical load by location on loading frame.*

Load Factor	Frame Number	Location on Loading Frame					
		A	B	C	D	E	F
1	1	*	117 (101%)	*	101 (80%)	75 (103%)	*
	2	*	111 (96%)	*	94 (75%)	91 (125%)	*
	3	*	33 (28%)	*	121 (96%)	64 (88%)	*
2	1	190 (93%)	236 (102%)	217 (96%)	225 (89%)	141 (97%)	113 (94%)
	2	203 (100%)	224 (97%)	220 (98%)	223 (88%)	210 (144%)	31 (26%)
	3	177 (87%)	223 (96%)	214 (95%)	228 (90%)	126 (86%)	112 (93%)
3	1	384 (94%)	354 (102%)	454 (101%)	395 (104%)	207 (96%)	245 (102%)
	2	410 (100%)	349 (100%)	472 (105%)	373 (99%)	291 (135%)	151 (63%)
	3	379 (93%)	337 (97%)	436 (97%)	332 (88%)	184 (85%)	246 (103%)
4	1	591 (97%)	464 (100%)	703 (104%)	565 (112%)	275 (94%)	313 (87%)
	2	636 (104%)	486 (105%)	788 (117%)	672 (133%)	369 (126%)	198 (55%)
	3	565 (92%)	457 (98%)	677 (100%)	440 (87%)	244 (84%)	322 (89%)
5	1	824 (101%)	762 (131%)	742 (82%)	531 (84%)	352 (96%)	491 (102%)
	2	**	491 (85%)	**	**	474 (130%)	497 (104%)
	3	821 (101%)	597 (97%)	917 (102%)	514 (82%)	371 (102%)	143 (30%)

*No load applied at load Factor 1.
**No reading due to debonded strain gages.

Tables 1 and 2 compare the measured loads from the strain gauge readings with the theoretical values expected.

The deflection/rotational plots (Figs. 15 and 16) show considerable departures from the computer model. This may be understood as showing the effect of the cracked section behavior of the actual steel reinforced panel over the uncracked, unreinforced behavior as assumed in the computer model.

Since the span-depth ratio is low, about 5.5, the spandrel panel is expected to be stiff approaching a thin deep beam. At service loads (load factor of one), the span deflection ratio is estimated to be more than 4000. This estimate is based on the assumption that the change in deflection between a load factor of one to two is the same as from zero to one. Such an assumption is required because the actual deflection of the specimen under its own gravity loading is necessarily unknown: it is conservative to use the calculation made. Even at a load factor of five, the vertical midspan deflection was only about 4/10ths of an inch or a span-deflection ratio of 725.

When similar analysis is made of the rotations at service loads (load factor of one), the rotation is about 0.001 rad. The horizontal movement of the tip of the sloping sill relative to the juncture of fascia and soffit is estimated to be 1/20th of an inch. At a load factor of five, this movement becomes about 1 in. It is apparent from these results that, relatively, the spandrel panel displays greater flexibility in torsion than in flexure as is to be expected of a deep, thin, open C-section. This observation becomes very significant when it is noted that the flexural span is approximately three times the torsional span, the wind supports being at the third points where the bottom horizontal tie was never installed.

Despite continuous inspection during loading and a thorough check of all specimen surfaces after the test, no visible cracking was noted of the brick and mortar materials. The total absence of cracking at the higher load factors was unexpected.

Investigation Summary

From the code review and the finite-element analysis, it was expected that the panel would fail in excessive principal stresses due to torsion. Even at a load factor of five, no brittle cracking was observed. The joint action between the steel reinforcing and the brick masonry proved to have a greater capacity than either the BIA or ACI Codes predicted. In fact, the ultimate behavior was surprisingly ductile-like. The authors conclude that both code requirements and analytical results are conservative. While the results proved to show satisfactory behavior for the problem at hand, it does suggest that code criteria may be more conservative than needed.

Hung-Liang Chen[1] *and Surendra P. Shah*[2]

Test of Model Masonry Single Pier Under Dynamic Shaking and Quasistatic Cyclic Loading

REFERENCE: Chen, H.-L. and Shah, S. P., **"Test of Model Masonry Single Pier Under Dynamic Shaking and Quasistatic Cyclic Loading,"** *Masonry: Materials, Design, Construction, and Maintenance, ASTM STP 992,* H. A. Harris, Ed., American Society for Testing and Materials, Philadelphia, 1988, pp. 145-165.

ABSTRACT: In order to improve seismic design of masonry structures, research is needed in the area of masonry response to seismic loading. One of the purposes of this program is to investigate the difference of the structural response of masonry single pier when in-plane shear force is applied dynamically and slowly. In addition, modeling techniques for brick masonry were studied.

The experimental results are reported for the structural behavior of two brick masonry piers under dynamic shaking and slowly applied cyclic loading. Each specimen was an unreinforced model pier having a height-to-width ratio of 1 with fixed-end conditions. The model was constructed of one fourth scale brick and mortar composed of fine sand, lime, and cement. Compressive tests of model mortar cubes, prisms, and one square panel were done to understand the modeling techniques of brick masonry. It was found that the piers under dynamic shaking are characterized by localized failure and abrupt stiffness degradation. However, the peak strength was lower under dynamic shaking. This tentative, preliminary conclusion should be verified with further tests.

KEY WORDS: modeling, masonry, bricks, mortar, prisms, single pier, square panel, cyclic loading, dynamic loading, shaking, compressive test, scale factor, shear

Although masonry is one of man's oldest and most common building materials, it has probably remained the least understood. The structural behavior in particular is not well known, and very little work has been done in studying the dynamic behavior of masonry structure.

Historically, masonry has shown poor resistance against earthquakes. After an extensive review of the literature, it was concluded that shear walls penetrated by numerous window openings, which were the components of multistory masonry buildings, were most frequently damaged in past earthquakes [*4,5,12*]. These structural components can be identified as the piers and the spandrel beams of a shear wall [*4*].

A testing fixture was designed in order to study the pier behavior under dynamic shaking and statically applied cyclic lateral loads (Fig. 1). Since experimental programs on large-scale specimens are expensive and difficult to conduct, model masonry piers having a height-to-width ratio of 1 were chosen. Each specimen was an unreinforced model pier with fixed-end conditions. One-fourth scale model brick and model mortar composed of fine sand, lime, and cement were used for model construction.

In this paper the experimental investigation is reported for the structural behavior of model

[1]Research assistant, Department of Civil Engineering, Northwestern University, Evanston, IL 60201.

[2]Professor, Department of Civil Engineering, and Director, Center for Concrete and Geomaterials, Northwestern University, Evanston, IL 60201.

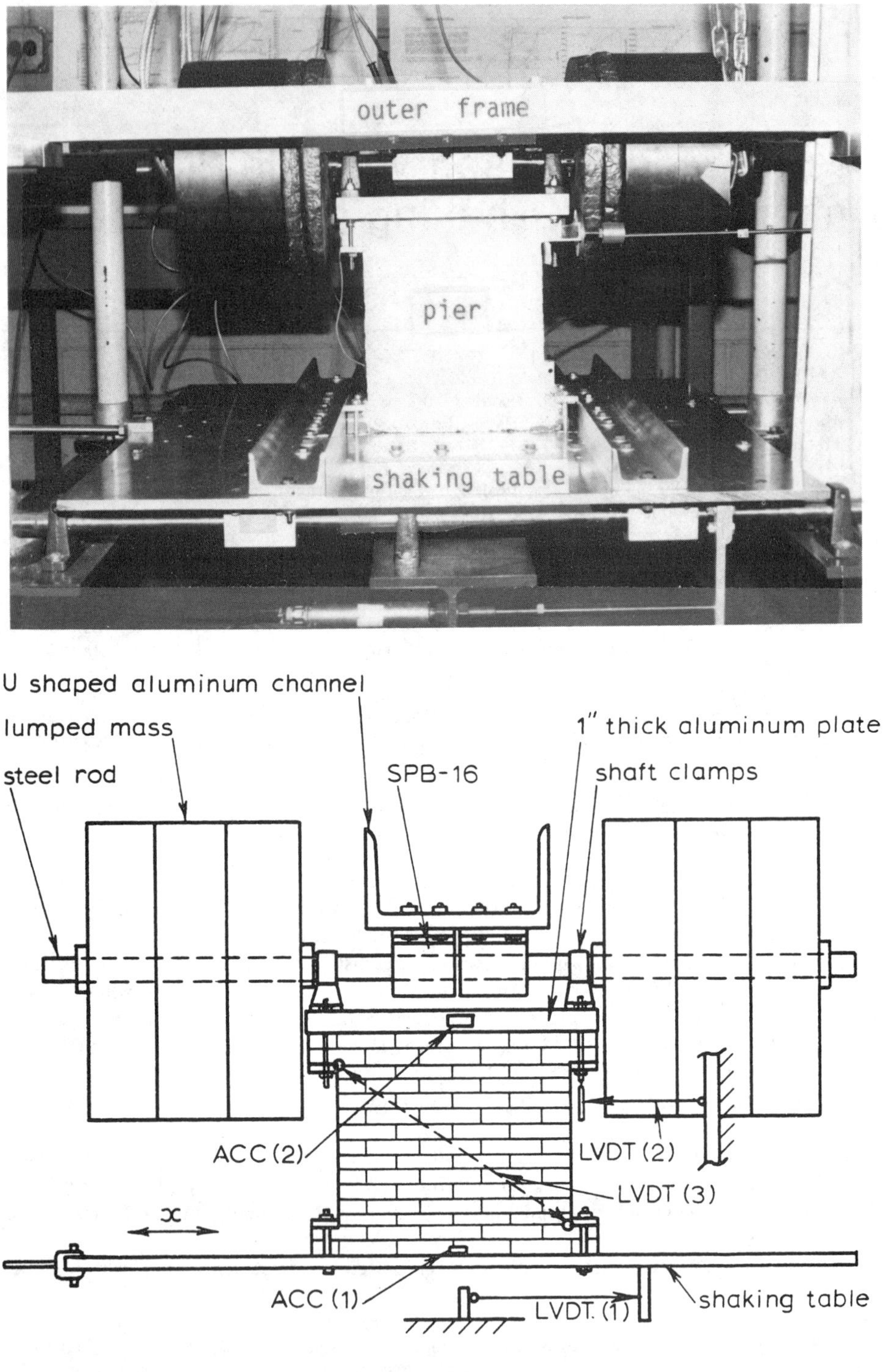

U shaped aluminum channel
lumped mass
steel rod
SPB-16
1" thick aluminum plate
shaft clamps
ACC (2)
LVDT (2)
LVDT (3)
x
ACC (1)
LVDT (1)
shaking table

FIG. 1—*Test setup.*

masonry single pier under dynamic shaking and slow-rate cyclic loading. The variable included in the program is the rate of loading. The results are presented in the form of hysteresis loops, envelope curves, and stiffness degradation. Simple analytical models are developed in order to check the shaking facilities and to understand the behavior of the pier under dynamic shaking. The modeling technique was checked by comparing the static behavior of the model pier with a prototype pier.

Objectives and Scope

In order to improve seismic design of masonry structures, research is needed in the area of masonry response to seismic loading. One of the purposes of this program was to investigate the difference of the structural response of masonry single piers under dynamic shaking and slow-rate cyclic loading. To our knowledge, this difference is not yet well understood, although some theoretical as well as experimental studies have been undertaken [*3,5,7–9,11,13,14*].

Modeling techniques for brick masonry have been studied by some researchers [*1,10*]. It is recognized that the influences of scale factor and material properties are the main difficulties in extrapolating the model performance to the actual case. The behavior of approximately 1/4-scale specimens was chosen for examination. In order to compare the material properties, compressive tests of model mortar cubes, prisms, and of a square masonry panel were done. A more detailed examination of scale effects will be possible in the future using the data obtained in this research. The scope was limited to model brick masonry units. Other masonry units, such as model concrete blocks and reinforced masonry assemblages, could form part of the basis for continued studies in the future.

The applied loadings were induced as horizontally in-plane shear forces with approximately 276 kPa (40 psi) vertically bearing stress at the fixed ends. Dynamic analysis was performed, based on linear assumption, to calculate the response of the model pier under dynamic shaking and to check the loading facilities. The calculations compared with the experimental results showed that the dynamic shaking setup was satisfactory and that the dynamic response of the model pier can be predicted provided the stiffness remains constant during loading.

Test Specimens

Material Properties

The specimens used to determine the material properties are shown in Fig. 2, and Table 1 shows the mechanical properties of the materials used in the construction of the piers. The 1/4-scale (50.8 by 25.4 by 14.3 mm) model bricks were made by Belden Brick Co. and were donated by Ramm Brick Inc. for support of this research. The brick was made from shale and was classified as Grade SW according to ASTM Specification for Facing Brick (Solid Masonry Units Made from Clay or Shale) (C 216-81). The test results following ASTM Method of Sampling and Testing Brick and Structural Clay Tile (C 67-81) are shown in Table 2A. The area of the holes is about 12.5% of the gross area.

The mortar was simulated as Type N, that is, Cement : Lime : Sand = 1 : 1 : 6 by volume. Portland Type I cement and Type S hydrated lime were used. Finer sand was used in constructing the model mortar. The sand gradation is shown in Table 2B. Compressive strength of 50.8-mm (2-in.) cubes for model mortar is also listed in Table 1. Three samples of cubes were taken at the same time when constructing the pier specimens. The sampling, curing, and testing of cubes followed ASTM Test Method for Compressive Strength of Hydraulic Cement Mortars (Using 2-in. or 50-mm Cube Specimens) (C 109-77).

Six prisms for uniaxial compression tests and one square panel for diagonal tension tests were constructed. Three of the six prisms were three-stack bond (48.4 by 50.8 by 25.4 mm) prisms,

1/4 scale model brick

square panel

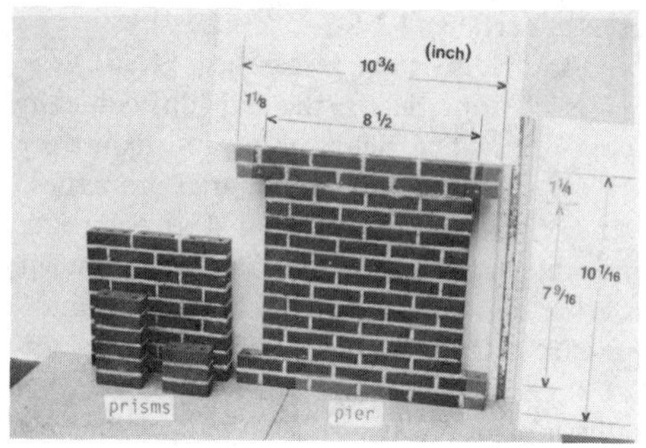

prisms and single pier

diagonal tensile test

FIG. 2—*Specimens.*

TABLE 1—*Material properties of model masonry.*[a]

Specimen, age	Mortar Type N, 28 Days	3-Stacks Prism, 28 Days	6-Stacks Prism, 31 Days	Square Panel, 61 Days		
				Ultimate load, kN	A, cm^2	$\sigma_o{}^b = 0.734\ P/\sqrt{2A}$, MPa
Compressive strength, MPa	4.96	47.82	47.23			
	5.58	44.81	43.78			
	5.37	36.41	40.68			
Average, MPa	5.31	43.03	43.90	8.01	40.0	1.04
Codes	ASTM C 270-84[c]	ASTM E 447-84				

NOTE: 1 in. = 25.4 mm, 1 lb = 4.448N, 1 ksi = 6.895 MPa, 1 lb/in. = 0.175 N/mm.

[a]Based on gross area.

[b]Frocht, M. M., 1931.

[c]ASTM Specifications for Mortar for Unit Masonry (C 270-84).

TABLE 2A—*Material properties of model brick.*[a]

Absorption	Saturation Coefficient	Initial Rate of Absorption, grams/min. 194 cm^2	Compressive Strength, MPa	Code
2.2% (24 h immersion)	0.88	2.1	138.20	ASTM C 216-81, Grade SW
2.5% (5 h boiling)				

[a]The values listed are the average of three samples.

TABLE 2B—*Mortar sand gradation.*

Sieve No.	Passing %, Model Mortar	Passing %, ASTM C 144[a]
30	100	40 to 75
50	70	10 to 35
100	20	2 to 15
200	0	. . .

[a]ASTM Specification for Aggregate for Masonry Mortar (C 144-84).

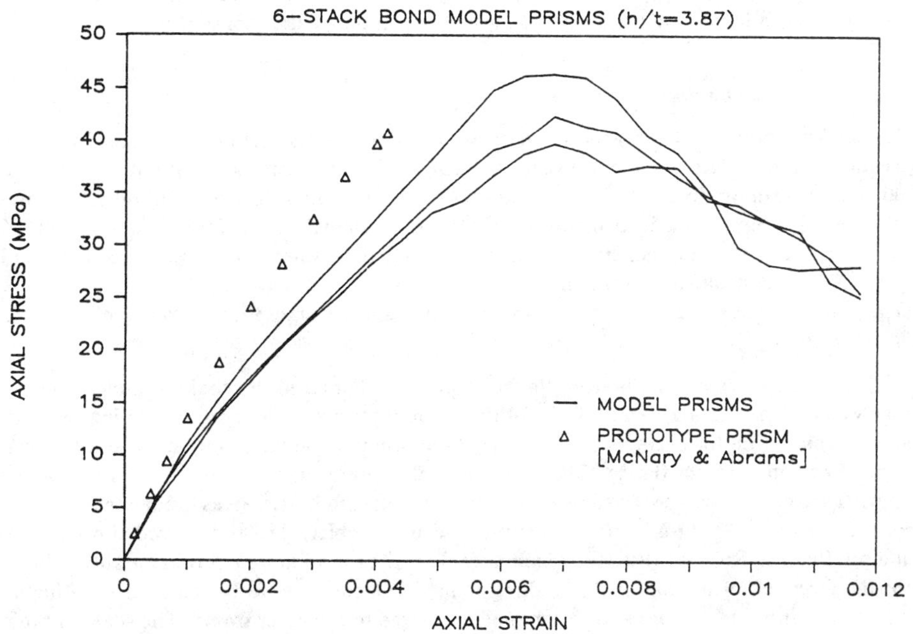

FIG. 3—*Stress-strain curves for model prisms.*

and the other three were six-stack bond (98.4 by 50.8 by 25.4 mm) prisms. These prisms were constructed, prepared, and basically tested according to ASTM Test Methods for Compressive Strength of Masonry Prisms (E 447-84). The thickness of mortar joint was approximately 2.8 ± 0.3 mm. All the compression tests were performed on an MTS loading machine at a loading rate of about 4448 N/min (1000 lb/min). The stress-strain curves for the three tests of six-stack bond prisms are shown in Fig. 3. The modulus of elasticity for the model masonry was obtained from these results. It is noted that the vertical cracks were usually formed before peak load and the prisms always failed due to the splitting of bricks. The stress-strain curves of the model prisms are compared with a test result of a prototype prism from Ref *6*.

The values of compressive strength of mortar, brick, prism, and square panel of larger scale specimens tested by other investigators are compared with the model specimens in Table 3. A direct comparison with the prototype and the model is not possible because of the variations in mortar strengths, brick strength, and test parameter involved in the prism test and square panel tests. However, considering these inherent variations, it can be said that the model prism strength and the model diagonal tension strength are reasonable when compared to the typical prototype values.

The prisms and the square panel were cured under the same normal atmospheric conditions as the piers. The ages of the specimens are listed on Table 1.

Model Piers

The model pier specimen is composed of two parts: top and bottom flanges and a square panel at the center. The weight of the pier was 30.7 N (6.9 lb). The overall dimensions of the model pier are shown in Fig. 2. The thickness of mortar joint is about 2.8 ± 0.3 mm. The thickness of the pier is 25.4 mm. The piers were constructed vertically by an experienced mason employing careful leveling techniques. Water was added to the mortar until it became workable. Two unreinforced ungrouted model piers were constructed at the same time. One pier was tested one month later. The other one was tested four months after construction.

Test Equipment and Setup

The shaking table used can generate a horizontal motion of single degree of freedom. The drive unit is an MB electromagnetic exciter model C5HE. It has a sinusoidal frequency range of 2500 Hz, a maximum force of 1779 N (400 lb), and a maximum displacement of ±12.7 mm (±0.5 in.). The excitation instrument used to produce the vibration signal is a WAVETEK Model 185 sweep function generator. The data acquisition system includes an IBM PC and ISAAC 2000 data acquisition system.

There were two test setups: one for dynamic shaking and the other for slow-rate cyclic loading (Fig. 1):

1. *Dynamic shaking*—The bottom flange of pier was bolted to the shaking table. The top flange was bolted with a 25.4-mm-(1-in.)-thick aluminum plate. Prior to the bolting process, epoxy was placed on the surfaces between the aluminum plate and the top flange and between the table and the bottom flange. The 1-in.-thick aluminum plate was bolted with two shaft clamps. The two clamps were fixed to a 1-in.-diameter steel rod. There was lumped mass on the steel rod, providing 1423-N (320-lb) vertical load on the pier. The rod passed through two Thomson Bearing Rollers (SPB-16), which only allow the rod to move freely along the X-direction (the direction of motion of the shaking table). The rollers were bolted with a U-shaped aluminum channel at the top, and the channel was fixed to the outer frame. The shaking table and the outer frame applied fixed-fixed end conditions for the pier.

Figure 1 also shows the instrumentation for measurements. Shaking table displacement was measured by LVDT(1), and the displacement at the top of the pier was measured by LVDT(2).

TABLE 3—*Comparison with prototype.*

Specimen	Compressive Strength, MPa [Reference]			
	[2]	[6]	[15]	Model (1/4)
Brick, (mm × mm × mm)[a]	122.4 (196 × 91 × 57)	101.7	99.8	138.2 (51 × 25 × 14)
Mortar	10.1 (Type N)	13.7 (Type N)	Not reported[c]	5.3 (Type N)
Prism (h/t)	38.5 (4.89)	32.5	22.0	43.9 (3.87)
Square Panel,[b] (mm × mm × mm)[a]	. . .	. . .	0.90 (1220 × 1220 × 94)	1.04 (155 × 160 × 25)

[a]Dimensions.
[b]Diagonal tensile strength.
[c]Conventional low-strength mortar (C:L:S = 1:1:4.5).

The LVDT(3) measured the diagonal displacement of the specimen. The working range for LVDT(1) and (2) was ±12.7 mm (0.5 in.) and for LVDT(3) was ±2.54 mm (0.1 in.). The accelerometer ACC(1) measured the acceleration of the table. The acceleration at the top of the pier was measured by ACC(2).

2. *Slow-rate cyclic loading*—There is only one change of the testing setup than that for the dynamic shaking. Four shaft collars (steel clamps) were used to fix the steel rod against the bearing rollers. Therefore, the top of the pier was not allowed to move and the outer frame was working as a reaction frame at this moment. The horizontal in-plane force was applied from the bottom of the pier by displacing the table along the X-direction.

One load cell with capacity ±4448 N (1000 lb) was placed between the exciter and the shaking table in order to measure the horizontal inputting force. The LVDT(1), (2), and (3) were at the same places as described in the dynamic shaking, and a LVDT(4) was added for measuring the vertical displacement of the pier.

Testing Procedure

Two testing procedures are described below:

1. *Dynamic shaking test:* The five stages (S1 to S5) during this test are reported in Table 4. At each stage, the motion history of the table was prescribed by the function generator. Sinusoidal cycles of motion at a specified displacement amplitude of the table were set through the entire test. The measured displacement of table was slightly different at different frequencies. The duration between each stage was about 15 min to 1 h. After each stage, the pier was visually inspected and the crack pattern was identified and photographed.

2. *Slow-rate cyclic loading test:* There were 18 stages (R1 to R18) of loading. The pier was subjected to a series of increasing displacement amplitude controlled in-plane shear forces. Each stage of loading consisted of two sinusoidal cycles of loading at a specified exciter displacement amplitude. The specified amplitude was gradually increased stage by stage. The full sequence of loading was applied at a frequency of 0.01 Hz. Each stage had a duration of 5 to 20 min.

The test was terminated at Stage R18 because of the capacity of the exciter. The shear strength of the pier at this stage was still in the ascending part. The test of the same pier by dynamic shaking was continued the following day in order to obtain more information about the model.

All of the tests were carried out under an essentially constant vertical load of 1423 N (320 lb). The additional vertical force developed during the cyclic displacing will be discussed in the presentation of test results.

TABLE 4—*Testing procedures.*

Pier, age	Loading Stage	Frequency, Hz	Cycles per Stage	Sampling Rate, points/s	Initial Stiffness, N/mm
No. 1, 30 days	S1	0.5	5	200	. . .
	S2	1	5	200	1064
	S3	7	5	350	1456
	S4	3	20	300	780
	S5	4	20	350	986
No. 2, 120 days	R1 ~ R18	0.01	2	2	834

Test Results

Mode of Failure

The mode of failure under dynamic loading is shown in Fig. 4. Flexural mode of failure was observed for both piers. This is characterized by horizontal cracks going through the toes of the pier. Shear slips along the horizontal cracks were observed during the dynamic shaking test and the permanent deformation of the measured relative displacement was mainly due to the slips. The pier was considered to fail when the horizontal cracks were completely formed, and further increase in inertia force only produced more slips. It is noted that the failure under dynamic shaking is characterized by localized cracking. Such failure has already been reported by other researchers [*16*].

Pier No. 1

The time histories of the relative lateral displacement is plotted in Fig. 5. The relative lateral displacement was computed from the difference between the lateral deflection at the top of the pier and the table displacement. For the dynamic shaking test, the acceleration at the top produces the inertia force necessary to displace the specimen. The acceleration measurements are also shown in Fig. 5.

The average shear stress (the inertia force divided by the net area) versus the relative displacement for Pier No. 1 is plotted in Fig. 6. In Stage S1, the shaking frequency was low (0.5 Hz), and the pier with its heavy lumped weight (1423 N) was moving as a rigid body. The measured top acceleration is about zero at Stage S1. Prior to Stage S3 there were no cracks. At Stage S3, a frequency of 7 Hz was applied and failure occurred. Two main flexural cracks and shear slips of the specimen were observed. A permanent deflection occurred as seen in Fig. 5, Stage S3. The LVDTs were zeroed again after Stage S3. At Stage S4, no further damage of the pier was found

FIG. 4—*Mode of failure.*

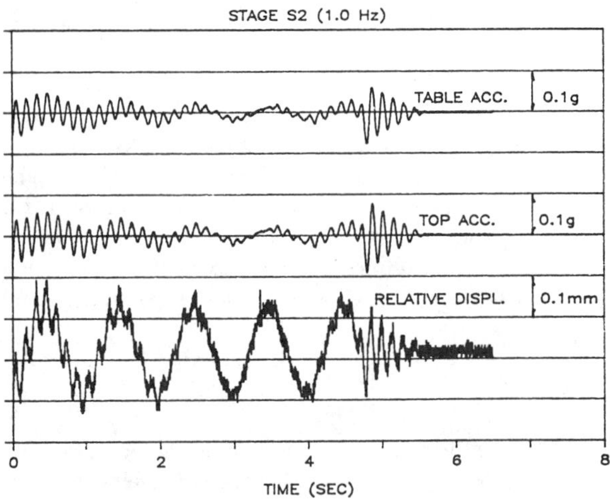

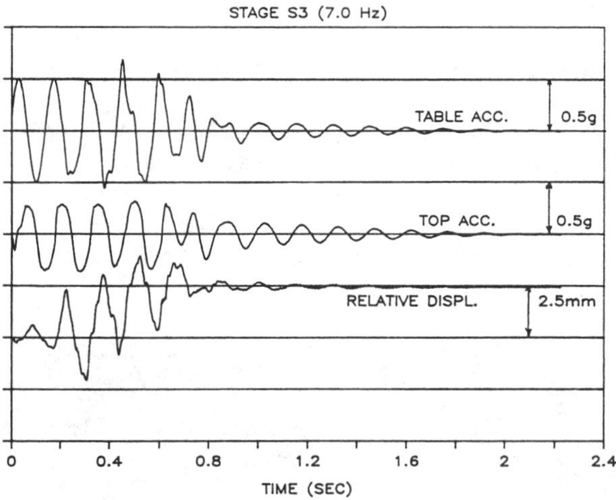

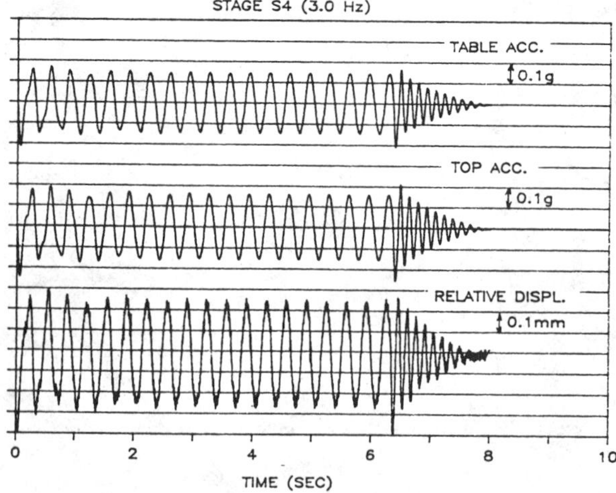

FIG. 5—*Time histories of test results.*

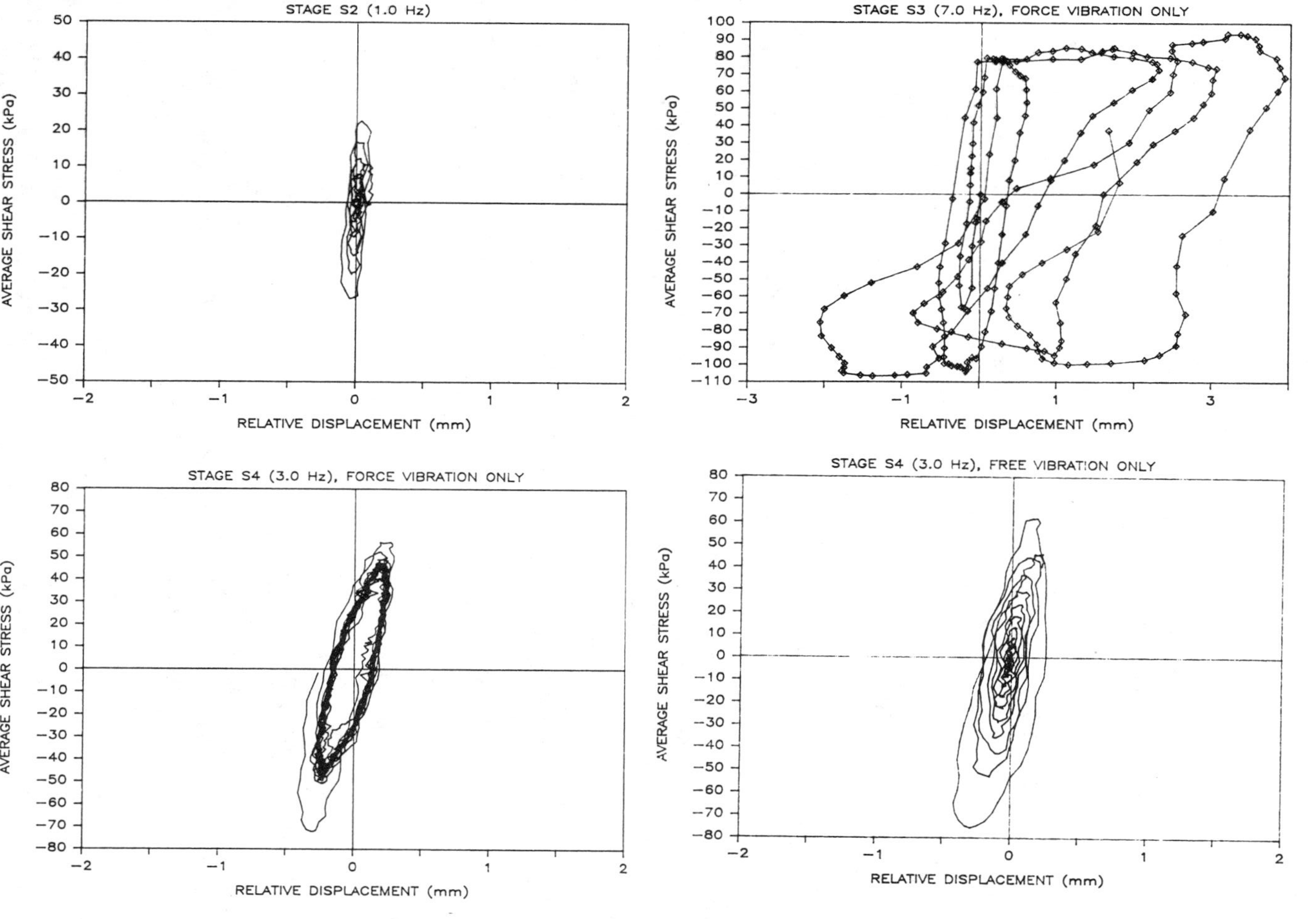

FIG. 6—*Test results under dynamic shaking (Pier No. 1).*

and the hysteretic loop went back to the origin at rest. During Stage S3 loading, the stiffness of the pier degraded 27% (as indicated by the difference in the observed initial stiffness during Stages S4 and S2). At Stage S5, the pier had further slips and crushing, and splitting at the top right and left corners was observed. No cracks appeared in the diagonal direction and the pier became unstable under shaking. Therefore, tests were stopped after this stage. It is interesting to note that even after severe damage during Stage S3, Pier No. 1 could still sustain a 3-Hz, 0.22-g shaking at Stage S4 without any further damage. As shown in Fig. 6, the damaged pier behaved as a rigid body. It vibrated around the origin under the inertia forces and returned back to the origin at rest.

Pier No. 2

The envelope (the maximum load of each stage) of the hysteresis is shown in Fig. 7. The individual hysteretic loop of Stage R5, R13, and R18 is also shown in Fig. 7. The average shear stress is calculated as the applied horizontal force divided by the net area. Each stage contains two cycles. The pier behaved linearly up to R5. At R13 the nonlinearity was observed because of cracking. The plots of shear stress versus vertical displacement are shown in Fig. 8. The nonlinearity of the vertical displacement shows the opening of the flexural cracks.

The reading of the LVDT(4) (vertical displacement) was recorded when the lumped load was put on the pier. There was concern that the reaction frame was not rigid enough and the constancy of the vertical load was not maintained. From the measured displacements we found that the maximum change of the average of vertical force is about −222 N (−50 lb), 15% of total vertical load), and we noted that the reading of vertical LVDT went back to the same reading after each stage. Furthermore, the maximum rotation of the reaction frame (Stage R18) is only 0.0075 rad. This influence of the fixed-end condition is negligible. Therefore, the rigidity of the reaction frame is confirmed and the vertical load can be assumed approximately constant during the tests.

The dynamic shaking of Pier No. 2 was started with small increments of acceleration, but no new crack was found. The complete opening of a new horizontal crack (other than the cracks at R18) was found after a 4-Hz, 0.25-g excitation. It is noted that under dynamic shaking the cracks may be very unstable. Even though the increment of acceleration was small, the displacement jump was very noticeable. There was no diagonal crack. The measured vertical displacement was due to the opening of horizontal cracks at the toe.

Discussion of Test Results

Static Response Comparison with a Prototype

The experimental results of the model pier under slowly applied cyclic loading is compared with the test results of a prototype pier, HCBR-11-2 [*3*]. The specimen HCBR-11-2 is an unreinforced partial grouted brick masonry single pier with fixed-ended conditions and height-to-width ratio of 1. For comparison purposes, the stiffness of the model pier, K_m was multiplied by a scale factor, F_r according to similitude requirements as follows:

$$F_r = \frac{K_p}{K_m} = K_r f_r \qquad (1)$$

where K_p is the stiffness of prototype pier and K_r is the theoretical initial stiffness ratio of prototype pier to model pier due to difference of dimension and elastic modulus. The calculated values are listed in Table 5. Since the mortar compressive strength and the bearing stress were found to be important factors influencing the behavior of the pier [*5*], we chose these two

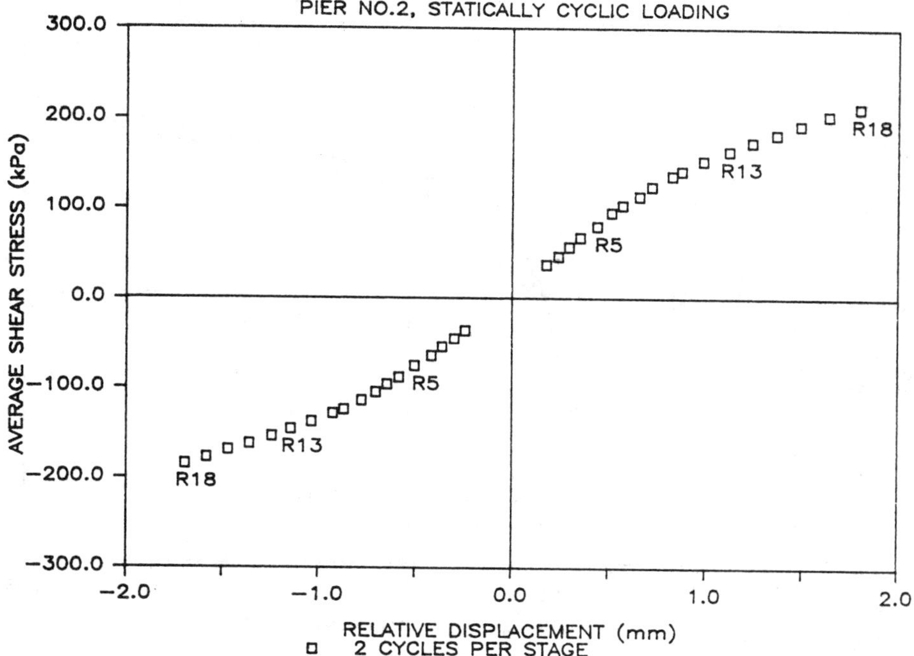

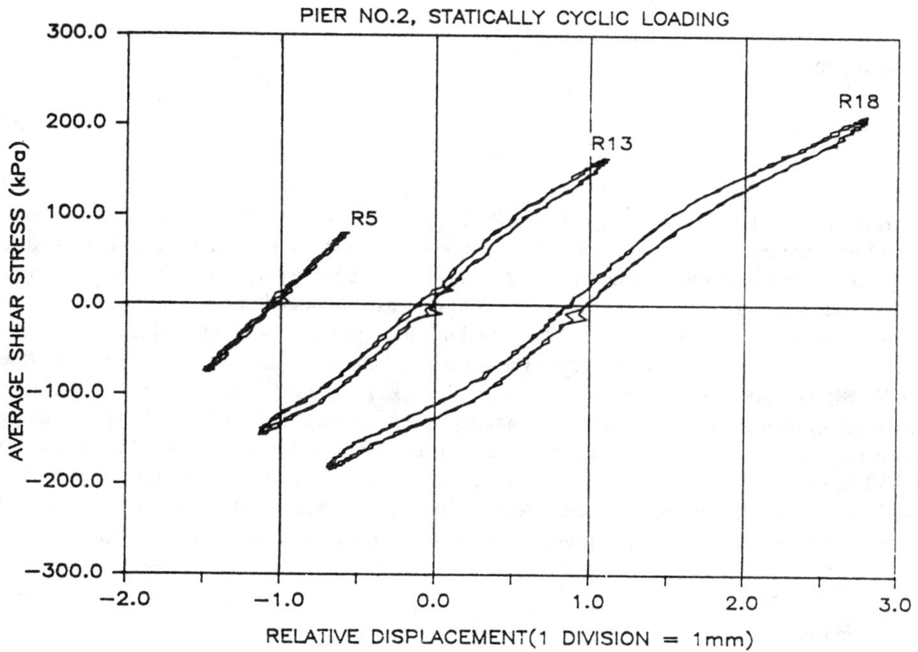

FIG. 7—*Test results under statically cyclic loading (Pier No. 2).*

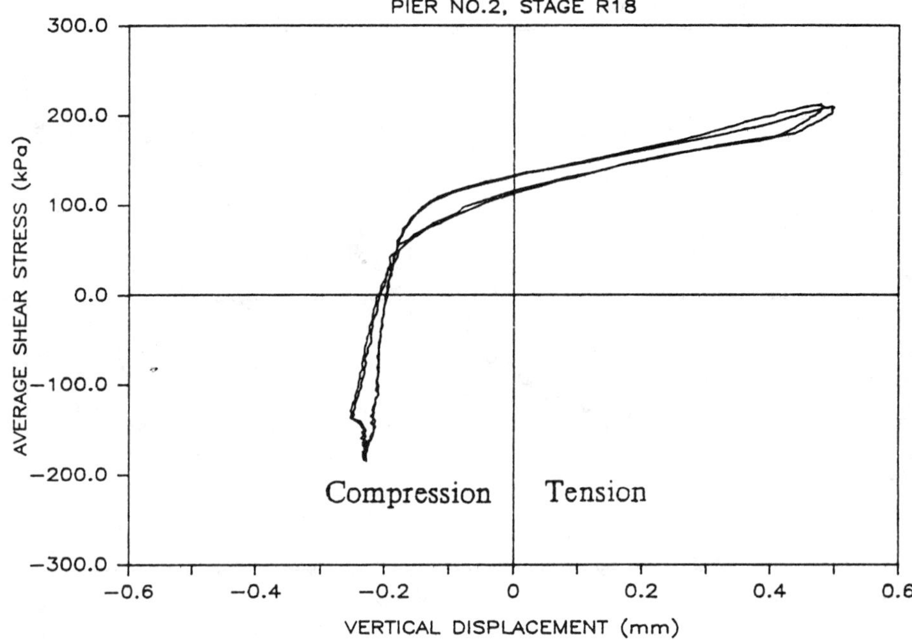

FIG. 8—*The opening of the flexural cracks.*

parameters and assumed a proportional relation as a first approximation for the current comparison,

$$f_r = \sigma_{Mr} \sigma_{Br} \tag{2}$$

where σ_{Mr} is the ratio of the compressive strength of mortar between prototype and model and σ_{Br} is the ratio of the initial bearing stress of prototype pier to that of model pier. The comparison of load-deflection diagram, stiffness degradation diagram, and equivalent damping ratio diagram are shown in Fig. 9. The comparison shows an adequate correspondence between prototype and model. However, because the model pier was still loaded on the ascending part, the comparison is only meant to illustrate the trend of the response. The validity of the model cannot be completely established until the overall loading behavior is simulated. Moreover, unlike the HCBR-11-2 specimen, there were no diagonal cracks in the model pier. This may be because of the relatively lower bearing stress. The model pier experienced about 276 kPa (40 psi) bearing stress during the test, but the stress on HCBR specimen increased from 730-kPa (106-psi) to 2793-kPa (405-psi) bearing stress when the ultimate load was reached. It is noted that during the first few loading stages, it is reasonable to assume the bearing stress remained as initial bearing stress for the HCBR specimen. This assumption was used in the comparison.

Dynamic Response

In order to understand the dynamic response of the model masonry and to check the dynamic shaking setup, the following analytic models were developed. As shown in Fig. 10, a Single

TABLE 5—*Calculation of scale factor.*

Specimen	L, mm	D, mm	t, mm	I, 10^6 mm^4	A (net), 10^3 mm^2	E, MPa	ν	K_o,[a] kN/mm	σ_M, MPa	σ_B, kPa	f_r	F_r
Prototype HCBR-11-2	1422.4	1219.2	187.3	20126.9	121.8	16893	0.15	382.4	26.5	730		
Model	192.0	215.9	25.4	15.5	4.8	11480	0.15	77.2	5.3	276	13.2	65.2
Ratio	7.4	5.7	7.4	1299	25.4	1.47	1	4.95	4.99	2.64		

[a] $k_o^{-1} = \dfrac{L^3}{12\,EI} + 1.2\dfrac{L}{AG},\ G = \dfrac{E}{2(1+\nu)}$.

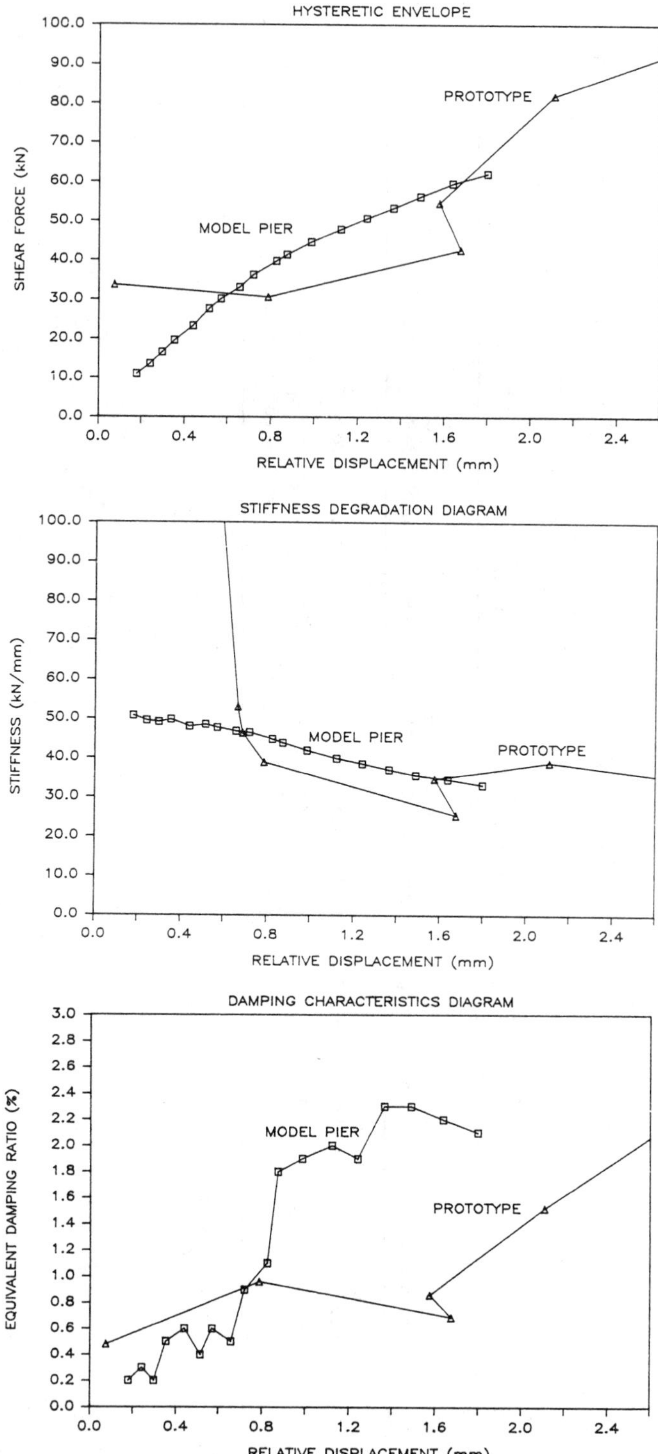

FIG. 9—*Static results comparison with prototype.*

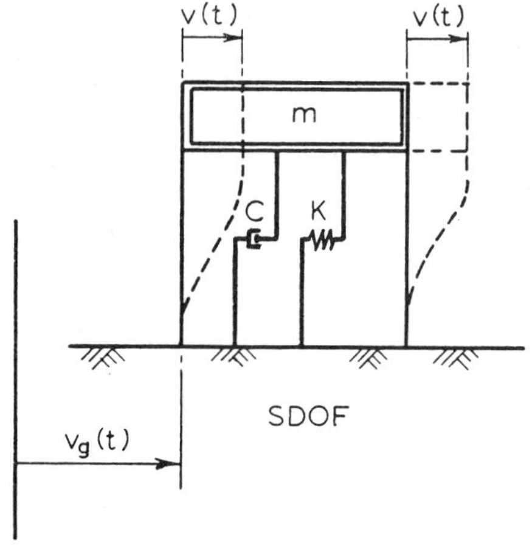

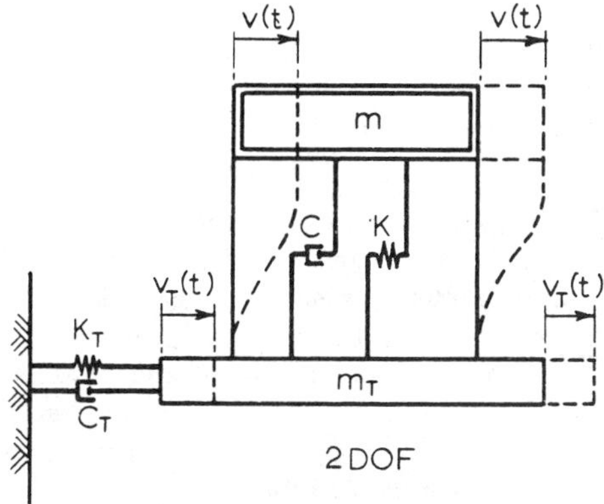

FIG. 10—*SDOF and 2DOF.*

Degree of Freedom (SDOF) model was used for the analysis of the response under dynamic shaking. The equation of motion is

$$m\ddot{v} + c\dot{v} + Kv = -m\ddot{v}_g(t) = P_o \sin \bar{\omega}t \tag{3}$$

For an underdamped system, the steady state response can be shown as

$$v(t) = \frac{P_o}{K} D \sin(\bar{\omega}t - \phi) \tag{4}$$

where

m = the lumped mass,
c = the damping coefficient,
K = the stiffness,
ω = the shaking frequency,
P_o = the amplitude of shaking,
D = the dynamic magnification factor, and
ϕ = the phase angle.

Assuming that the system is linear, we let K be equal to the initial stiffness of the pier, damping ratio equal to 5%, and the loading function equal to the table excitation. These data are shown in Table 6. It was also noted that the dynamic magnification factor was not influenced significantly by the assumed damping ratio at the currently applied loading frequency.

The free vibration response was simulated by a 2 Degree of Freedom (2DOF) system as shown in Fig. 10. This was necessary since the shaking table did not stop right after the force vibration. It was assumed that the shaking table had the damping coefficient c_T and spring constant K_T. The equations of motion are as follows:

$$m_T \ddot{v}_T + c_T \dot{v}_T + K_T v_T - c\dot{v} - Kv = 0 \tag{5}$$

$$m(\ddot{v} + \ddot{v}_T) + c\dot{v} + Kv = 0 \tag{6}$$

where m_T is the mass of the shaking table. The solution of the equations of motion is of the form

$$\underset{\sim}{v} = \underset{\sim}{A} \sin(\bar{\omega}t - \phi) \tag{7}$$

We end up with the frequency equation of the system as

$$(-m_T\omega^2 + c_T\omega + K_T)(-m\omega^2 + c\omega + K) - (c\omega + K)\, m\omega^2 = 0 \tag{8}$$

The spring constant of the shaking table was measured while the damping ratio was assumed equal to 5%. Table 6 shows the input data and the calculated frequency of the first mode. Since the response of the pier is of main interest, we use the measured table acceleration to calculate the time history of the relative displacement of the pier. The comparisons with the experimental results are shown in Fig. 11. It was noted that in Stage S2 and Stage S4 the pier behaved linearly under dynamic shaking and the calculated results are in good agreement with the experimental results. But in Stage S3, as a result of damage, the stiffness of the pier varied during loading (Fig. 6). Thus the assumption of a constant stiffness could not predict the measured response. For further investigation of the dynamic behavior of this stage, a model constructed with a nonlinear spring is suggested. However, more experiments should be conducted to achieve this aim.

Difference Between Dynamic Response and Static Response

By comparing the hysteretic behavior of the model pier under dynamic shaking (Fig. 6) with the behavior of the pier under static loading (Fig. 7), it is seen that the conventional quasistatic cyclic test results may not be able to represent the dynamic properties of the masonry pier. The use of an envelope curve constructed from statically cyclic tests may lead to an unconservative estimation of the ultimate strength of the masonry pier when dynamic response is concerned. By comparing Stage S3 (Pier No.1) with Stage R18 (Pier No.2), and assuming Pier No.1 and Pier No.2 are identical, it can be noted that the ultimate load of the pier under dynamic shaking was 118 lb (816 N; 7 Hz, 0.37 g). This value is lower than the maximum load (230 lb; 1586 N) applied statically during Stage R18. When Pier No.2 was subjected to dynamic shaking after the static test, it failed at a 4 Hz, 0.25 g (80 lb; 552 N) excitation.

It should be noted that a frequency of 7 Hz applied to the model pier during Stage S3 corresponds to a frequency of 7/4 = 1.75 Hz applied to a full-scale pier. Thus it is likely that masonry structures subjected to dynamic loading during a strong earthquake may exhibit a behavior similar to the model pier tested here. Additional model and full-scale dynamic tests should be conducted to verify this preliminary tentative conclusion.

TABLE 6—*Calculation of dynamic response.*

Pier No. 1	SDOF									2DOF				Measured	
Stage	m, kg	K, N/mm	$\omega = \sqrt{K/m}$, rad/s	$\bar{\omega}$, rad/s	ξ, %	P_o, N	$\beta = \bar{\omega}/\omega$	θ, 10^{-3} rad	D	m_T, kg	K_T, N/mm	ξ_T, %	ω_1, rad/s	ω_1, rad/s	ξ, %
S2	145	1064	85.7	6.3	5	106	0.073	7.4	1.005	23	252	0.05	37.1	44.1	4.9
S3	145	1456	100.2	44.0	5	996	0.439	54.3	1.237	23	252	0.05	38.0	45.2	7.0
S4	145	780	73.4	18.8	5	191	0.257	27.5	1.070	23	252	0.05	36.0	42.1	4.6

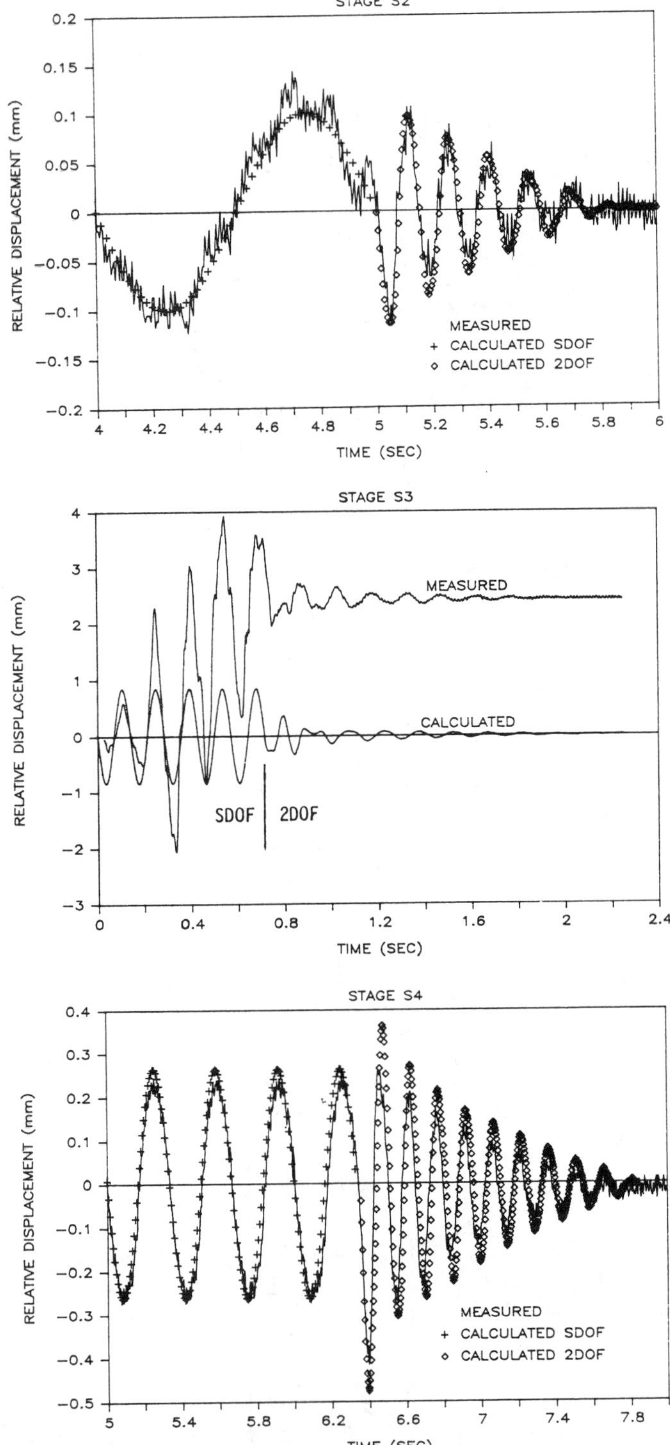

FIG. 11—*Calculation of dynamic response.*

Summary

As a preliminary study, the possibility of using the model brick masonry to study the static behavior as well as dynamic behavior of single pier was explored. The difference of the masonry response under static loading and dynamic loading was observed, and simple analytical models were developed in order to check the shaking facilities and to understand the behavior of the pier under dynamic shaking. The strength of model prisms and square panels were found in good correlation with the prototype prisms and square panels. The possibility of using model piers to reproduce the behavior of prototype piers was examined by considering a scale factor. The model piers under dynamic shaking were characterized by localized flexural failure. However, the peak strength was lower under dynamic shaking.

Acknowledgment

Acknowledgment is made to C. Ostrander from Masonry Advisory Council for his valuable suggestions. Ramm Brick Inc. is gratefully acknowledged for providing the model brick, and we thank J. Gilstrom from Ramm for providing us with very useful information. R. Nudd from Bricklayers Local 21 of Illinois is gratefully acknowledged for his construction of the model masonry specimens. The authors also want to acknowledge M. L. Wang (currently at the University of New Mexico) for his helpful suggestions and his contributions to the experimental setup that he made while he was research associate at Northwestern University.

References

[*1*] Benjamin, J. R. and Williams, H. A., "The Behavior of One-Story Brick Shear Walls," *Proceedings of ASCE, Journal of Structural Division,* Vol. 84, No. ST4, July 1958.

[*2*] Brown, R. H., "Prediction of Brick Masonry Prism Strength from Reduced Constraint Brick Tests," *Masonry: Past and Present, ASTM STP 589,* American Society for Testing and Materials, Philadelphia, 1975, pp. 171-194.

[*3*] Chen, S. J., Hidalgo, P. A., Moyes, R. L., Clough, R. W., and McNiven, H. D., "Cyclic Loading Tests of Masonry Single Piers, Vol. 2—Height to Width Ratio of 1," EERC 78-28, Earthquake Engineering Research Center, Berkeley, CA, December 1978.

[*4*] Mayes, R. L. and Clough, R. W., "A Literature Survey—Compressive, Tensile, Bond and Shear Strength of Masonry," EERC 75-15, Earthquake Engineering Research Center, Berkeley, CA, 1975.

[*5*] Mayes, R. L. and Clough, R. W., "State-of-the-Art in Seismic Shear Strength Masonry—An Evaluation and Review," EERC 75-21, Earthquake Engineering Research Center, Berkeley, CA, 1975.

[*6*] McNary, W. S. and Abrams, D. P., "Mechanics of Masonry in Compression," *Journal of Structural Engineering,* American Society of Civil Engineers, New York, April, 1985.

[*7*] Mengu, Y. and McNiven, H. D., "A Mathematical Model of Masonry for Predicting Its Linear Seismic Response Characteristics," EERC 79-04, Earthquake Engineering Research Center, Berkeley, CA, 1979.

[*8*] Priestley, M. T. N. and Bridgeman, D. O., "Seismic Resistance of Brick Masonry Walls," *Bulletin of the New Zealand National Society for Earthquake Engineering,* Vol. 7, No. 4, 1974.

[*9*] Qamaruddin, M., Arya, A. S., and Chandra, B., "Dynamic Response of Multi-Storied Brick Buildings," *Earthquake Engineering and Structural Dynamics,* Vol. 13, March-April 1985, pp. 135-150.

[*10*] Samarasinghe, W., Page, A. W., and Hendry, A. W., "Behavior of Brick Masonry Shear Walls," *The Structural Engineer,* Vol. 59B, No. 3, September 1981.

[*11*] Stafford Smith, B., Carter, C., and Choudhury, J. R., "The Diagonal Tensile Strength of Brickwork," *The Structural Engineer,* Vol. 48, No. 6, June 1970.

[*12*] Stephen, R. M., Hollings, J. P., Bouwkamp, J. G., and Jurukovski, D., "Dynamic Properties of an Eleven Story Masonry Building," EERC 75-20, Earthquake Engineering Research Center, Berkeley, CA, 1975.

[*13*] Sucuoglu, H., Mengi, Y., and McNiven, H. D., "A Mathematical Model for the Response of Masonry Walls to Dynamic Excitations," EERC 82/24, Earthquake Engineering Research Center, Berkeley, LCA, 1982.

[*14*] Williams, D. W., "Seismic Behavior of Reinforced Masonry Shear Walls," Ph.D. thesis, University of Canterbury, Christchurch, New Zealand, 1971.

[*15*] Yokel, F. Y. and Fattal, S. G., "Failure Hypothesis for Masonry Shear Walls," *Journal of the Structural Division,* Vol. 102, No. ST3, American Society of Civil Engineers, New York, March 1976.

[*16*] Gulkan, P., Mayes, R. L., and Clough, R. W., "Shaking of Single-Story Masonry Houses", Vol. 1, 2 and 3, EERC 79/23-79/25, Earthquake Engineering Research Center, Berkeley, CA, 1979.

Construction

Clayford T. Grimm[1]

Statistical Primer for Brick Masonry

REFERENCE: Grimm, C. T., "**Statistical Primer for Brick Masonry,**" *Masonry: Materials, Design, Construction, and Maintenance, ASTM STP 992,* H. A. Harris, Ed., American Society for Testing and Materials, Philadelphia, 1988, pp. 169–192.

ABSTRACT: Methods of sampling and statistical data reduction applicable to brick masonry are reviewed. Techniques for plotting histograms and probability density functions for normal and log normal distributions are discussed. Criteria for establishing sample size and specification requirements are enumerated. Binary operations with variables are illustrated. The concept of structural reliability is introduced. Seventeen log-normal graphs of various brick masonry properties are included.

KEY WORDS: absorption, brick, strength (compressive) (flexural), cracks, expansion (freezing) (moisture) (thermal), joint thickness variation, masonry, modulus of elasticity, mortar, sampling, statistics

Sampling

The purpose of materials testing is to formulate generalizations about the characteristics of those materials. To determine absolutely the mean strength of a large quantity of material, it would be necessary to test all of the material, which may be impractical, virtually impossible, or even self-defeating. Accordingly, some of the material may be tested from which results inferences may be drawn about the strength of all the material. Valid tests of a sample give results which are certain only for the sample, but those results permit conclusions of varying certitude about the material from which the sample was taken [*11*].

The material from which a sample is taken is called a *lot,* that is, a quantity of material which, insofar as is practical, consists of a single type, grade, class, size, finish (texture), and composition produced by a single source by the same process and under practically the same conditions. The size of a lot is sometimes standardized; for example, ASTM Method of Sampling and Testing Brick and Structural Clay Tile (C 67-83) [2] establishes 250,000 brick as a lot.

A *specimen* is an individual piece of material from a lot. A *sample* is a number of specimens. Valid conclusions about a lot can be drawn by testing a sample only if the sample is representative of the lot. For that reason the method of selecting specimens must be unbiased. Bias is avoided by selecting a *random sample,* that is, every specimen of the sample has an equal chance of being selected in every trial. A sample selected haphazardly, without a conscious plan, is not a random sample. It is a haphazard sample. It is virtually impossible to draw a sample at random by exercise of human judgement alone. The proper use of an artificial or mechanical device of selecting a random sample is necessary. Specimens may be selected by assigning a number to each unit or small group of units in the lot and using a table of random numbers [*11*] or an electronic random number generator to select a number of specimens. For any given set of conditions there are usually several possible sampling plans, all valid, but differing in speed, simplicity, and cost.

[1]Consulting architectural engineer and senior lecturer in architectural engineering, University of Texas at Austin, Austin, TX.

The degree of certitude with which conclusions are drawn from test results increases with the number of randomly selected specimens in the sample. The required number of specimens depends on: (1) the variability in the characteristic to be tested; (2) the permissible difference between the tested mean value and the mean value of the lot; and (3) the degree of confidence with which one wishes to state what that difference may be.

Histogram

Having collected a random sample and performed the tests, the first step in data analysis is often the preparation of a *histogram,* that is, a plot of the frequency of occurrence versus selected *class intervals* (i) of the random variable. The number of intervals can be any convenient number, usually not less than six and increasing with the amount of data. Consider the 25 data points on the tensile strength of a mortar in psi given in Table 1.

Note in Table 1 that the minimum value is 496 and the maximum if 598. The *range* (R) is the difference between the maximum and minimum. If there are 7 class intervals, there are about R/7 or 15 units per class interval, that is, $i = 15$.

The frequency distribution is given in Table 2. Note for example that the class interval is from and including 495 to but not including 510. In Table 2 the number 525 occurs in the 525 to 540 class interval rather than in the 510 to 525 class interval. The histogram is shown in Fig. 1.

Mean and Variance

The sample *mean* ($\bar{x}$) is defined as the sum of the individual data (Σx) divided by the number of data points (n), $\bar{x} = (\Sigma x)/n$. $\bar{x}$ is an estimate of the *lot mean* (μ).

The *residual* is the difference between the sample mean ($\bar{x}$) and an individual value (x), that is, $(x - \bar{x})$.

The sample variance (s^2) is the quotient of the sum of squares of the residuals and the number of data (n) less one. s^2 is an estimate of the *lot variance* (σ^2).

$$s^2 = (n - 1)^{-1} \sum_{x=0}^{x=n} (x - \bar{x})^2 \tag{1}$$

TABLE 1—*Sample data.*

518, 560, 538, 577, 544, 525, 574, 528, 534, 538, 554, *598,* 579, 524, 544, 555, 567, 550, 570, 542, 556, *496,* 541, 562, 511

TABLE 2—*Frequency distribution.*

Interval	Data	Frequency
495 to 510	496	1
510 to 525	511, 518, 524	3
525 to 540	525, 528, 534, 538, 538	5
540 to 555	541, 542, 544, 544, 550, 554	6
555 to 570	555, 556, 560, 562, 567, 570	6
570 to 585	574, 577, 579	3
585 to 600	598	1

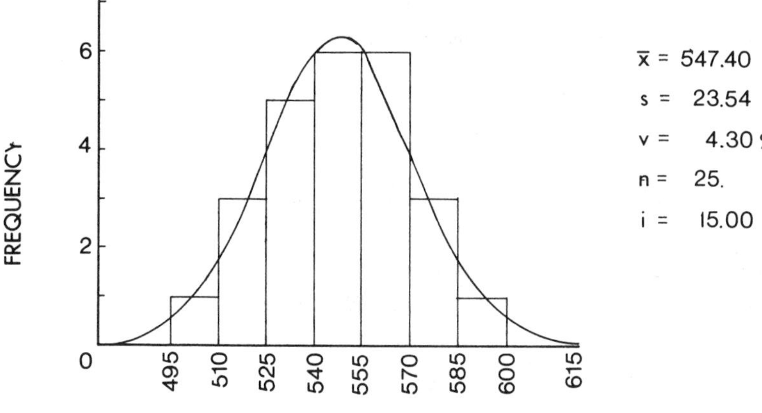

FIG. 1—*Histogram.*

The *sample standard deviation* (s) is the square root of the variance (s^2). It is an estimate of the standard deviation of the lot (σ).

The *sample coefficient of variation* (v) is the sample standard deviation divided by the sample mean, $v = s/\bar{x}$ or expressed as a percentage, $v\% = 100\,s/\bar{x}$. Typical coefficients of variation (v) for some samples of materials are given in Table 3.

From the data in Table 1, $n = 25$, $\bar{x} = 547.4$, $s^2 = 554.13$, $s = 23.54$, and $v = 4.3\%$.

Normal Distribution

As the total number of observations approaches infinity and the class interval approaches zero, the histogram approaches a continuous curve, that is, a frequency distribution curve referred to as a *probability density function*. The integration of the equation for that curve between specified limits is a *probability function*. Although there are several types of such functions, an axiom of probability theory, the *central limit theorem,* states: "Under very general conditions, as the number of variables in the sum becomes large, the distribution of the sum of

TABLE 3—*Typical coefficients of variation in some masonry properties,* v, %.

Material	See Figure No.	Low	Normal	High
	BRICK			
24-h cold water absorption	5	<4.9	4.9 to 7.8	>7.8
Saturation coefficient	7	<2.9	2.9 to 5.3	>5.3
Initial rate of absorption, suction	8	<13.1	13.1 to 21.3	>21.3
Compressive strength	10	<7.0	7.0 to 12.0	>12.0
	BRICK MASONRY			
Compressive strength[a]	12	<3.4	3.4 to 5.7	>5.7
Mortar head joint thickness	18	<17.0	17.0 to 21.0	>21.0
Mortar bed joint thickness	19	<11.9	11.9 to 15.5	>15.5

[a]Laboratory specimens.

random variables will approach the normal distribution." The normal distribution function is also known as the *normal probability density function*.

$$f(x) = (2\pi)^{-1/2} \exp - (x - \bar{x})^2/2s^2 \tag{2}$$

where

$f(x)$ = the normal probability density function for a sample, that is, the height of an ordinate erected at a distance $x - \bar{x}$ from the mean.

It is convenient to express the deviation of x from the mean ($\bar{x}$) in terms of s, so that $z = (x - \bar{x})/s$. Thus, the density function for the standard normal curve is

$$f(z) = (2\pi)^{-1/2} \exp - (z^2/2) \tag{3}$$

where

$f(z)$ = the normal probability density function in terms of z.

At the mean, $z = 0$ and $f(z) = (2\pi)^{-1/2}$ or 0.3989. In terms of a histogram the computed frequency is $nif(z)/s$. Although the data in Fig. 1 are not symmetrically arranged, we will assume by virtue of the central limit theorem that the data would be normally distributed given more data. Accordingly, for the histogram in Fig. 1 at $x = 518$, $x - \bar{x} = 518 - 547.4$ or -29.4, $z = (x - \bar{x})/s$ or $-29.4/23.54$ or -1.2489, from Eq 3, $f(z) = 0.1829$, and the frequency is $25 \times 15 \times 0.1829/23.54$ or 2.91. The computed normal frequency at the midpoint of each interval in the histogram in Fig. 1 is given in Table 4. With those data the normal curve may be drawn on the histogram. The accuracy of such a plot increases as the number of class intervals increases.

The probability that x will fall between x_1 and x_2 can be determined by expressing x in terms of z and integrating Eq 3 between z_1 and z_2.

$$P(z) = \int_{-\infty}^{z} f(z)dz = (2\pi)^{-1/2} \int_{-\infty}^{z} e^{-(z^2/2)}dz \tag{4}$$

TABLE 4—*Normal curve fit.*

Intervals (i)		Observed Frequency, Table 2 Data	Interval Midpoint	Deviation at Interval Midpoint from x: Residual, $x - \bar{x}$	Deviation at Interval Midpoint from x: Number of Standard Deviations, z	Computed Frequency, Ordinate
(1)	(2)	(3)	(4) = [(1) + (2)]/2	(5) = (4) − $\bar{x}$	(6) = (5)/s	(7)
495	510	1	502.5	−44.9	−1.91	1.03
510	525	3	517.5	−29.9	−1.27	2.84
525	540	5	532.5	−14.9	−0.63	5.21
540	555	6	547.5	0.1	−0.004	6.36
555	570	6	562.5	15.1	0.64	5.18
570	585	3	577.5	30.1	1.28	2.80
585	600	1	592.5	45.1	1.92	1.01

where

$P(z)$ = the normal probability function, that is, the area under the normal curve between $-\infty$ and z (see Table 5).

Equation 4 cannot be integrated in closed form and must be evaluated by numerical integration. Tabular values of $P(z)$ between $-\infty$ and z may be found in Table 5. The following polynomial approximation may be used for $P(z)$ between $-\infty$ and z [1].

$$P(z) \simeq 1 - 0.5(1 + 0.049867347z + 0.0211410061z^2 + 0.0032776263z^3 + 0.0000380036z^4 + 0.0000488906z^5 + 0.000005383z^6)^{-16} + 1.5 \times 10^{-7} \quad (5)$$

Thus, for example, from the data in Table 1 the probability that the tensile strength of the mortar sample will be greater than 533 psi is determined as follows

$$z = (x - \bar{x})/s = (533 - 547.4)/23.54 = -0.61$$

From Eq 5, $P(z) = 0.2710$ or $1 - P(z) = 0.7290$, which compares with $0.2291 + 0.5$ or 0.7291 from Table 5. Similarly, the probability that the tensile strength of a specimen lies between plus or minus one standard deviation of the sample mean, that is, between 523.96 and 571.04 psi, is 0.6826, that is, from Table 5 where $z = 1$, $P(z) = 0.8413$ and between $z = -1$ and $z = 1$, $P(z) = 2(0.8413 - 0.5)$ or 0.6826 (see Fig. 2). The shape of the normal curve is wider and flatter as the standard deviation increases, that is, as the data become more variable.

It is often necessary to know the value of z required for a given probability, $P(z)$, in which case refer to Table 5 or the following polynomial approximation, where $P(z)$ is the portion of the area under the normal curve between $-\infty$ and z [1]. Where $P(z) > 0.5$, $z = F$. Where $P(z) < 0.5$, $z = -F$.

$$F \simeq c + 4.5 \times 10^{-4} - (2.515517 + 0.802853c + 0.010328c^2) \times (1 + 1.432788c + 0.189269c^2 + 0.002308c^3)^{-1} \quad (6)$$

where

$P(z) \geq 0.5$: set $c = \{\ln[1 - P(z)]^{-2}\}^{1/2}$, and
$P(z) < 0.5$: set $c = \{\ln[P(z)]^{-2}\}^{1/2}$.

For example, if tests on representative samples from the national brick production give a mean compressive strength of 10 437 psi with a standard deviation of 3636 psi, what are the limits of the 20 percentiles of the sample; that is, 20% of the sample has a compressive strength lower than what value? What are the strength limits for the next higher 20 percentiles?

$P(z)$	z (Eq 6)	zs	$\bar{x} + zs$
0.2	−0.8419	−3061	7376
0.4	−0.2534	−921	9516
0.6	0.2534	921	11358
0.8	0.8419	3061	13495

TABLE 5—*Cumulative normal distribution—values of* $P_{(z)}$ [11].

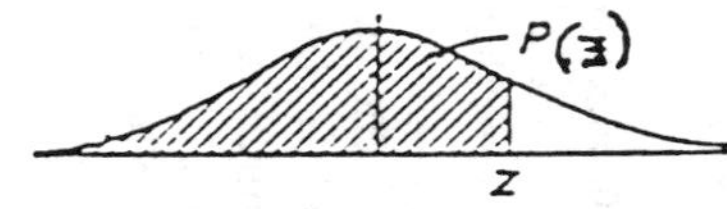

Values of P_z corresponding to z_p for the normal curve.
z is the standard normal variable. The value of P for $-z$ equals one minus the value of P for $+z$, e.g., the P for -1.62 equals $1 - .9474 = .0526$.

z	0.00	0.01	0.02	0.03	0.04	0.05	0.06	0.07	0.08	0.09
0.0	0.5000	0.5040	0.5080	0.5120	0.5160	0.5199	0.5239	0.5279	0.5319	0.5359
0.1	0.5398	0.5438	0.5478	0.5517	0.5557	0.5596	0.5636	0.5675	0.5714	0.5753
0.2	0.5793	0.5832	0.5871	0.5910	0.5948	0.5987	0.6026	0.6064	0.6103	0.6141
0.3	0.6179	0.6217	0.6255	0.6293	0.6331	0.6368	0.6406	0.6443	0.6480	0.6517
0.4	0.6554	0.6591	0.6628	0.6664	0.6700	0.6736	0.6772	0.6808	0.6844	0.6879
0.5	0.6915	0.6950	0.6985	0.7019	0.7054	0.7088	0.7123	0.7157	0.7190	0.7224
0.6	0.7257	0.7291	0.7324	0.7357	0.7389	0.7422	0.7454	0.7486	0.7517	0.7549
0.7	0.7580	0.7611	0.7642	0.7673	0.7704	0.7734	0.7764	0.7794	0.7823	0.7852
0.8	0.7881	0.7910	0.7939	0.7967	0.7995	0.8023	0.8051	0.8078	0.8106	0.8133
0.9	0.8159	0.8186	0.8212	0.8238	0.8264	0.8289	0.8315	0.8340	0.8365	0.8389
1.0	0.8413	0.8438	0.8461	0.8485	0.8508	0.8531	0.8554	0.8577	0.8599	0.8621
1.1	0.8643	0.8665	0.8686	0.8708	0.8729	0.8749	0.8770	0.8790	0.8810	0.8830
1.2	0.8849	0.8869	0.8888	0.8907	0.8925	0.8944	0.8962	0.8980	0.8997	0.9015
1.3	0.9032	0.9049	0.9066	0.9082	0.9099	0.9115	0.9131	0.9147	0.9162	0.9177

1.4	0.9192	0.9207	0.9222	0.9236	0.9251	0.9265	0.9279	0.9292	0.9306	0.9319
1.5	0.9332	0.9345	0.9357	0.9370	0.9382	0.9394	0.9406	0.9418	0.9429	0.9441
1.6	0.9452	0.9463	0.9474	0.9484	0.9495	0.9505	0.9515	0.9525	0.9535	0.9545
1.7	0.9554	0.9564	0.9573	0.9582	0.9591	0.9599	0.9608	0.9616	0.9625	0.9633
1.8	0.9641	0.9649	0.9656	0.9664	0.9671	0.9678	0.9686	0.9693	0.9699	0.9706
1.9	0.9713	0.9719	0.9726	0.9732	0.9738	0.9744	0.9750	0.9756	0.9761	0.9767
2.0	0.9772	0.9778	0.9783	0.9788	0.9793	0.9798	0.9803	0.9808	0.9812	0.9817
2.1	0.9821	0.9826	0.9830	0.9834	0.9838	0.9842	0.9846	0.9850	0.9854	0.9857
2.2	0.9861	0.9864	0.9868	0.9871	0.9875	0.9878	0.9881	0.9884	0.9887	0.9890
2.3	0.9893	0.9896	0.9898	0.9901	0.9904	0.9906	0.9909	0.9911	0.9913	0.9916
2.4	0.9918	0.9920	0.9922	0.9925	0.9927	0.9929	0.9931	0.9932	0.9934	0.9936
2.5	0.9938	0.9940	0.9941	0.9943	0.9945	0.9946	0.9948	0.9949	0.9951	0.9952
2.6	0.9953	0.9955	0.9956	0.9957	0.9959	0.9960	0.9961	0.9962	0.9963	0.9964
2.7	0.9965	0.9966	0.9967	0.9968	0.9969	0.9970	0.9971	0.9972	0.9973	0.9974
2.8	0.9974	0.9975	0.9976	0.9977	0.9977	0.9978	0.9979	0.9979	0.9980	0.9981
2.9	0.9981	0.9982	0.9982	0.9983	0.9984	0.9984	0.9985	0.9985	0.9986	0.9986
3.0	0.9987	0.9987	0.9987	0.9988	0.9988	0.9989	0.9989	0.9989	0.9990	0.9990
3.1	0.9990	0.9991	0.9991	0.9991	0.9992	0.9992	0.9992	0.9992	0.9993	0.9993
3.2	0.9993	0.9993	0.9994	0.9994	0.9994	0.9994	0.9994	0.9995	0.9995	0.9995
3.3	0.9995	0.9995	0.9995	0.9996	0.9996	0.9996	0.9996	0.9996	0.9996	0.9997
3.4	0.9997	0.9997	0.9997	0.9997	0.9997	0.9997	0.9997	0.9997	0.9997	0.9998

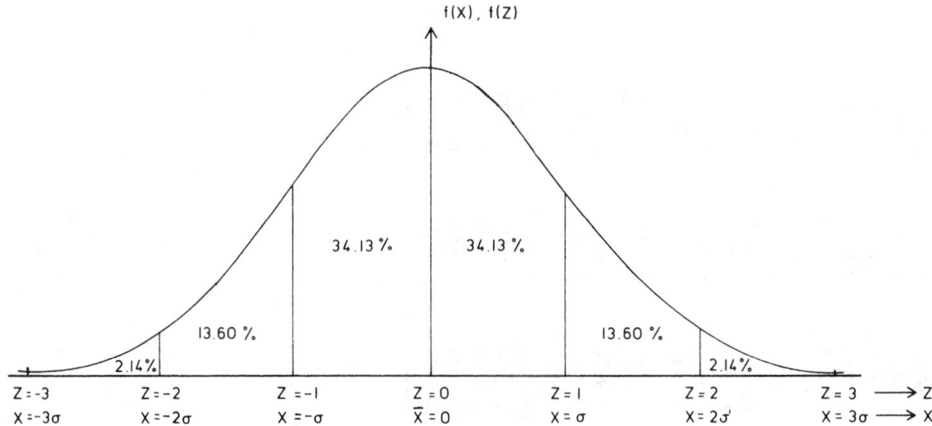

FIG. 2—*Areas under the normal probability curve.*

It is concluded that 20% of the national brick production sample has a compressive strength less than 7376 psi, that 20% is in the range of 7376 to 9516 psi, 20% is in the range of 9516 to 13 495 psi, and 20% have a compressive strength greater than 13 495 psi. It is important to realize that those conclusions are valid for the sample and not necessarily for the lot. Those conclusions are valid for the lot only if $\bar{x} = \mu$ and $s = \sigma$, that is, that the sample has the same distribution as the lot. That probably is not exactly true.

Sample Size

The distribution of the sample approaches that of the lot as the number of specimens increases. The deviation of the sample mean from the lot mean expressed as a percentage of the sample mean is the *sampling error,* $[\epsilon = 100(\mu - \bar{x})/\bar{x}]$. The number of specimens required for a sample is a function of: (1) the allowable sampling error; (2) the coefficient of variation of the sample; and (3) the confidence required that the allowable sampling error will not be exceeded [3].

$$n_r = (k\,v/\epsilon_a)^2 \tag{7}$$

where

n_r = required sample size to the next higher whole number,
k = a factor corresponding to a probability that the sampling error will not exceed ϵ_a (see Table 6),
v = coefficient of variation, and
ϵ_a = allowable sampling error, %.

The probability that the sampling error will not exceed ϵ_a is, of course, set at a low value. Approximate values of k for various levels of confidence are given in Table 6. An allowable sampling error (ϵ_a) is often set at 5%; for example, the true mean compressive strength of the brick in the lot may range from 7600 to 8400 psi, when the sample mean strength is 8000 psi. The use of Eq 7 requires an estimate of v, in which case Table 3 may be helpful.

Suppose we select a sample from a lot of what we believe is good quality-controlled brick to estimate the 24-h cold water absorption of the lot within 5% of the sample mean with a confi-

TABLE 6—k *Factors for various probabilities of excessive sampling error.*

Confidence Level, %	Approximate Probability of Excessive Error	Chance of Excessive Error	k
99.9999	0.000001	1 in 1 000 000	4.892
99.999	0.00001	1 in 100 000	4.417
99.99	0.0001	1 in 10 000	3.891
99.9	0.001	1 in 1000	3.291
99.7	0.003	3 in 1000	3.000
99.0	0.01	1 in 100	2.576
95.5	0.045	45 in 1000	2.000
95.0	0.05	5 in 100	1.960
90.0	0.10	1 in 10	1.645
70.00	0.20	1 in 5	1.282
80.0	0.30	3 in 10	1.037
60.0	0.40	2 in 5	0.842
50.0	0.50	1 in 2	0.674
40.0	0.60	3 in 5	0.524
30.00	0.70	3 in 10	0.385
20.0	0.80	4 in 5	0.253
10.0	0.90	1 in 10	0.127

dence level of 95%; that is, the probability is 0.05 that the sampling error will be exceeded. From Table 6, $k = 1.96$. From Table 3, $v = 6\%$ for the coefficient of variation in the 24-h cold water absorption of good quality-controlled brick. $\epsilon_a = 5\%$. Accordingly, from Eq 7, $n = (1.96 \times 6/5)^2$ or 5.5 or 6 to the next larger whole number.

Suppose further that when we test the 6 brick, we find that $\bar{x} = 20\%$ and $v = 10\%$. Apparently, what we thought was good quality brick is only fair brick by reference to Table 3. We may now determine the sampling error (ϵ) from a rearrangement of Eq 7.

$$\epsilon = k v n^{-1/2} \tag{8}$$

or

$$\epsilon = 1.96 \times 10(6)^{-1/2} = 8\%$$

Therefore, the mean 24-h water absorption of the lot might be between 18.4 and 21.6%. Since that range is intolerable (that is, 8% > 5%), more testing is necessary. We now set $v = 10$ and from Eq 7, $n = (1.96 \times 10/5)^2$ or 16 to the next higher whole number. Accordingly, we must test 10 more brick, that is, $16 - 6 = 10$.

If a national survey of the compressive strength of brick, consisting of 31 samples, indicates a mean strength of 10 437 psi with a standard deviation of 3636 psi, we can be 95% confident that the true national average compressive strength lies between 9157 psi and 11 717 psi, that is, $k = 1.96$, $v = 100 \times 3636/10\ 437 = 34.84\%$, and $n = 31$. Therefore, $\epsilon = 12.26\%$. Lower value $= 0.01\bar{x}(100 - \epsilon)$ or 9157, and upper value $= 0.01\bar{x}(100 + \epsilon)$ or 11 717.

Specification Versus Requirement

To be reasonably confident that the mean strength of a lot will exceed a specified mean strength, the required mean strength must be somewhat higher than the specified mean strength, that is, if the mean strength of the brick on a project equals the specified mean strength, half the brick is below the specified mean strength, if the strength is normally distributed. The required mean strength will be greater for brick having high variance in strength than for brick having low-strength variance. Also, the required mean strength will be higher if the mean strength is determined from a small sample than from a large one. The required mean strength is determined as follows:

$$\bar{x}_r = 100\,\bar{x}_s(100 - tvn^{-1/2})^{-1} \tag{9}$$

where

$\bar{x}_r$ = required mean value,
$\bar{x}_s$ = specified mean value,
v = coefficient of variation of the sample, %
t = factor depending on confidence that $\bar{x}_s < \bar{x}_r$ (See Table 7), and
n = number of specimens in sample.

For example, if you wish to be 90% confident that the mean strength of a lot of brick will exceed a specified mean compressive strength of 8000 psi, what mean strength would be required of a sample of eight bricks which had a coefficient of variation of 12%?

From Table 7, $t = 1.415$ and $\bar{x}_r = 100 \times 8000[100 - 1.415 \times 12(8)^{-1/2}]^{-1}$ or $\bar{x}_r = 8511$ psi. Similarly, the maximum required mean value can be determined as follows

$$x_{r\,\max} = 100\,\bar{x}_s(100 + tvn^{-1/2})^{-1} \tag{10}$$

If one wishes to be 99% confident that the mean 5-h boiling water absorption of a lot of brick will not exceed 17%, should brick be accepted on which the manufacturer offered test data indicating a mean 5-h boiling water absorption of 10% with coefficient of variation of 25% based on a sample of five bricks?

$$\bar{x}_{r\,\max} = 100 \times 17[100 + 3.747 \times 25(5)^{-1/2}]^{-1}$$

$$\bar{x}_{r\,\max} = 12\% > 10\% \text{ (ok)}$$

Under those conditions, brick having a 5-h boiling water absorption of less than 12% should be accepted.

Maximum and Minimum Properties

Given the mean ($\bar{x}$), the coefficient of variation (v), and the number of specimens tested (n), it can be stated with a given level of confidence (P_c) that a specified portion of a normally distributed lot (P_p) will exceed a given minimum value (x_l) or will not exceed a given maximum value (x_h) [*12*].

$$x_l = \bar{x}[1 - (Kv/100)] \tag{11}$$

$$x_h = \bar{x}[1 + (Kv/100)] \tag{12}$$

where

x_l = lower limit of x,
x_h = upper limit of x, and

v = coefficient of variation, %.

$$z_p = K - z_c[n^{-1} + 0.5K^2(n - 1)^{-1}]^{1/2} \tag{13}$$

where

z_p = number of standard deviations from the mean below which the values of x lie;
P_c = confidence probability; decimal and
z_c = the number of standard deviations from the mean below which the confidence probability lies.

Procedure:

Given x_l or x_h, $\bar{x}$, and v, determine K from Eq 11 or 12.
Given P_c, determine z_c from Eq 6 or Table 5.
Given K, z_c, and n, determine z_p from Eq 13.
Given z_p, determine P_p from Eq 5 or Table 5.

For example, a representative sample of nine mortar specimens has a mean sand/cement ratio of 3.04 with a coefficient of variation of 8.72%. We can be 99% confident what percent of the mortar has a sand/cement ratio of at least 3 (that is, 3 or greater).

From Eq 6, where $P_c = 0.99$, $z_c = 2.33$.
From Eq 11, where $\bar{x} = 3.04$, $v = 8.72$, and $x_L = 3$, $K = 0.1509$.
From Eq 13, where $K = 0.1509$, $n = 9$, and $z_c = 2.33$, $z_p = -0.6307$.
From Eq 5, where $z_p = -0.6307$, $P_p = 0.264$.

We can be 99% confident that at least 26.4% of the mortar sample has a sand/cement ratio greater than 3.

Binary Operations

The mean and the standard deviation of the sum, difference, product, or quotient of two normally distributed, independent variables may be determined as follows [*10*].

$$\mu_{x+y} = \mu_x + \mu_y \tag{14}$$

$$\sigma_{x+y} = [\sigma_x^2 + \sigma_y^2]^{1/2} \tag{15}$$

$$\mu_{x-y} = \mu_x - \mu_y \tag{16}$$

$$\sigma_{x-y} = [\sigma_x^2 + \sigma_y^2]^{1/2} \tag{17}$$

$$\mu_{xy} = \mu_x \mu_y \tag{18}$$

$$\sigma_{xy} = [(\mu_x \sigma_y)^2 + (\mu_y \sigma_x)^2 + (\sigma_x \sigma_y)^2]^{1/2} \tag{19}$$

$$\mu_{(x/y)} = \mu_x / \mu_y \tag{20}$$

$$\sigma_{(x/y)} = \mu_y^{-1}[(\mu_x \sigma_y)^2 + (\mu_y \sigma_x)^2]^{1/2}(\mu_y^2 + \sigma_y^2)^{-1/2} \tag{21}$$

$$\mu_{(x)}^{1/2} = [0.5(4\mu_x^2 - 2\sigma_x^2)^{1/2}]^{1/2} \tag{22}$$

$$\sigma_{(x)}^{1/2} = [\mu_x - 0.5(4\mu_x^2 - 2\sigma_x^2)^{1/2}]^{1/2} \tag{23}$$

TABLE 7—*Percentiles of the* t *distribution* [11].

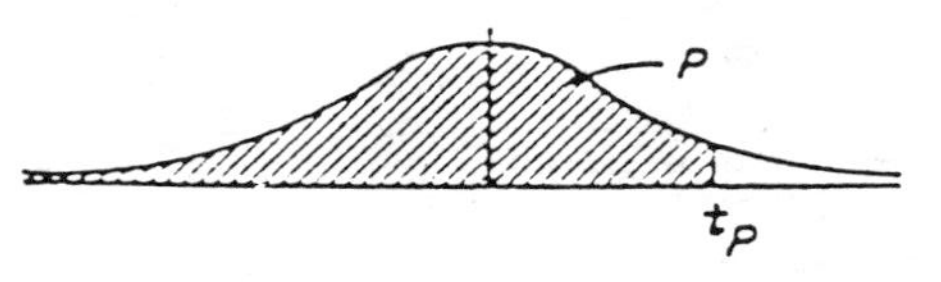

$m - 1$	$t_{.60}$	$t_{.70}$	$t_{.80}$	$t_{.90}$	$t_{.95}$	$t_{.975}$	$t_{.99}$	$t_{.995}$
1	0.325	0.727	1.376	3.078	6.314	12.706	31.821	63.657
2	0.289	0.617	1.061	1.886	2.920	4.303	6.965	9.925
3	0.277	0.584	0.978	1.638	2.353	3.182	4.541	5.841
4	0.271	0.569	0.941	1.533	2.132	2.776	3.747	4.604
5	0.267	0.559	0.920	1.476	2.015	2.571	3.365	4.032
6	0.265	0.553	0.906	1.440	1.943	2.447	3.143	3.707
7	0.263	0.549	0.896	1.415	1.895	2.365	2.998	3.499
8	0.262	0.546	0.889	1.397	1.860	2.306	2.896	3.355
9	0.261	0.543	0.883	1.383	1.833	2.262	2.821	3.250
10	0.260	0.542	0.879	1.372	1.812	2.228	2.764	3.169
11	0.260	0.540	0.876	1.363	1.796	2.201	2.718	3.106
12	0.259	0.539	0.873	1.356	1.782	2.179	2.681	3.055
13	0.259	0.538	0.870	1.350	1.771	2.160	2.650	3.012
14	0.258	0.537	0.868	1.345	1.761	2.145	2.624	2.977
15	0.258	0.536	0.866	1.341	1.753	2.131	2.602	2.947

16	0.258	0.535	0.865	1.337	1.746	2.120	2.583	2.921
17	0.257	0.534	0.863	1.333	1.740	2.110	2.567	2.898
18	0.257	0.534	0.862	1.330	1.734	2.101	2.552	2.878
19	0.257	0.533	0.861	1.328	1.729	2.093	2.539	2.861
20	0.257	0.533	0.860	1.325	1.725	2.086	2.528	2.845
21	0.257	0.532	0.859	1.323	1.721	2.080	2.518	2.831
22	0.256	0.532	0.858	1.321	1.717	2.074	2.508	2.819
23	0.256	0.532	0.858	1.319	1.714	2.069	2.500	2.807
24	0.256	0.531	0.857	1.318	1.711	2.064	2.492	2.797
25	0.256	0.531	0.856	1.316	1.708	2.060	2.485	2.787
26	0.256	0.531	0.856	1.315	1.706	2.056	2.479	2.779
27	0.256	0.531	0.855	1.314	1.703	2.052	2.473	2.771
28	0.256	0.530	0.855	1.313	1.701	2.048	2.467	2.763
29	0.256	0.530	0.854	1.311	1.699	2.045	2.462	2.756
30	0.256	0.530	0.854	1.310	1.697	2.042	2.457	2.750
40	0.255	0.529	0.851	1.303	1.684	2.021	2.423	2.704
60	0.254	0.527	0.848	1.296	1.671	2.000	2.390	2.660
120	0.254	0.526	0.845	1.289	1.658	1.980	2.358	2.617
∞	0.253	0.524	0.842	1.282	1.645	1.960	2.326	2.576

A unit of heat flow through a wall is one btu/h — ft^2 — °F. The rate through Material A has a mean value of 0.13 units with a standard deviation of 0.04 units. The mean rate through Material B is 1.13 units with a standard deviation of 0.1 units. What is the mean flow through both materials combined in parallel and what is the standard deviation of the mean flow? From Eqs 14 and 15

$$\bar{x} = 0.13 + 1.13 = 1.26 \text{ units}$$

$$s = (0.04^2 + 0.1^2)^{1/2} = 0.1077 \text{ units}$$

The gross area of a hollow masonry unit is 88.64 in.^{2} with a standard deviation of 4 in.^{2}. The net area of the unit has a mean value of 45 in.^{2} with a standard deviation of 2 in.^{2}. What is the mean cell area of the unit and its standard deviation? From Eqs 16 and 17

$$\bar{x} = 88.64 - 45.0 = 43.64 \text{ in.}^2,$$

$$s = (4^2 + 2^2)^{1/2} = 4.47 \text{ in.}^2$$

A pier has a mean width of 7.625 in. and a mean thickness of 11.6 in. The standard deviation in width and thickness is 0.17 in. What is the mean area of the pier and the standard deviation in the area? $\bar{A} = 7.625 \times 11.6 = 88.45$ in.^{2}. From Eq 18, $s = [(7.625 \times 0.17)^2 + (11.06 \times 0.17)^2 + (0.17)^4]^{1/2}$ or $s = 2.2839$ in.

If that short pier at failure carries a mean axial load of 30 kips, having a standard deviation of 5 kips, what is the mean unit compressive strength of the pier and what is its standard deviation? $\bar{x} = \bar{P}/\bar{A} = 30/88.45$ or 0.339 ksi. From Eq 21, $s = 88.45^{-1}[(30 \times 2.28)^2 + (88.45 \times 5)^2]^{1/2}[88.45^2 + 2.28^2]^{-1/2}$ or $s = 0.057$ ksi. What is the allowable stress, if we wish to be 95% confident that stress will not exceed strength? From Eq 6 where $P(z) = 0.95$, $z = 1.6458$. Therefore $P/A = 0.339 - (1.6458 \times 0.057)$ or $P/A = 0.245$ ksi. This is correct only if $\bar{x} = u$.

Reliability

Consider Fig. 3. $f(y)$ is the frequency distribution curve for strength of a structural member, and $f(x)$ is the frequency distribution curve for the stress in the same member. When stress equals strength, failure occurs. Reliability (R) is the probability that strength will exceed stress. R may be determined from Eq 5 or Table 5 where [*10*]:

$$z = (\bar{x}_r - \bar{x}_s)(s_s^2 + s_r^2)^{-1/2} \tag{24}$$

where

$\bar{x}_r$ = mean strength, psi,
$\bar{x}_s$ = mean stress, psi,
s_s = standard deviation in stress, psi, and
s_r = standard deviation in strength, psi.

For example, the mean flexural strength of a particular type of masonry is 135 psi with a standard deviation of 33 psi. The flexural stress in the masonry is 25 psi with a standard deviation of 9 psi. What is the reliability of the system? $z = (135 - 25)(33^2 + 9^2)^{-1/2}$ or $z = 3.22$. From Table 5 or Eq 5, $R = 0.9994$. There is one chance in about 1667 that failure will occur.

What is an acceptable level of probability of being killed by structural failure? The appropriate value of R varies with the type of material, load, and structure.

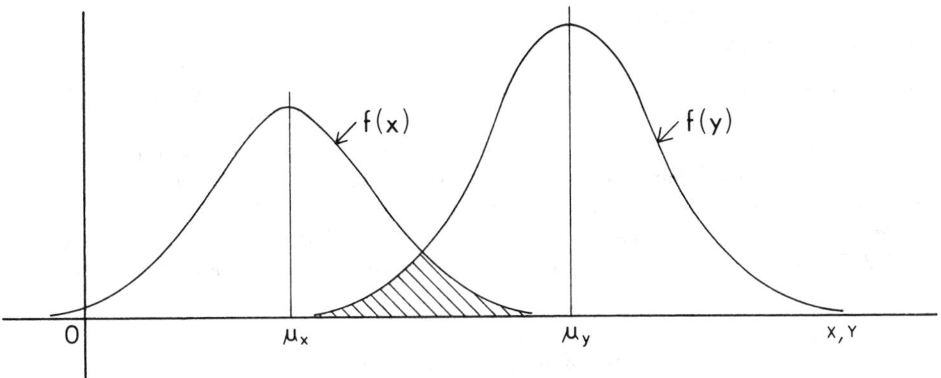

FIG. 3—*Applied stress versus strength.*

Log-Normal Distribution

The frequency distribution is often not symmetrically arranged but is skewed to the right or left. When that happens the mean, mode, and median are not coincident as they are in normally distributed data. The *mode* is the value of the observation that occurs most frequently. It is at the peak value in a frequency distribution. The *median* in a set of observations is the middle observation when the observations are arranged in order of magnitude, that is, half the data are greater than and less than the median.

Many types of data are not symmetrically (normally) distributed. For example, if the mean price of a house is $100,000, a house selected at random is much more likely to cost $200,000 that it is to cost zero. In such cases when the coefficient of variation exceeds 30%, it is often helpful to use the *log-normal distribution,* that is, the natural logarithm of the variable is normally distributed. The log-normal density function is determined as follows in which v is expressed as a decimal:

$$f(x) = (2\pi)^{-0.5}(\beta x)^{-1}\exp[-(\ln x - \alpha)^2/2\beta^2] \tag{25}$$

$$\alpha = \ln[\bar{x}(1 + v^2)^{-0.5}] \tag{26}$$

$$v = s/\bar{x} \tag{27}$$

$$\beta = [\ln(1 + v^2)]^{0.5} \tag{28}$$

where

$z = (\ln x - \alpha)\beta^{-1}$:

$$f(z) = (2\pi)^{-0.5}\beta^{-1}\exp[-(z^2/2) - \beta z - \alpha] \tag{29}$$

$$y = nif(z) \tag{30}$$

where

y = the histogram ordinate at any value of x,
n = number of data in the histogram,

i = histogram interval, and
v = coefficient of variation, decimal.

For a given probability that x will not be exceeded, values of z are determined from Eq 6 or Table 8.

A national survey of 30 samples of brick were tested for initial rate of absorption (suction), grams of H_2O per 30 in.^{2}/min. The mean value was 23.7 with a standard deviation of 31.3. The test data were distributed as indicated in Table 9.

From those data, $\bar{x} = 23.7$, $s = 31.3$, $v = 31.3/23.7$ or 1.32 or 132%, $\alpha = 2.6607$, and $\beta =$ 1.00473. This histogram in Fig. 9 is drawn with intervals of 6 for 31 samples (that is, $i = 6$ and

TABLE 8—*Probability that* x *will not be exceeded.*

Probability, %	z
1	−2.3267
5	−1.645
10	−1.28
20	−0.842
25	−0.674
30	−0.524
33.3	−0.4317
40	−0.253
Mode	$-\beta$
50 (median)	0
Mean	$\beta/2$
60	0.253
66.7	0.4317
70	0.524
75	0.674
80	0.842
90	1.28
95	1.645
99	2.3267
99.5	2.575
99.9	3.10

TABLE 9—*Log-normal fit of brick suction.*

Interval (i)	Observed Frequency	Interval Midpoint	Computed Frequency (Ordinate)
(1) (2)	(3)	(4)	(5)
0 to 6	7	3	7.3
6 to 12	5	9	7.4
12 to 18	6	15	4.9
18 to 24	4	21	3.3
24 to 30	1	27	2.2
30 to 36	3	33	1.6
36 to 42	0	39	1.2
42 to 48	2	45	0.9
48 to 54	2	51	0.7

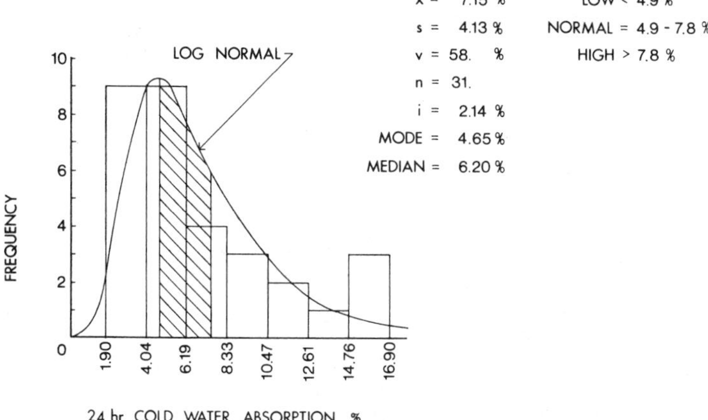

FIG. 4—*Twenty-four-hour cold water absorption of brick* [4].

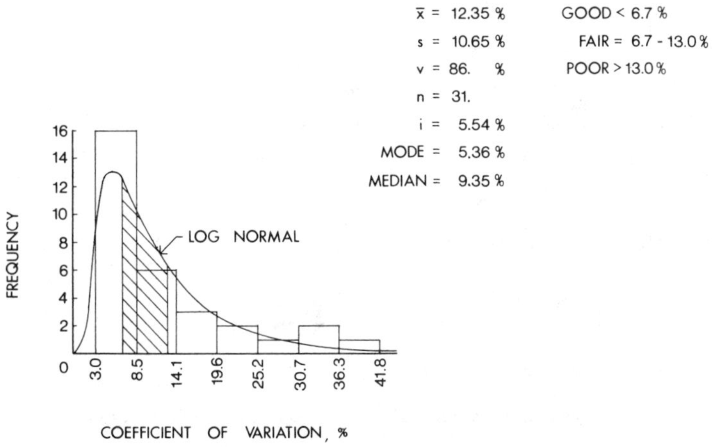

FIG. 5—*Coefficient of variation of 24-h cold water absorption of brick* [4].

$n = 31$). Since $z = (\ln x - \alpha)\beta^{-1}$, $x = \exp(\beta z + \alpha)$. For example where $P(z) = 80\%$, from Table 8, $z = 0.842$, from Eq 29, $f(z) = 0.3938(1.00473)^{-1} \exp[-(0.842^2/2) - (1.00473 \times 0.842) - 2.6607]$ or $f(z) = 0.008325$, and from Eq 30, $y = 0.008325 \times 31 \times 6$ or $y = 1.55$ and $x = \exp[(0.842 \times 1.00473) + 2.6607]$ or $x = 33.3$. The ordinates to the distribution curve are recorded in Table 9.

Application

The principles of statistics as discussed above have been applied to some published test data on masonry. Histograms and log-normal probability density functions have been plotted in Figs. 4 thru 21. In those figures, "high" values are in the upper 33 percentile, low values are in the lower 33 percentile, and normal values are in the middle third of the distribution.

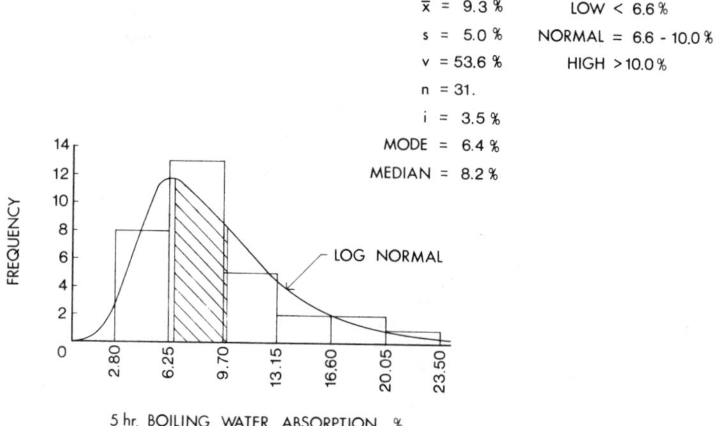

FIG. 6—*Five-hour boiling water absorption of brick* [4].

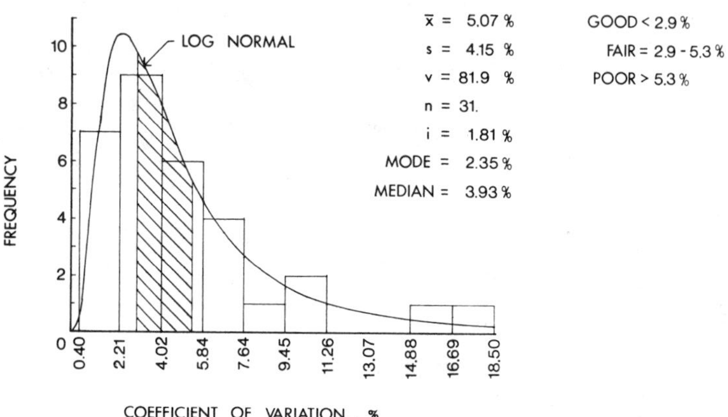

FIG. 7—*Coefficient of variation in the mean saturation coefficient of brick* [4].

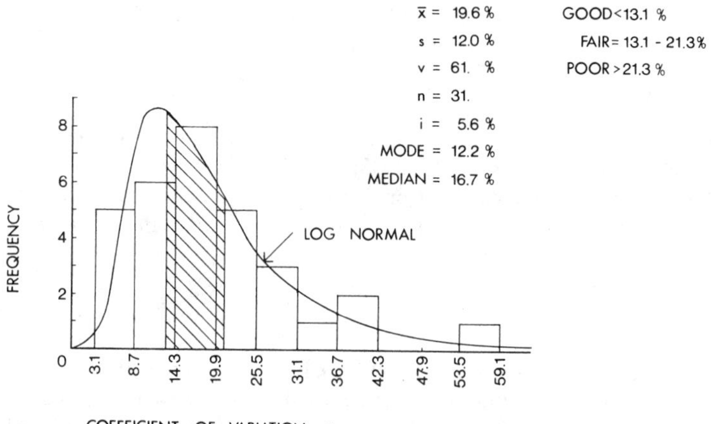

FIG. 8—*Coefficient of variation in the mean initial rate of absorption of brick* [4].

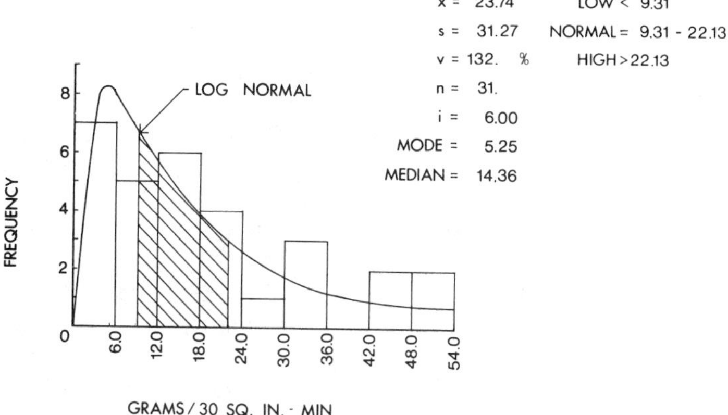

FIG. 9—*Initial rate of absorption of brick* [4].

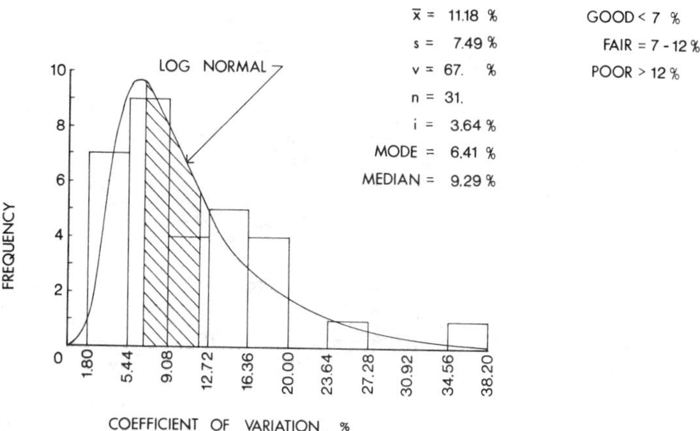

FIG. 10—*Coefficient of variation in brick compressive strength* [4].

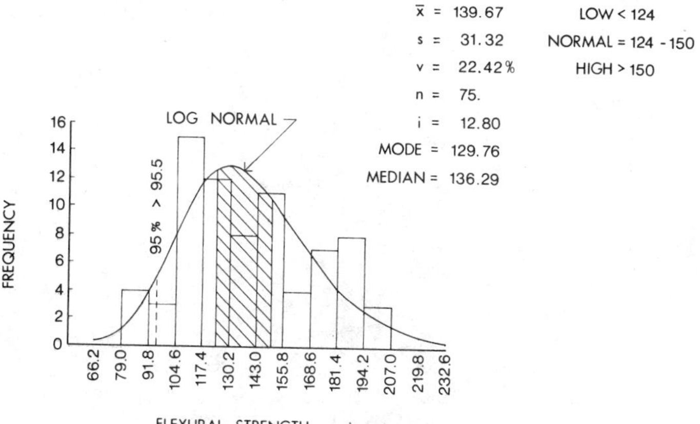

FIG. 11—*Flexural strength of brick masonry wall panels Type S mortar, inspected workmanship, psi* [9].

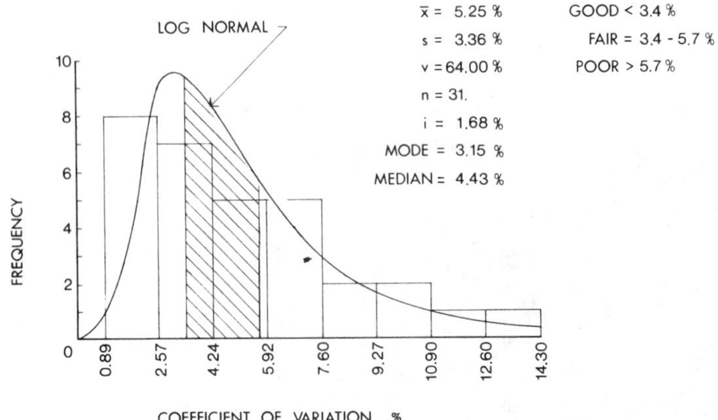

FIG. 12—*Coefficient of variation in compressive strength of laboratory-prepared brick masonry prisms* [4].

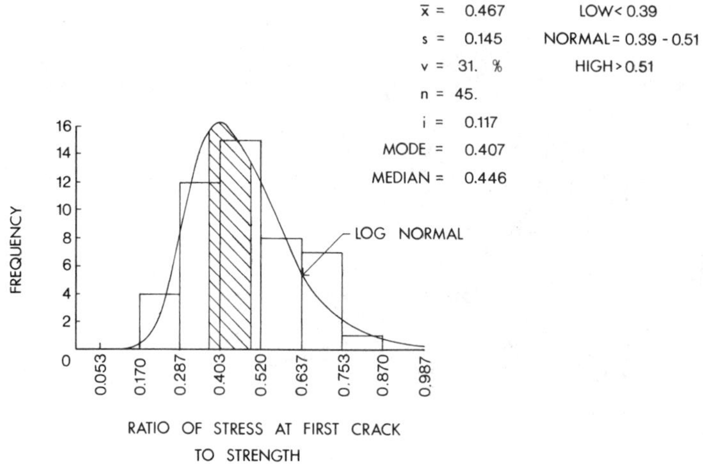

FIG. 13—*Ratio of stress at first crack to ultimate strength of brick masonry in compression* [5].

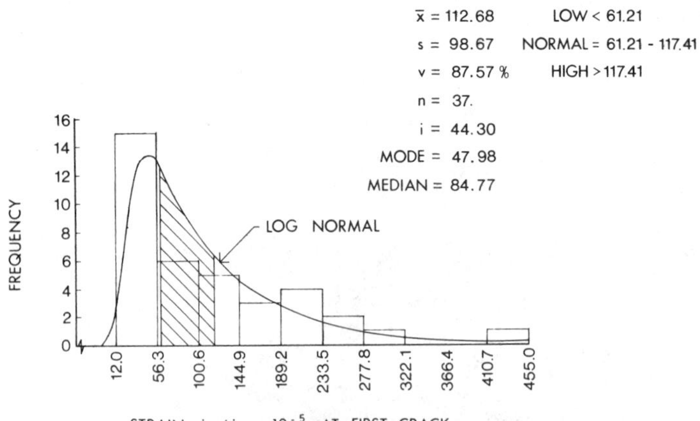

FIG. 14—*Brick masonry prisms; longitudinal strain at first crack; three brick strengths with three mortar types* [5].

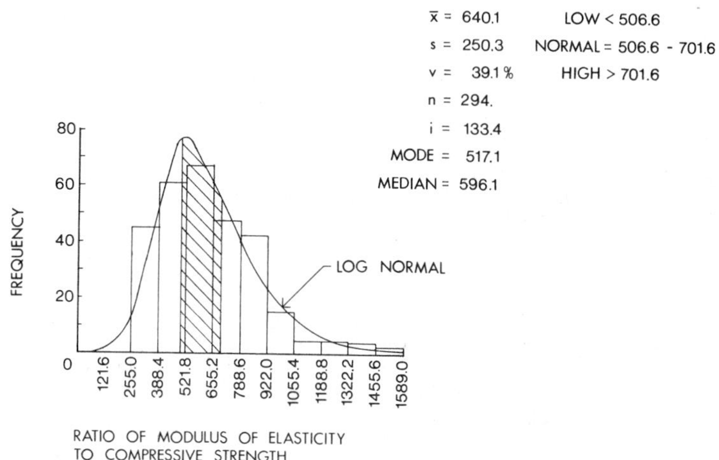

FIG. 15—*Ratio of modulus of elasticity to compressive strength of brick masonry prisms, dimensionless* [6].

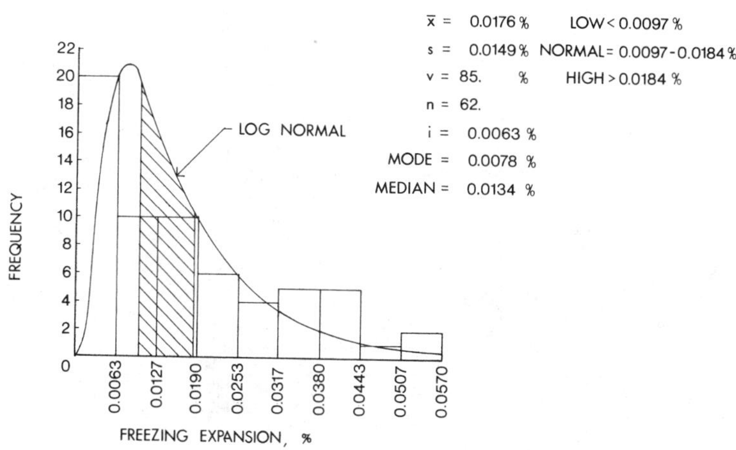

FIG. 16—*Freezing expansion of brick* [7].

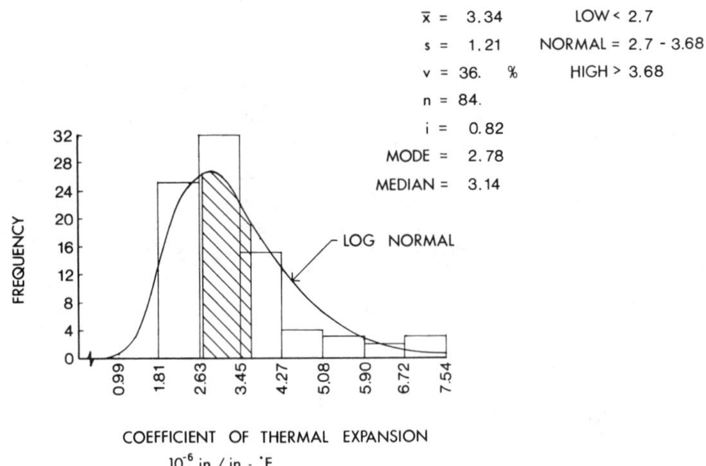

FIG. 17—*Coefficient of thermal expansion of brick (10^{-6} in./in. °F)* [7].

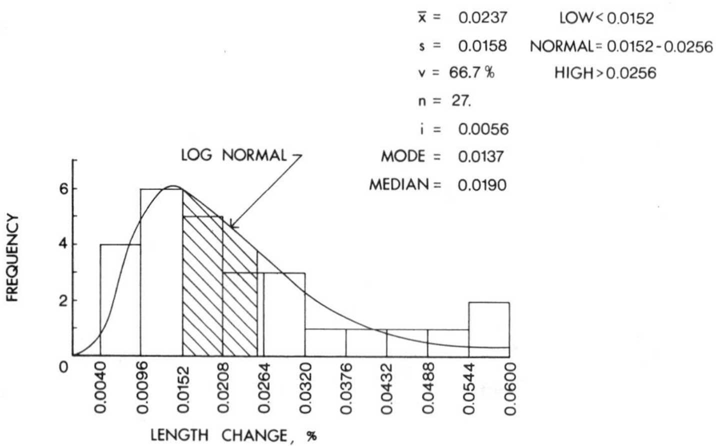

FIG. 18—*Brick moisture expansion at 17 months* [13].

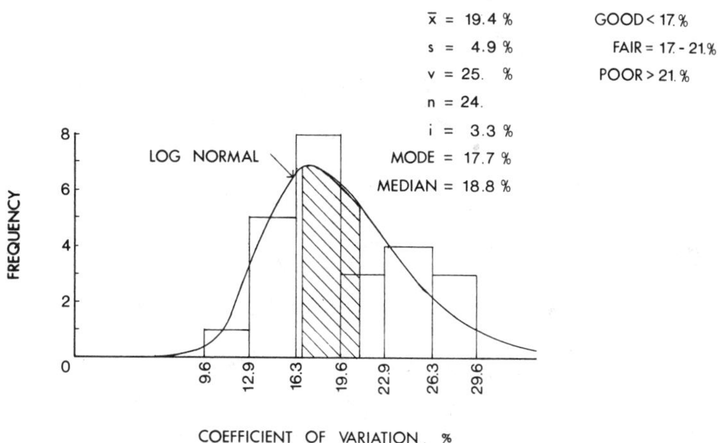

FIG. 19—*Coefficient of variation in head joint thickness in brick masonry* [8].

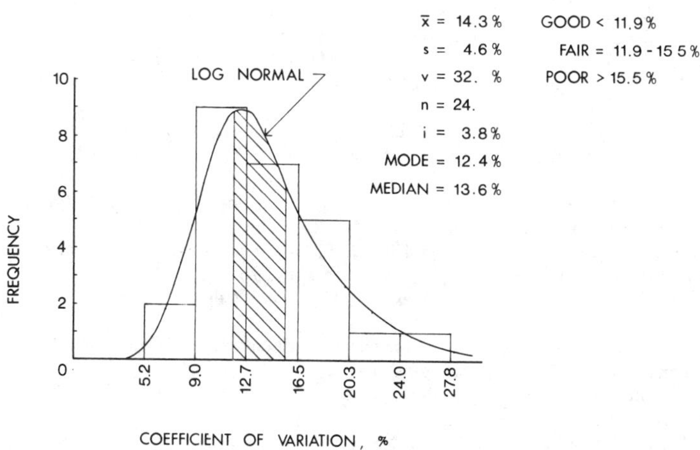

FIG. 20—*Coefficient of variation in bed joint thickness in brick masonry* [8].

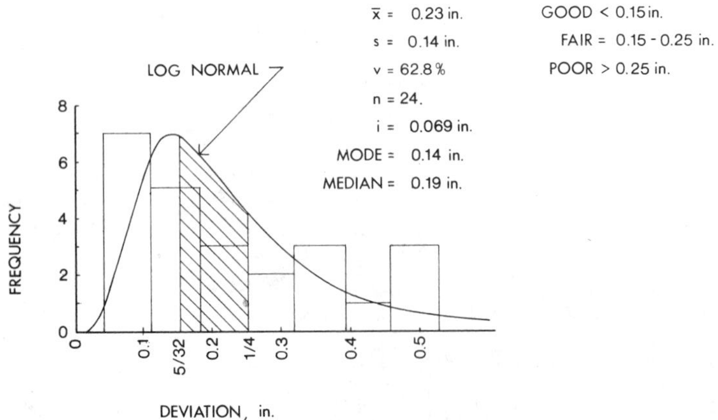

FIG. 21—*Deviation of masonry head joint from vertical in 6-ft height of brick wall* [8].

References

[*1*] Abromourtz, M. and Stegum, I. A., *Handbook of Mathematical Functions,* NBS Applied Science Series 55, Superintendent of Documents, U.S. Government Printing Office, Washington, DC, Nov. 1964.

[*2*] ASTM Method of Sampling and Testing Brick and Structural Clay Tile (C 67-84), *1985 Annual Book of ASTM Standards,* ASTM, Philadelphia, PA, 1985, pp. 49–58.

[*3*] ASTM Recommended Practice for Choice of Sample Size to Estimate the Average Quality of a Lot or Process (E 122-72), *1985 Annual Book of ASTM Standards,* American Society for Testing and Materials, Philadelphia, PA, 1972.

[*4*] *National Testing Program, Progress Report No. 1, Small Scale Specimen Testing,* Brick Institute of America, Herndon, VA, Oct. 1964.

[*5*] Grimm, C. T. and Fok, C.-P., "Brick Masonry Compressive Strength at First Crack," *Masonry International,* University of Edinburgh, Edinburgh, Scotland, Vol. 1, No. 2, July 1984, pp. 18–23.

[*6*] Grimm, C. T., "Elastic Modulus of Clay Brick Masonry," *Proceedings of International Symposium on Reinforced and Prestressed Masonry,* University of Edinburgh, Scotland, 26–28 Aug. 1984, pp. 225-252.

[*7*] Grimm, C. T., "Probabilistic Design of Expansion Joints in Brick Cladding," *Proceedings: 4th Canadian Masonry Symposium,* University of New Brunswick, Freducton, NB, Canada, June 1986, pp. 553-568.

[*8*] Grimm, C. T., "Some Brick Masonry Workmanship Statistics," *Journal of the Construction Division,* American Society of Civil Engineers, New York, NY, March 1988.

[*9*] Gross, J. G., Dickers, R. D. and Grogan, J. C., *Recommend Practice for Engineered Brick Masonry,* Brick Institute of America, Reston, VA, Nov. 1969, pp. 257-262.

[*10*] Haugen, E. B., *Probabilistic Approaches to Design,* John Wiley & Sons, Inc., New York, NY, 1968, p. 123.

[*11*] Natrella, M. G., *Experimental Statistics,* NBS Handbook 91, Superintendent of Documents, U.S. Government Printing Office, Washington, DC, 1966.

[*12*] Hald, A., *Statistical Theory with Engineering Applications,* John Wiley & Sons, New York, NY, 1965, pp. 303-316.

[*13*] Young, J. E. and Brownell, W. E., "Moisture Expansion of Clay Products," *Journal of The American Ceramic Society,* Columbus, OH, 1 Dec. 1959, Vol. 42, No. 12, pp. 571-581.

John H. Matthys,[1] *James T. Houston,*[2] *and Alireza Dehghani*[3]

An Investigation of an Extended Plastic Life Mortar

REFERENCE: Matthys, J. H., Houston, J. T., and Dehghani, A., **"An Investigation of an Extended Plastic Life Mortar,"** *Masonry: Materials, Design, Construction, and Maintenance, ASTM STP 992,* H. A. Harris, Ed., American Society for Testing and Materials, Philadelphia, 1988, pp. 193–219.

ABSTRACT: The use of an extended plastic life mortar which can be stored for a prolonged period of time without losing its desirable properties can be beneficial to the masonry industry in several ways.

This paper presents the findings of a two-phase study related to an extended plastic life mortar (ready mix mortar) using a single brand of set-reducing additive. Phase 1 was optimization of a final mix in terms of setting characteristics and compressive strength. Mixes were made which examined the variables of dosage rates, cement content, lime content, and fly ash content. Each mix was characterized by compressive strength, air content, water retention, and plasticity. Several mix designs were generated for achieving particular strength and setting characteristics.

Phase 2 compared one of the generated extended plastic life mortars to a standard portland cement lime Type N mortar as defined in ASTM Specification for Mortar for Unit Masonry (C 270) with respect to mechanical properties. The properties examined included: compressive strength, bond strength, freeze thaw, corrosion, shrinkage, efflorescence, and compressive strength and modulus of elasticity of masonry prisms.

KEY WORDS: mortar, ready mixed, retarded strength, freeze-thaw resistance, corrosion, shrinkage, efflorescence

The commercialization of extended plastic life (EPL) mortars as a ready-mixed product available to the masonry industry is a relatively new concept in the United States. At this time several chemical admixtures have been developed and are being promoted for use in producing ready-mixed EPL mortars. The marketing of such products in the United States has developed quite rapidly and has been spurred on by such factors as prior usage in non-U.S. construction markets and desires of inventors, entrepreneurs, and major companies to be first in the market place for obvious reasons. The incentive of the potential user is to achieve a promised or implied net savings over the conventional job-mixed mortars.

Agencies and individuals who participate in the development of specifications and standards for the use of construction materials in the United States are traditionally conservative in their approach to that task, and that stance is certainly proper in these writers' points of view. Public safety and long-term performance issues as well as the more common and obvious short-term physical properties of new products must be adequately assessed prior to public use.

In the case of EPL mortars the authors have found little published research data from U.S. sources; foreign sources were not reviewed. In addition, there is currently no recognized U.S. specification or standards specifically applicable to the evaluation and use of EPL mortars. The

[1]Professor, Civil Engineering Dept., University of Texas at Arlington, Arlington, TX 76013.
[2]Consulting engineer, Constructive Engineering, 1711 S. Parkway Blvd., Mesquite, TX 75149.
[3]Student, Civil Engineering Dept., University of Texas at Arlington, Arlington, TX 76013.

current, generally referenced U.S. specification for mortar for unit masonry is ASTM C 270, and EPL mortars do not comply with regard to duration of use provisions and in some cases may exceed the air content limitations. Consequently, there is a definite need for both thorough research and specification development to address the subject of EPL mortars for use in masonry construction in the United States. The market testing for such products has produced sufficient results to motivate both the technology and marketing interests in the masonry industry to work together in producing a cost-effective product with dependable short-term and long-term performance properties.

There are at least six basic parameter study areas related to the proper evaluation of construction materials composed of filler aggregates bound together in a matrix containing portland cement and other related binder components.

1. Physical and chemical properties, variability, and compatibility of each binder component.
2. Compatibility of the binder components with the aggregate filler of the product matrix.
3. The effects of proportioning the binder components upon the performance of both the plastic and hardened phase of the product matrix.
4. The influence of the batching and mixing cycle and plastic phase history upon short-term and long-term performance of the product matrix.
5. The influence of environmental factors, especially temperature, upon the plastic phase and in-place performance of the product matrix.
6. The long-term performance of the product matrix with emphasis on chemical and physical stability in those environmental conditions where the product will actually be used.

Other areas of needed study for new products relate to considerations of constructability, that is, the influence of construction variables upon the performance of the in-place product. The emphasis of this study was to evaluate selected materials parameters only.

Phase 1: Early Screening Tests of EPL Mortars

Research on EPL mortars conducted at the Construction Research Center at The University of Texas at Arlington has been selectively reported in this paper. One experimental chemical admixture proposed for use in producing EPL mortar for unit masonry construction was evaluated with respect to selected performance criteria. Preliminary testing using a wide variety of controlled test parameters was conducted to provide an opportunity to develop testing procedures and to become familiar with the general characteristics of EPL mortars (see Figs. 1–8). This screening work, termed Phase 1, provided an opportunity to make several general observations concerning data trends for the parameters studied. This work is considered very preliminary by the authors and will not be reported in detail since the process of optimizing testing procedures was not deemed to be complete and only a minimum of replicate testing was completed at this writing date. However, the authors have made a number of significant observations regarding EPL mortar performance which may be helpful in the planning of future-related research by these authors and others. These observations and data presented here are *not known* to be characteristic of commercially available EPL mortars now in use in the U.S. *nor* are such relationships *implied* by these authors. As previously stated, the research discussed in this paper is limited to one experimental admixture used to produce a variety of EPL mortar mixes under laboratory-controlled conditions. A summary of general observations made to date is as follows:

1. The mix proportions and binder types had a significant effect upon the mortar setting characteristics and upon the retempering procedures required to maintain plasticity beyond about 4 h and up to 72 h of age. (A 5-min mixing cycle was standardized for all work done to date.)

FIG. 1—*Mortar batch mixing.*

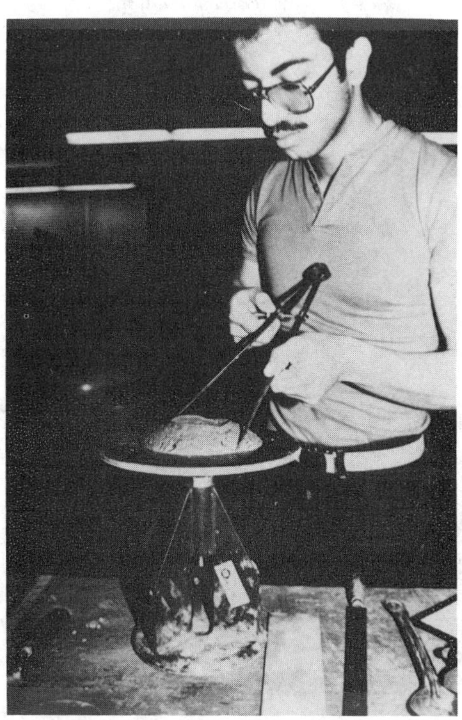

FIG. 2—*Flow test.*

FIG. 3—*Water retention test.*

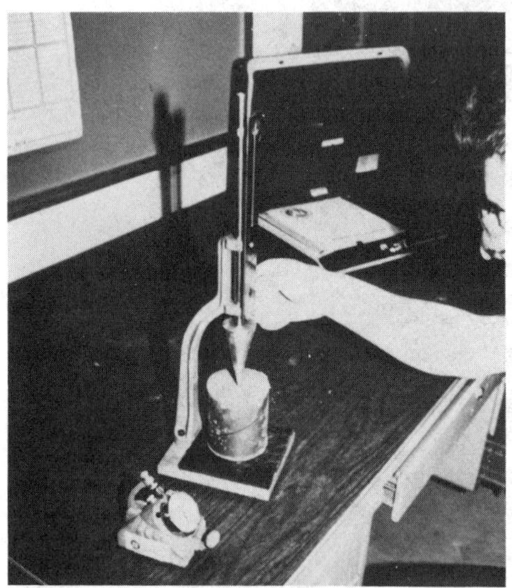

FIG. 4—*Penetrometer test.*

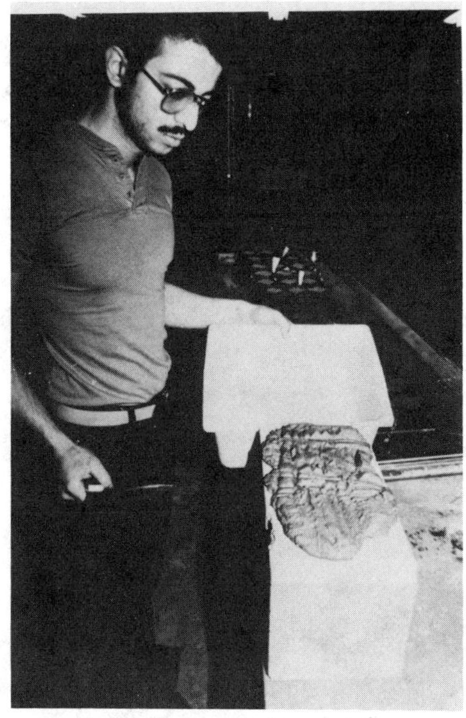

FIG. 5—*Masonry unit suction effect on mortar.*

FIG. 6—*Two-inch by two-inch cube compression specimen.*

FIG. 7—*Containerized EL mortar.*

FIG. 8—*Compression cube specimens.*

2. Early screening tests suggest that EPL mortars may be significantly subject to the effects of specimen size, shape, and curing regimen.

3. Strength retrogression from 7 to 28-day air-cured tests was observed in a significant number of EPL mortar mixes, especially when the mixtures were kept plastic for 24 h or more. It should be noted that compressive strength retrogression did not occur in any moist-cured specimens.

4. Almost all EPL mortar mixes had air contents exceeding 12%, with many exceeding 20%.

5. For all combinations of binders (cement only, cement plus fly ash, and cement plus lime)

and the use of EPL admixtures in varying dosages, a severe reduction in compressive strength was noted when comparing cubes made immediately after initial mixing and those made after maintaining mortar plasticity for 24 to 72 h. Strength reductions typically ranged from 50 to 90+% for 28-day tests.

6. Increasing cement content and varying EPL admixture dosage were only marginally successful in improving strengths of mortars aged 24 h or more.

7. Increasing fly ash content from 20 to 50% of cementitious material weight produced a significant reduction in strengths of cubes molded immediately after initial mixing, but much less effect was seen in strengths of EPL mortars kept plastic for 24 to 72 h.

8. In general, it is felt that EPL mortar testing is quite sensitive to testing procedure, both in procedure design and operator technique. The influence of the binder systems employed, including the chemical admixtures coupled with the uncharacteristically long set time extension for portland cement-based binders, has resulted in a hardened mortar product which would be expected to be "testing techniques sensitive." As a result, the authors recommend that those responsible for test and standards development should give due consideration to the potential of this product to display significant sensitivity to field production and field construction variables as well as laboratory test procedures.

Tables 1 and 2 have been included in this paper for general reference only in order to illustrate the relative magnitude of cube strengths for selected variations in mortar mix proportions.

Phase 2: Comparisons of Extended Plastic Life Mortar to a Conventional Type N Mortar

Based on Phase 1 study, several final types of mortar mixes were generated based on setting times and 5.08-cm cube compression strength values. These mixes, when quantified on compression strength requirements (initial mixing specimens only), would be classified as Type N, Type S, or Type M mortars. With regard to relative mix economy, these EPL mortars varied from 12% less to 11% more expensive than a standard Type N proportion specification portland cement lime mortar.

Since the majority of all mortars used in masonry construction would probably be Type N, it was felt that a comparison of some general physical properties between a Type N EPL mortar and a Type N ASTM C 270 portland cement lime mortar would be of interest. It was also decided to examine a few properties of assemblages constructed with these mortars. Mortar prop-

TABLE 1—*Mortar batch properties*[a] *(mixes with cement and sand only).*

Mix No.	≈C.F., kg/m^3	Cement, kg	Water, kg	Sand, kg	EPL Admixture	% Air	Water Retention	Flow	W/C
EPM5A	0.98	2.67	1.70	14.20	170	25.8	85	127	0.63
EPM5B	0.98	2.67	1.52	14.20	283	27.9	86	127	0.57
EPM5C	0.98	2.67	1.50	14.20	425	25.6	84	134	0.56
EPM6A	1.21	3.31	1.67	14.20	170	23.8	86	129	0.51
EPM6B	1.21	3.31	1.40	14.20	283	27.6	91	129	0.42
EPM6C	1.21	3.31	1.0	14.20	425	25.2	80	129	0.30
EPM7A	1.44	3.93	1.73	14.20	170	22.9	86	133	0.44
EPM7B	1.44	3.93	1.69	14.20	283	25.3	90	130	0.43
EPM7C	1.44	3.93	1.0	14.20	425	22.6	73	134	0.25

[a]Batch weights [kg saturated surface dry (SSD)] flow = % (ASTM Specification for Flow Table for Use in Tests of Hydraulic Content C 230-80); air = % by volumetric method (ASTM Test Method for Air Content of Hydraulic Cement Mortar C 185-80); EPL admixture = dosage rate (g/45.3 kg total cementitious material); water retention = % (ASTM Methods for Physical Testing of Quicklime, Hydrated Lime, and Limestone C 110); W/C = water cement ratio (by weight).

TABLE 2—*Air-cured cube strengths (kPa) for cement only EPL mortars.*

EPL Admixture Dosage Rate	fc—2 Days			fc—7 Days			fc—28 Days		
	170 g/45.3 kg	283 kg	425 g	170 g	283 g	425 g	170 g	283 g	425 g
EPM5 (A, B, C) 0 h	2799	2847	69	6315	5536	985	6364	6336	1158
Suspension times, 24 h	751	1537	103	1399	2847	1385	1227	2951	1454
48 h	434	696	234	723	1854	1110	799	1744	2213
72 h	337	344	544	503	655	1434	475	613	1868
EPM6 (A, B, C) 0 h	543	1716	144	8018	8370	1179	8811	5171	1399
Suspension times, 24 h	813	1889	75	1744	3516	799	1847	2640	613
48 h	537	2020	13	1006	3330	420	1054	2061	317
72 h	137	468	144	475	1068	351	455	723	455
EPM7 (A, B, C) 0 h	7212	1765	110	11163	7067	923	8453	7694	1165
Suspension times, 24 h	2592	3171	124	4481	1385	627	3937	4474	544
48 h	854	2558	75	1413	4212	730	889	4019	675
72 h	337	1089	27	558	2261	475	475	1641	365

NOTE: EPM = extended plastic mortar; fc = cube compressive strength.

erties evaluated included freeze-thaw resistance, corrosion performance, shrinkage behavior, and efflorescence characteristics. Assemblage properties examined included compressive strength, compression stiffness, flexural bond strength, and assemblage shrinkage. The tests involving extended plastic life mortar used specimens constructed using three different suspension time intervals: initial mixing, 6 to 7-h suspension, 24-h suspension except as noted.

Materials

Masonry Units—Two types of clay brick were selected for construction of assemblages as shown in Fig. 9. The net cross-sectional area was approximately 78% of the gross area, classifying them as solid units. Initial rate of absorption tests and compression tests were conducted on the units according to ASTM Method of Sampling and Testing Brick and Structural Clay Tile (C 67). Test results and unit dimensions are given in Table 3.

Type I brick was a medium compressive strength and medium suction unit. Type II brick was a high compressive strength and low suction unit.

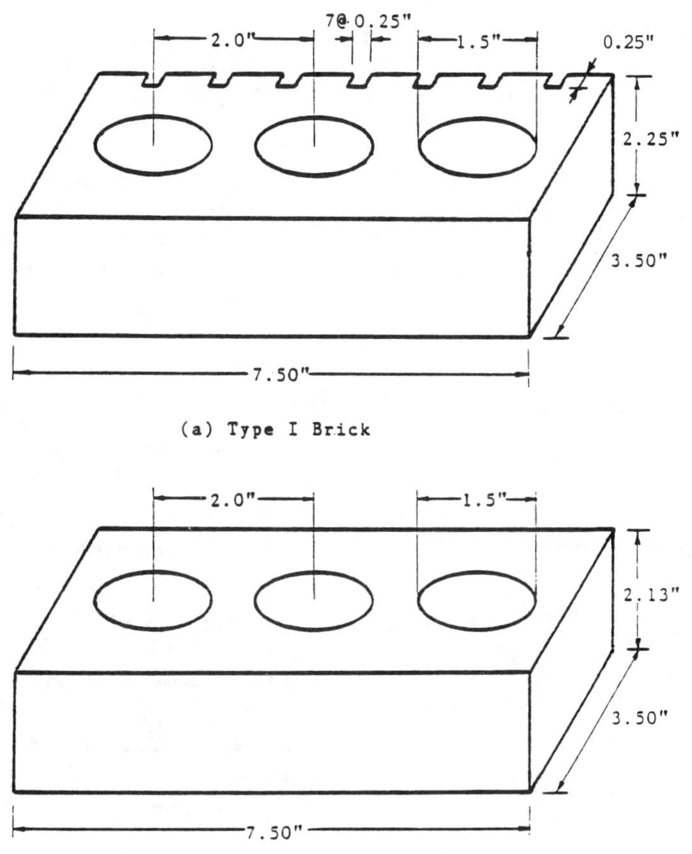

FIG. 9—*Masonry units.*

TABLE 3—*Physical properties of masonry units.*

Type	Height, cm	Width, cm	Length, cm	Net Area, cm^2	Gross Area, cm^2	Initial Rate of Absorption, g/min/193 cm^2	Compressive Strength, kPa
I[a]	5.71	8.89	19.05	132	169	12.3	72 983
II	5.41	8.89	19.05	134	169	2.32	145 794

[a]Average of five tests.

Mortar

The conventional Type N portland cement lime mortar used was based on the proportion specification. Water was added to produce a flow of 130 ± 5%. All conventional mortar was used within a 2½-h period and exhibited average 28-day moist-cured cube compressive strength of 7949 kPa.

Assemblage Tests

Flexural Bond Test—Two stack bonded prisms five units high, fully bedded, with concave tooled mortar joints were built by a qualified mason (Fig. 10). Prisms were air cured in the laboratory environment for testing at 3, 7, and 28 days using the bond wrench. Figures 11 and 12 show the flexural bond strength with age for prisms constructed with Type I and II units, respectively.

For the medium absorption (Type I brick) prisms, the extended plastic life mortar specimens for initial mixing had 3-day bond strengths comparable to that of the conventional Type N mortar, a 40 and 50% higher strength after 7 and 28 days, respectively. The 7-h suspension mortar specimens had a 50% lower 3-day strength but about the same 7 and 28-day strength as

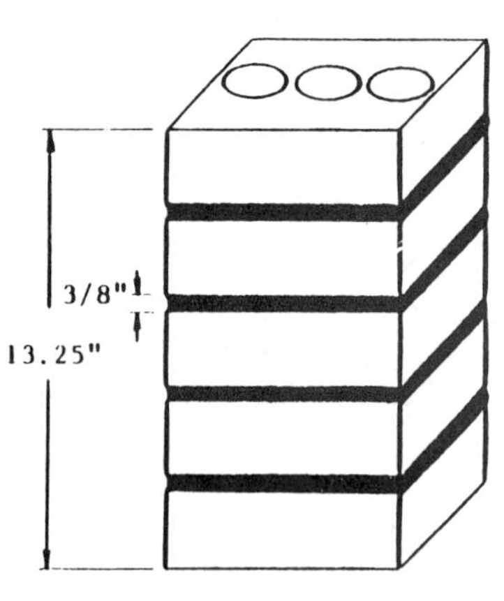

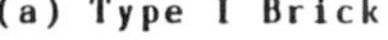

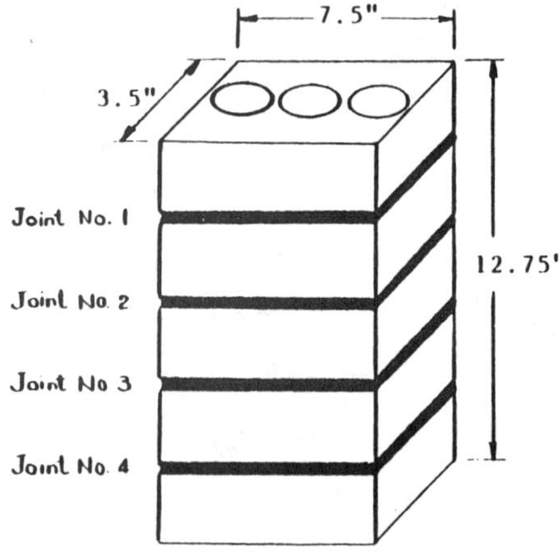

FIG. 10—*Prism specimens for bond and compression test.*

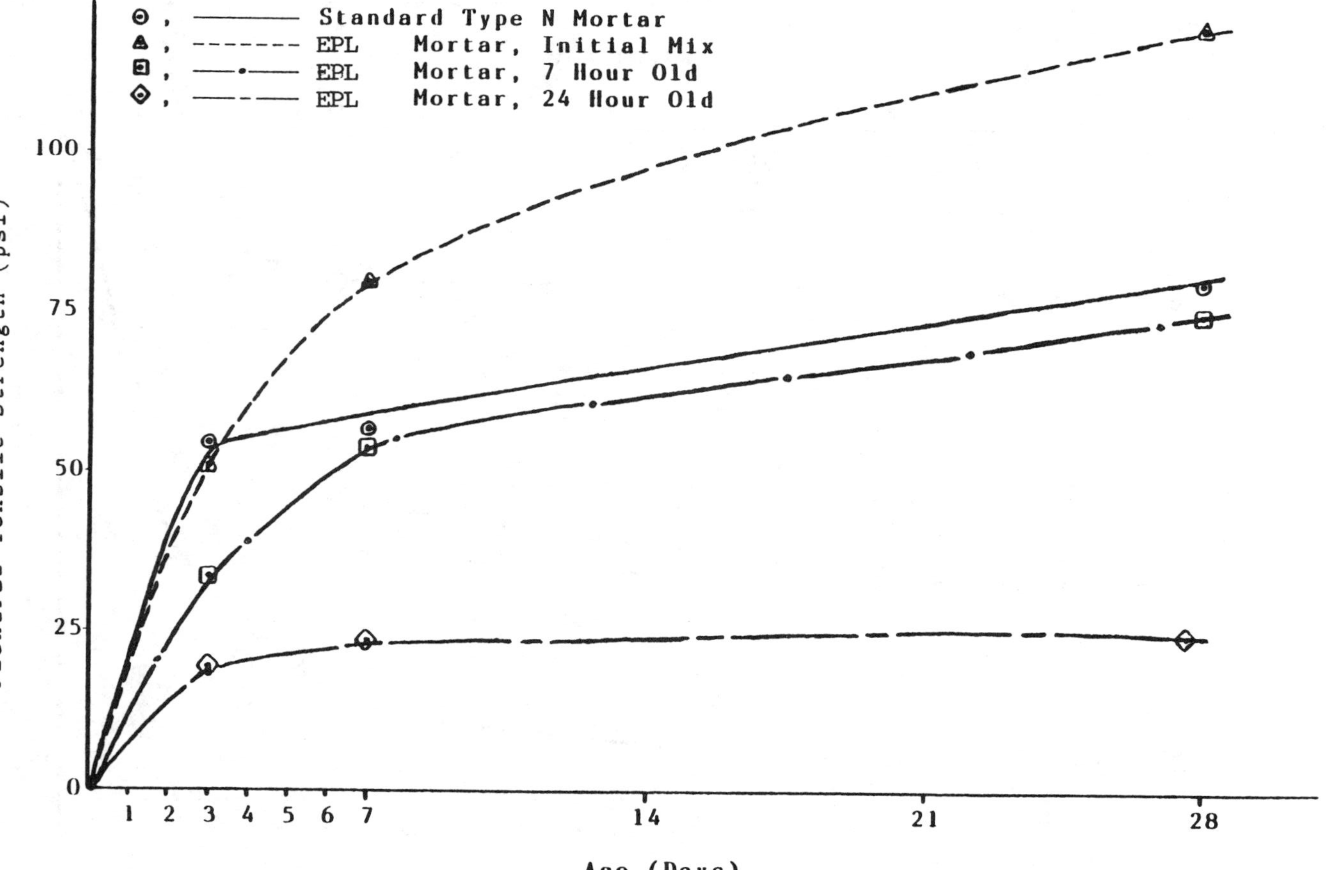

FIG. 11—*Flexural tensile strength of masonry bonds with time, Type I brick.*

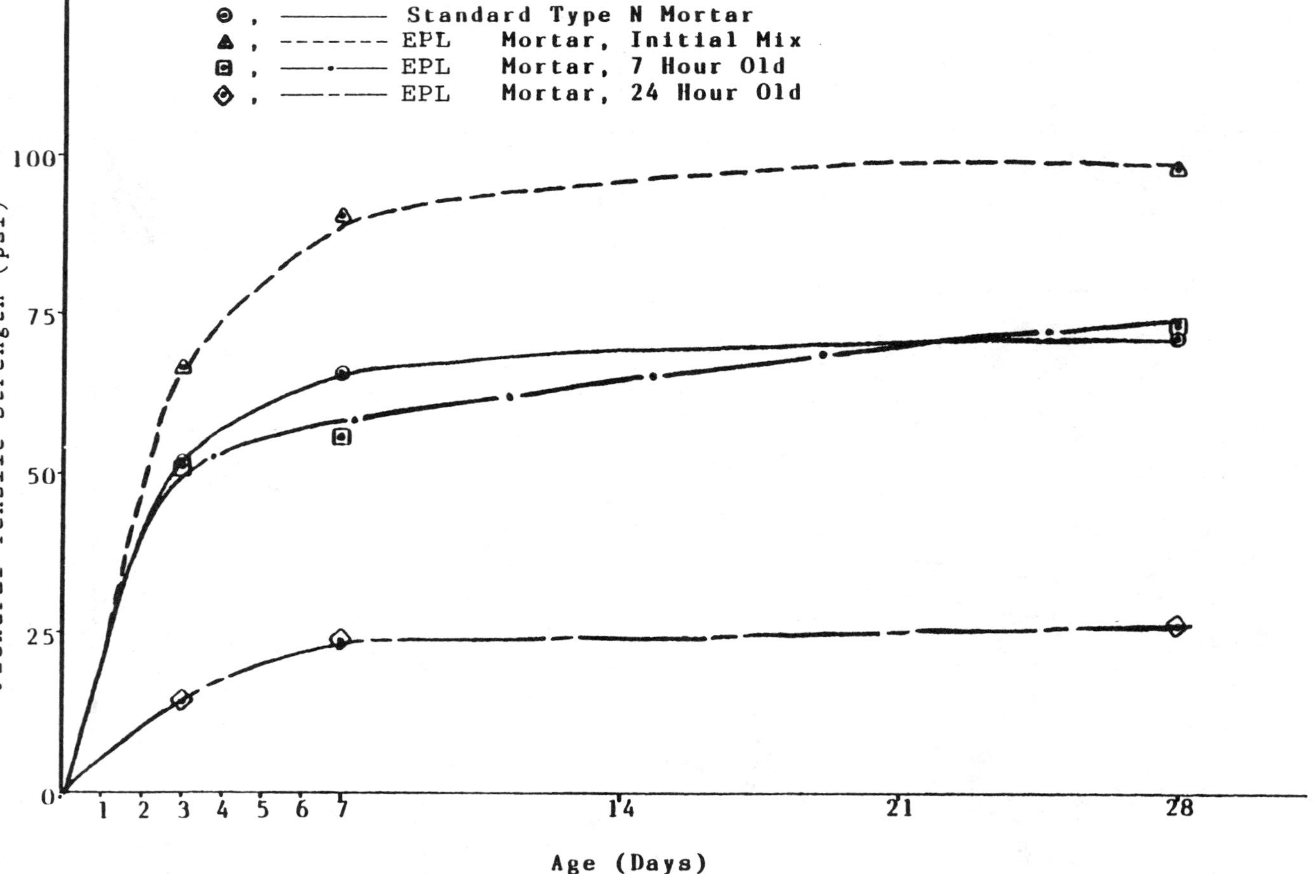

FIG. 12—*Flexural tensile strength of masonry bonds with time, Type II brick.*

conventional Type N mortar. The retempered 24-h extended mortar suspension specimens exhibited 3-day, 7-day, and 28-day strengths about 60% to 70% less than that for standard Type N mortar.

For the low absorption (Type II brick) prisms both mortars exhibited about 7% lower flexural strength compared to the medium absorption specimens. The comparison of the two mortars at 3, 7, and 28 days are basically the same as for the Type I brick specimens.

Compression Prisms

As for the flexural bond study, two stack bonded prisms were constructed for examining compression strength of masonry assemblages for the two brick types at 3, 7, and 28 day ages. The specimens were capped and tested according to ASTM Test Methods for Compressive Strength of Masonry Prisms (E 447) (see Figs. 13 and 14). All specimens were air cured in the laboratory.

FIG. 14—*Compression prism testing.*

Test results for medium absorption (Type I brick) prisms are given in Fig. 15. Three-day strength and 28-day strength for initial mixing for extended plastic life mortar was equivalent to the Type N standard. The prisms for 7-h and 24-h suspension periods, exhibiting lower strength at all ages, gave 28-day strength at 20% and 40% less than those of standard mortar, respectively.

Test results for the low absorption (Type II brick) prisms are given in Fig. 16 and exhibited a lower overall strength as compared to Type I brick. With the exception of the 7-h suspension period strength at 28 days, the values are basically the same as for Type I brick.

Modulus of Elasticity in Compression—To determine the initial modulus of elasticity in compression, the two brick types were used as in the flexural bond and compression test to build stack bonded prisms using standard Type N portland cement lime mortar and the extended plastic life mortar at 0, 7, and 24-h suspension periods. For each test two prisms were constructed and tested after 28 days of lab curing. To measure strains, a transducer was placed at the center line on both sides of the prism's wide face with a 20.32-cm gage length. Test data curves are shown in Figs. 17 and 18. Modulus values are tabulated in Table 4. The prisms with the higher strength units were stiffer.

The extended plastic life mortar selected for comparison purposes was the most economical of the four final mixes generated in Phase 1. It contained only portland cement, admixture, and sand. This mortar showed good workability and, according to the mason, was easy to work with. It remained quite plastic for 24 h while maintaining workability. Without retempering, this mix hardened in 35 h. The EPL mortar initial flow was in the order of 125%. Flows at 6-h suspension were about 105%. Flows at 24-h suspension were about 55%. All suspended mortar was retempered to approximately the same initial flow prior to construction of specimens. The mortar set time in assemblages was approximately 30 min for the high IRA brick (Type I) and 1½ h for low IRA (initial rate of absorption) brick (Type II). The mortar had an average air content of 21.3% and water retentivity value of 78.6%. The average 28-day moist-cured strength of mortar was 6338, 7239, and 4343 kPa for 5.08-cm cubes made initially, 6 h, and 24 h after initial mixing, respectively. The drop in mortar strength for 24-h suspension was associated with retempering.

It should be noted that although both mortars were mixed to approximately a 130% flow, the extended plastic life mortar showed excellent workability at flows as low as 80%. The color of cubes made from the extended plastic mortar was slightly darker than the standard mortar cubes. However, in assemblages, both mortars approached a similar shade of color after a few days.

Masonry Assemblage Shrinkage—To examine the shrinkage of mortar when used in masonry assemblages, brass gage plugs were epoxyed to two faces of the 28-day compression test stacked bonded specimens. The gage length was then measured at time intervals of 1, 4, 7, 14, 21, and 28 days using a dial micrometer conforming to ASTM C 246. The length changes were then

TABLE 4—*Tabulated values of modulus of elasticity.*

Mortar Mix	Modulus of Elasticity, Avg of Two Samples, kPa	
	Type I Brick	Type II Brick
Type N	9 653 000	14 755 300
EPL mortar		
Initial mix	11 514 650	14 410 550
After 7 h	8 274 000	13 721 050
After 24 h	393 000	468 860

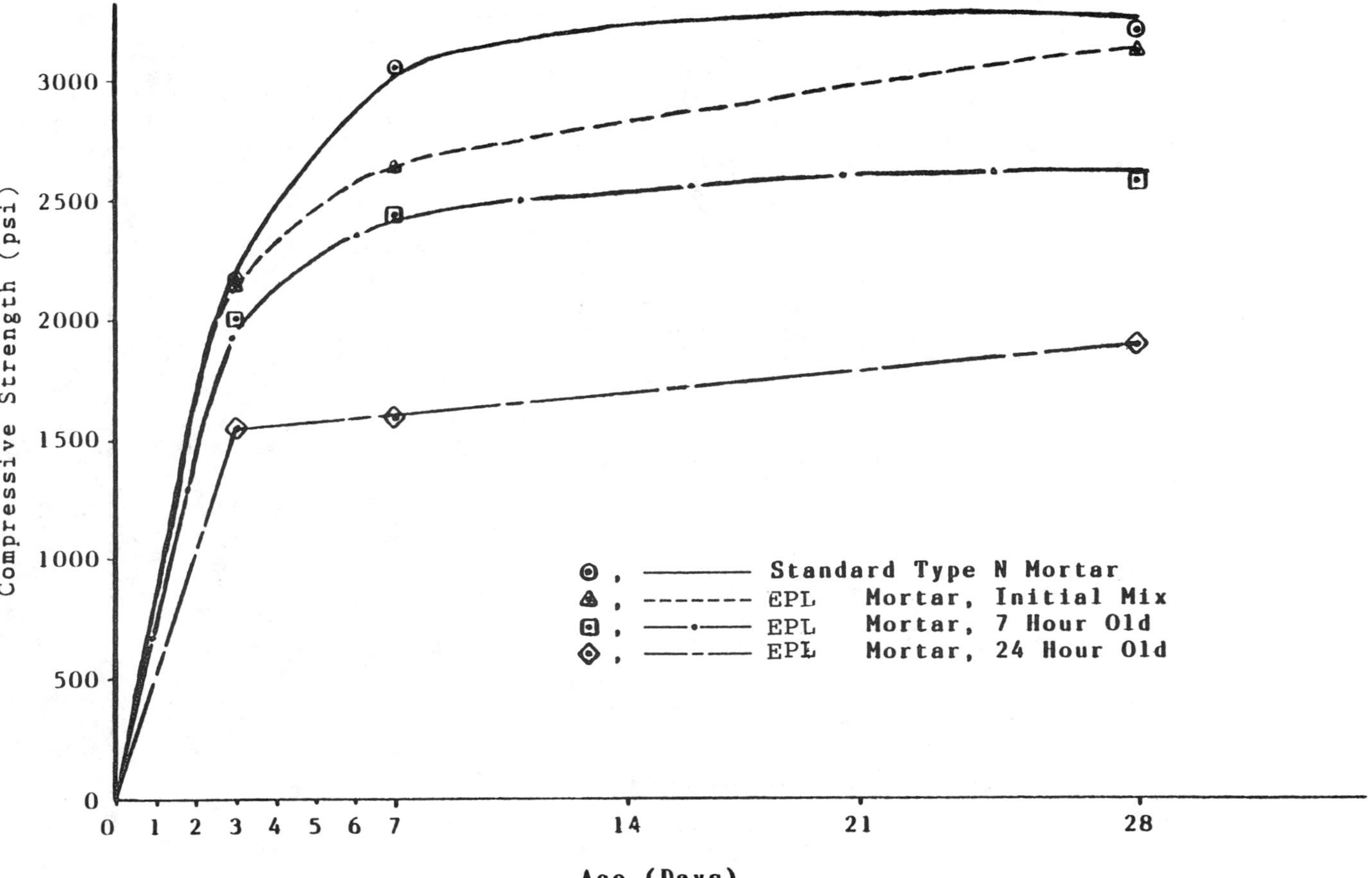

FIG. 15—*Compressive strength of masonry prisms with time, Type I brick.*

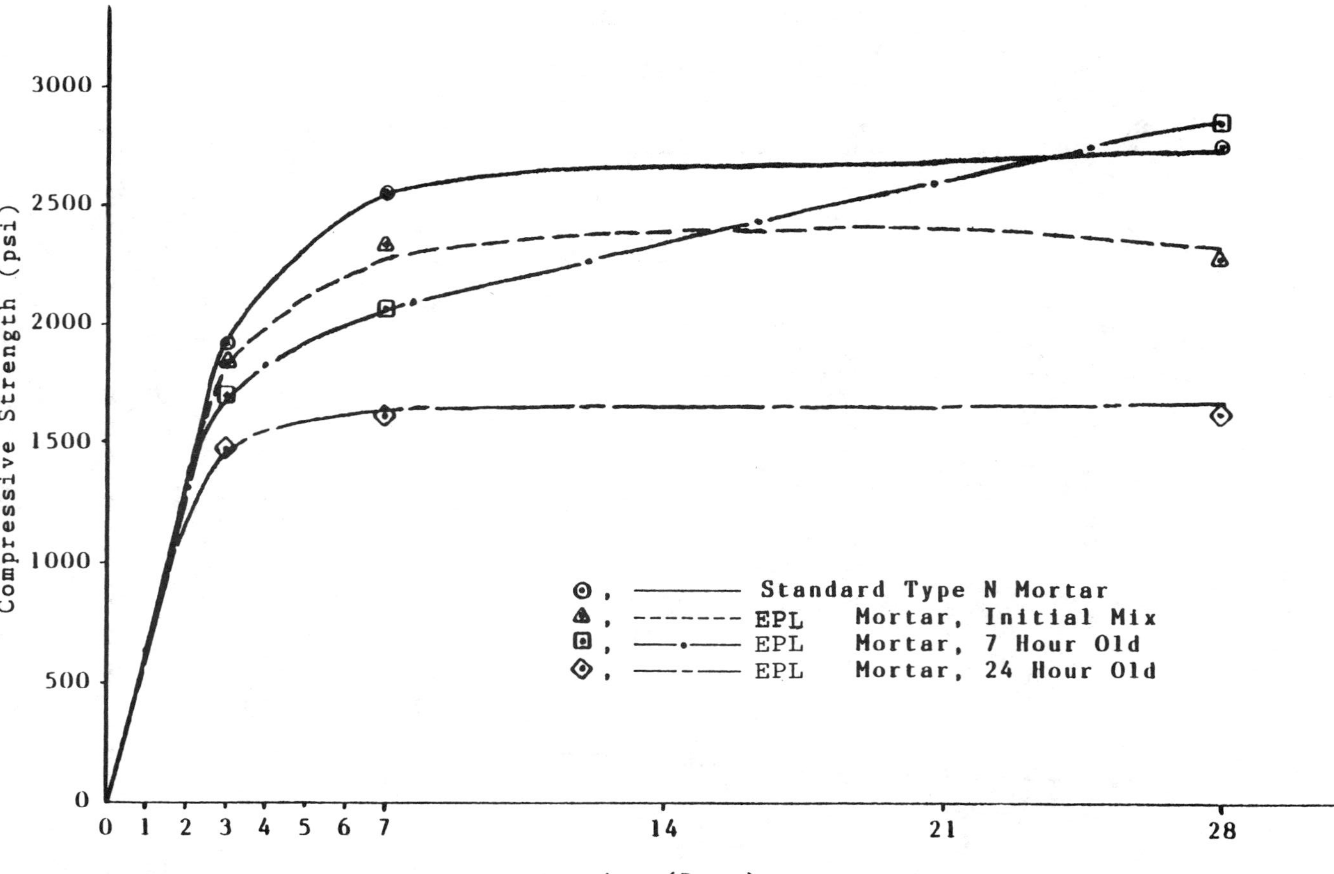

FIG. 16—*Compressive strength of masonry prisms with time, Type II brick.*

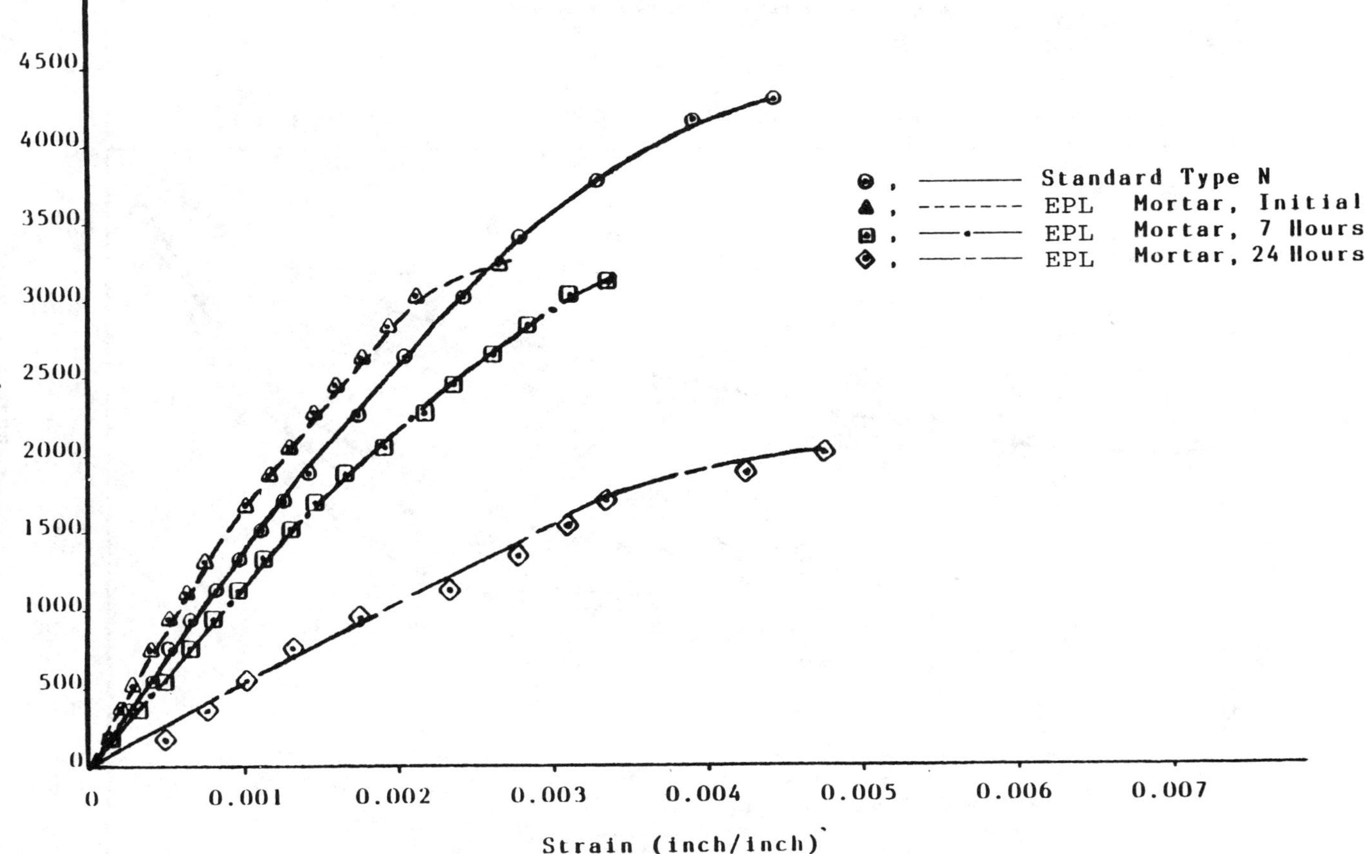

FIG. 17—*Stress-strain curve for compression prisms (Type I brick).*

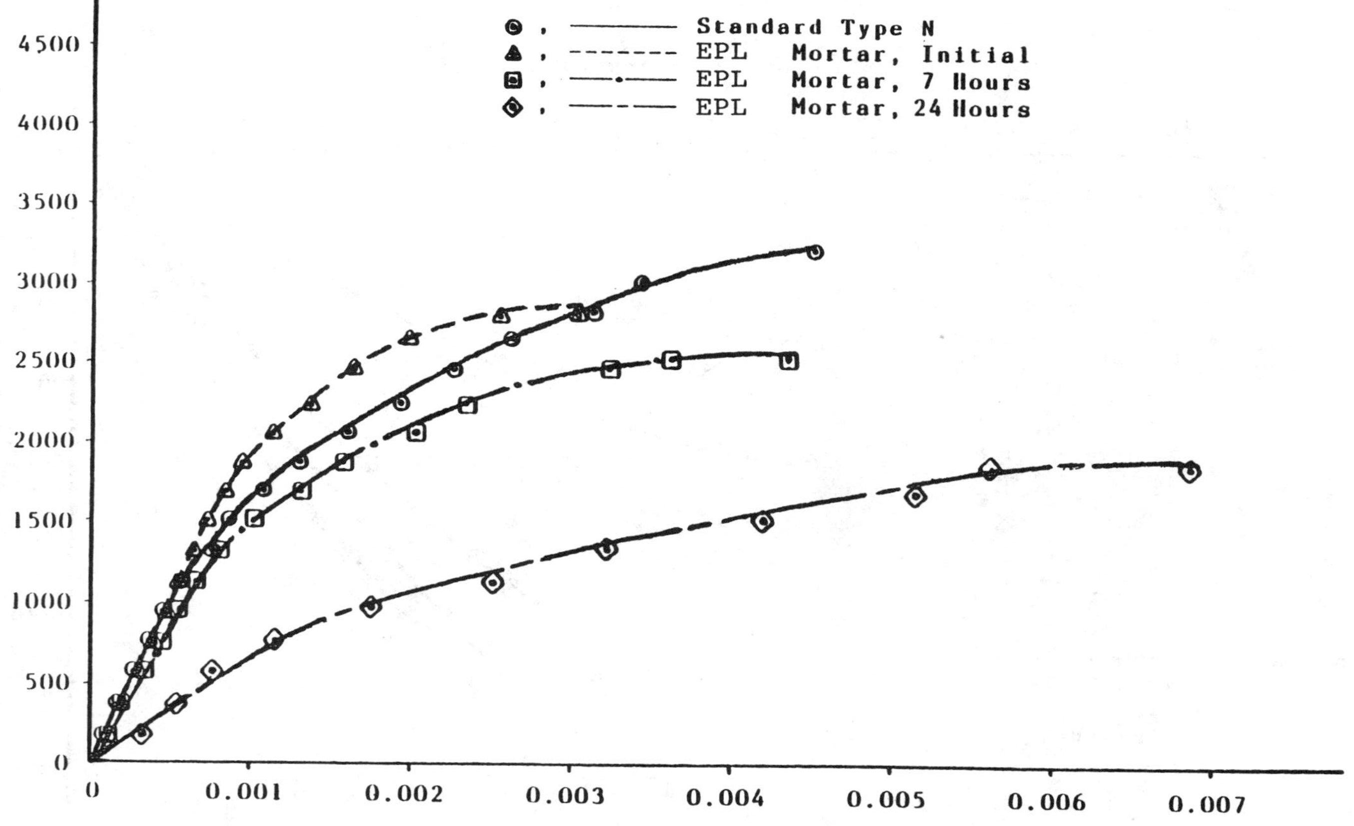

FIG. 18—*Stress-strain curve for compression prisms (Type II brick).*

calculated using the initial one-day observation length as the reference. The test data is shown in Figs. 19 and 20. Specimens were air cured in the laboratory.

As shown, the Type I brick prisms made with initial mixed extended plastic life mortar had less shrinkage than that of Type N portland cement lime mortar. The 7-h suspended mortar had the least shrinkage while the 24-h suspended mortar had the most. The results were basically the same for the Type II brick prisms, except that the amount of shrinkage was about 50% higher. The low absorption brick masonry assemblages exhibited similar shrinkage to that of the mortar shrinkage specimens. The high absorption brick masonry assemblages exhibited a smaller shrinkage than that of the mortar shrinkage specimens.

Freeze Thaw Resistance—To examine the resistance of the mortars to alternate freezing and thawing, 5.08-cm mortar cubes were made for the conventional Type N portland cement lime mortar and the comparison extended plastic life mortar. (Here only initially mixed cubes were evaluated.) Cubes were air cured for 28 days and their dry weights recorded. The samples were then subjected to alternate freezing and thawing cycles in accordance with ASTM C 67.

After nine cycles of freezing and thawing, it appeared the conventional standard mortar had lost a considerable amount of material from the portion of the cube in contact with water (see Fig. 21). The extended plastic life mortar specimens did not appear to have lost any weight. All cubes were dried and weighed. The data are given in Table 5.

The conventional mortar cubes, having exceeded the 3% weight limit of ASTM C 67, were considered to have failed. The extended plastic life mortar cubes had not lost any weight; therefore, the freezing and thawing procedure was continued. At the end of 50 cycles, these cubes were still undamaged (see Fig. 22).

Efflorescence Test—A decision was made to examine the conventional and extended plastic life mortar with regard to efflorescence. Standard 5.08-cm cubes were made for conventional Type N mortar and the extended plastic life mortar for 0 and 24-h suspension times. For each mix, six cubes were made, air cured in laboratory for 28 days, and divided into two three-cube sets. At the end of the curing period they were dried and tested according to ASTM C 67. One set of each specimen group was placed in the drying room, and the second set was placed in distilled water for seven days. Both sets were dried in a drying oven for 24 h and examined for changes due to efflorescence.

The results showed that the cubes made from standard Type N and the extended plastic mortar (0 suspension time and 24-h suspension time) had effloresced on the cube portion in the water. Figure 23 shows the extended life mortar cubes; Fig. 24 shows the standard Type N mortar cubes.

Shrinkage Characteristics—To compare the shrinkage characteristic of the mortars, two mortar specimens for Type N standard mortar and for the extended plastic life mortar (0, 7-h, and 24-h suspension times) were cast and tested according to ASTM Test Method for Length Change of Hardened Cement Mortar and Concrete (C 157) and ASTM Specification for Apparatus for Use in Measurement of Length Change of Hardened Cement Paste, Mortar, and Concrete (C 490). The specimens were mortar prisms 2.54 cm^2 and approximately 28.57 cm in length. Figures 25 and 26 show the specimens along with the reference bar. The samples were cured in a moist room for 48 h, then cured in lime-saturated water until they reached an age of 28 days. An initial observation of length was made after seven days of curing, this length to serve as the reference length. The second observation of length was made at the end of the 28-day curing period. The specimens were then stored in the laboratory environment with more length observations being made after total air storage periods of 4, 7, 14, and 28 days, 8 weeks, and 16 weeks. The results are shown in Fig. 27.

After 28 days of moist curing there was a slight increase in length of extended plastic life mortar samples for 0 and 7-h suspension times. The 24-h extended mortar and standard Type N mortar showed a decrease. In the air-curing period, all samples decreased in length; most of the shrinkage occurred in the first 14 days. Very little change occurred after 28 days in air with basically no change after 56 days.

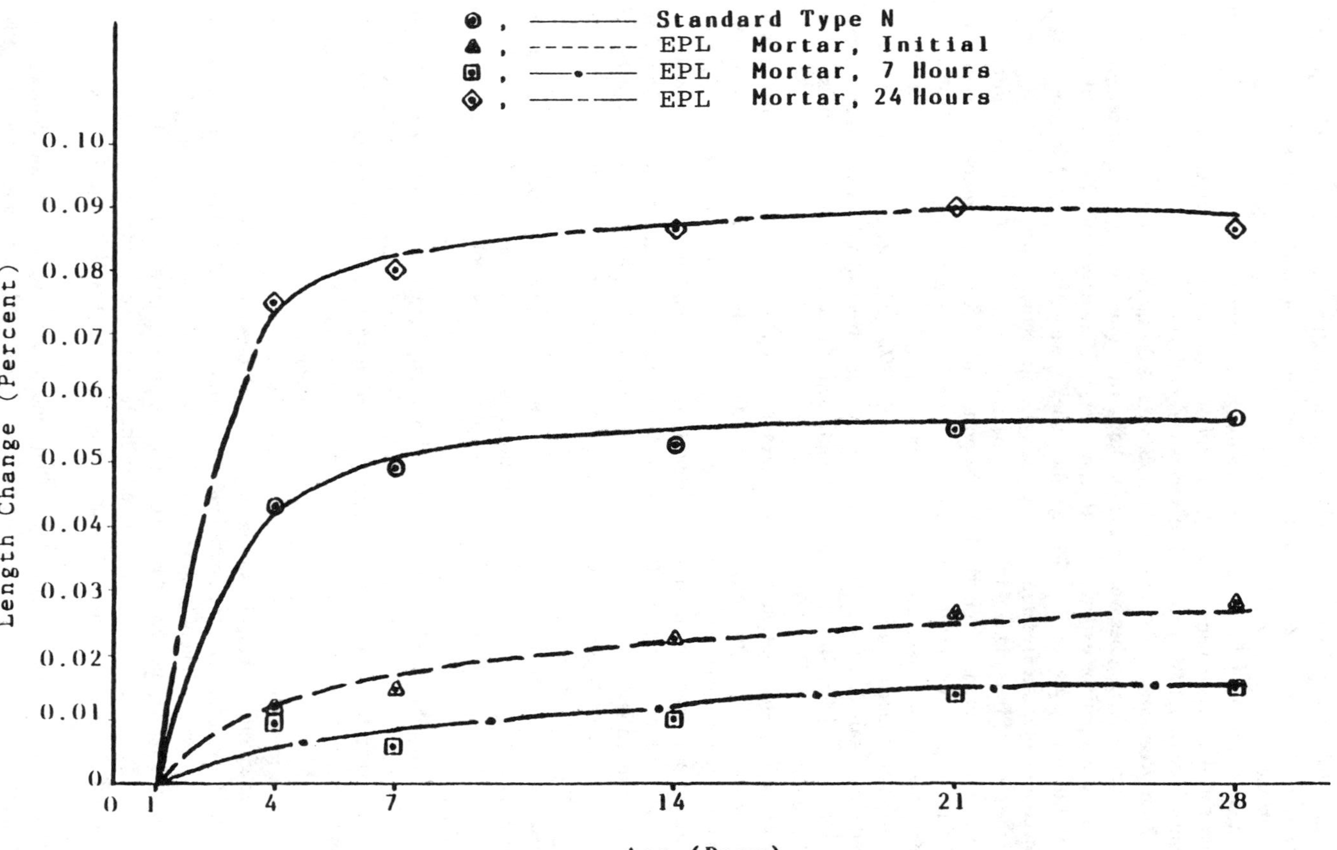

FIG. 19—*Drying shrinkage of masonry prisms (Type I brick).*

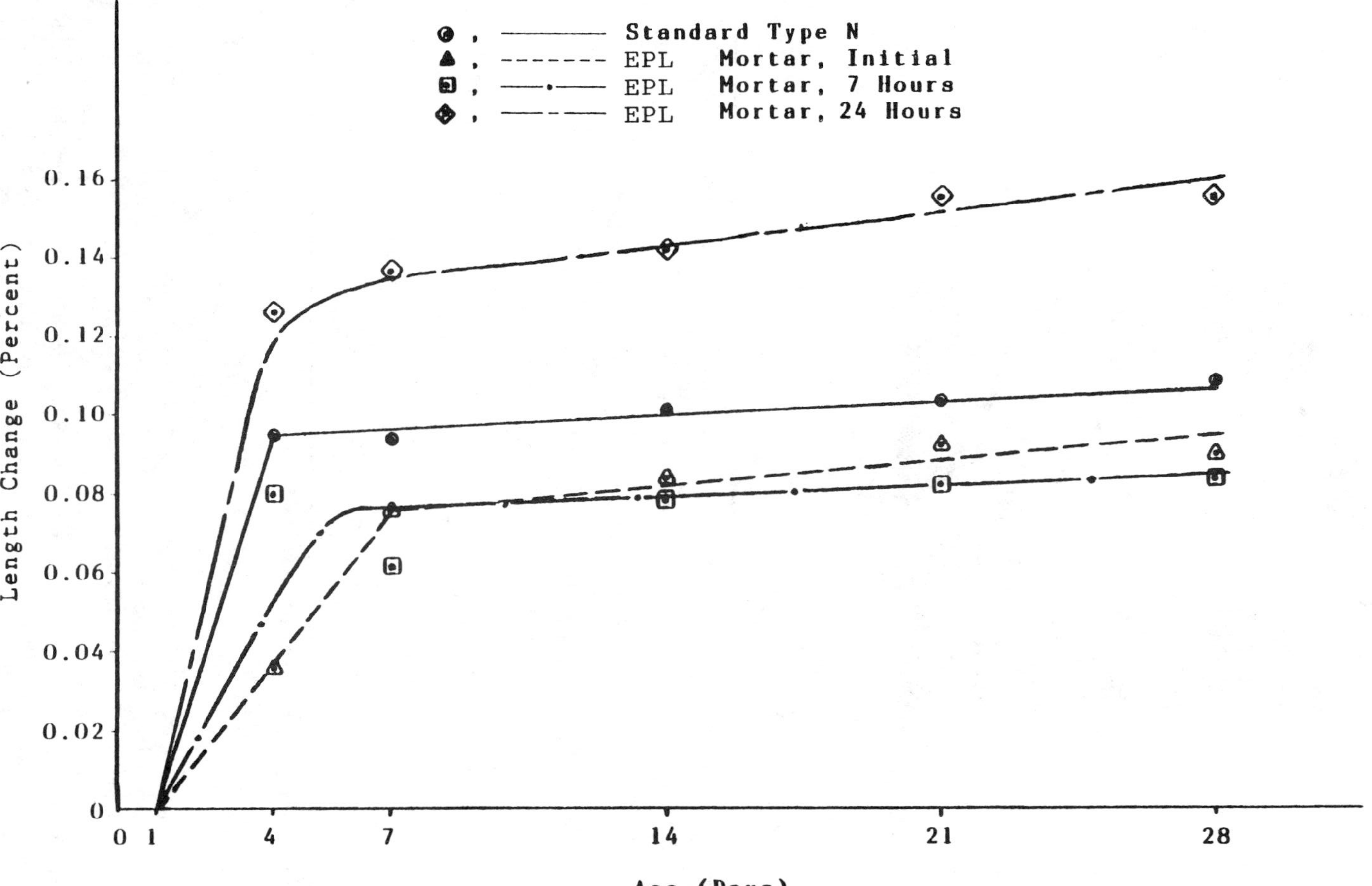

FIG. 20—*Drying shrinkage of masonry prisms (Type II brick).*

FIG. 21—*Freeze thaw—9 cycles, standard Type N mortar.*

TABLE 5—*Freeze thaw test results.*[a]

Mix No.	Sample No.	Initial Weight, g	Weight After Cycles, g[a]	Weight Loss, %	Avg, %
Standard Type N	1	243	236	2.9	
	2	243	235	3.3	
	3	241	233	3.3	3.5
	4	242	236	2.5	
	5	242	233	3.7	
	6	241	228	5.4	
EPL	1	218	218	0	
	2	217	217	0	
	3	217	217	0	0
	4	218	218	0	
	5	217	217	0	
	6	225	225	0	

[a]9 cycles for Standard Type N mortar, 50 cycles for EPL.

After 16 weeks the total percent length change of Type N standard mortar specimens was 0.147%. The extended plastic mortar specimen for 0, 7, and 24-h suspension times measured 0.120, 0.140, and 0.143%, respectively.

Corrosion—An attempt was made to examine the corrosion resistance characteristics of an ASTM Type N mortar and the extended plastic life mortar for 0 suspension time. Mortar samples were made using two different specimen sizes, namely 5.08 cm wide by 5.08 cm high by 30.48 cm long and 10.16 cm wide by 10.16 cm high by 30.48 cm long (see Figs. 28 and 29). A No. 3 steel rod was placed in the center of each specimen for either an embedded condition or an

FIG. 22—*Freeze thaw—50 cycles, EPL mortar.*

FIG. 23—*Efflorescence—EPL mortar.*

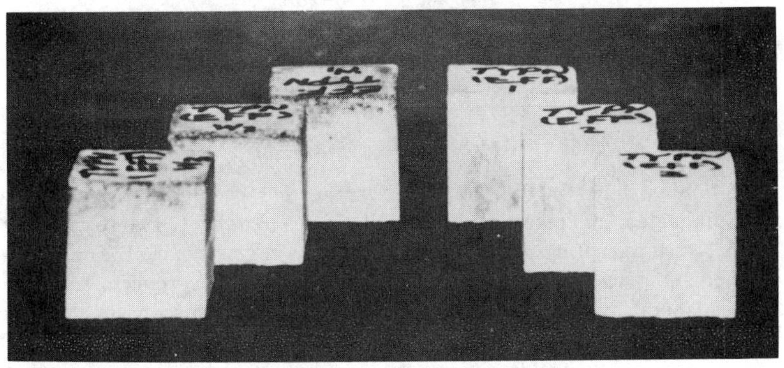

FIG. 24—*Efflorescence—standard Type N mortar.*

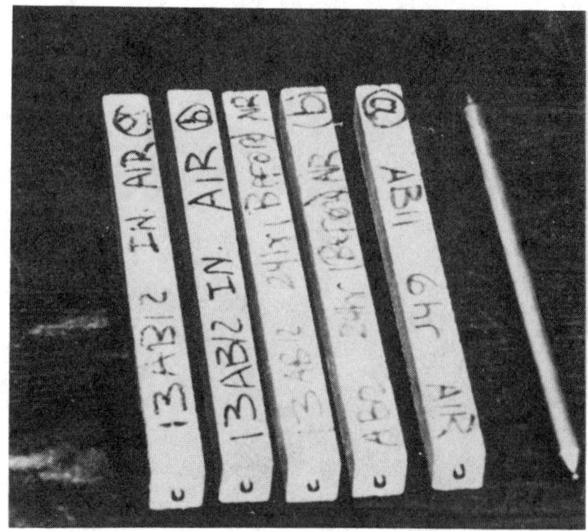

FIG. 25—*Mortar shrinkage bars.*

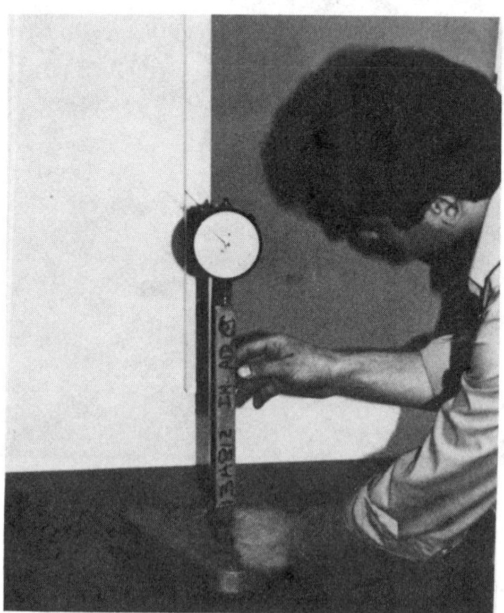

FIG. 26—*Shrinkage measurements.*

exposed condition as shown in Fig. 30. Three surface bar conditions were examined: zinc plated, cadmium plated, and regular steel without surface treatment. Two specimens per mortar type per bar surface coating per exposure condition were constructed. The No. 3 rod was initially weighed and placed inside the molds in the middle of the specimen. All samples were air cured in the laboratory for 28 days and then moved to the moist room. Each specimen was then checked at 3, 6, and 9 months for extent of corrosion. At the end of each specified period the rods were taken out of the molds, cleaned, and weighed. No definitive trends could be established from this data.

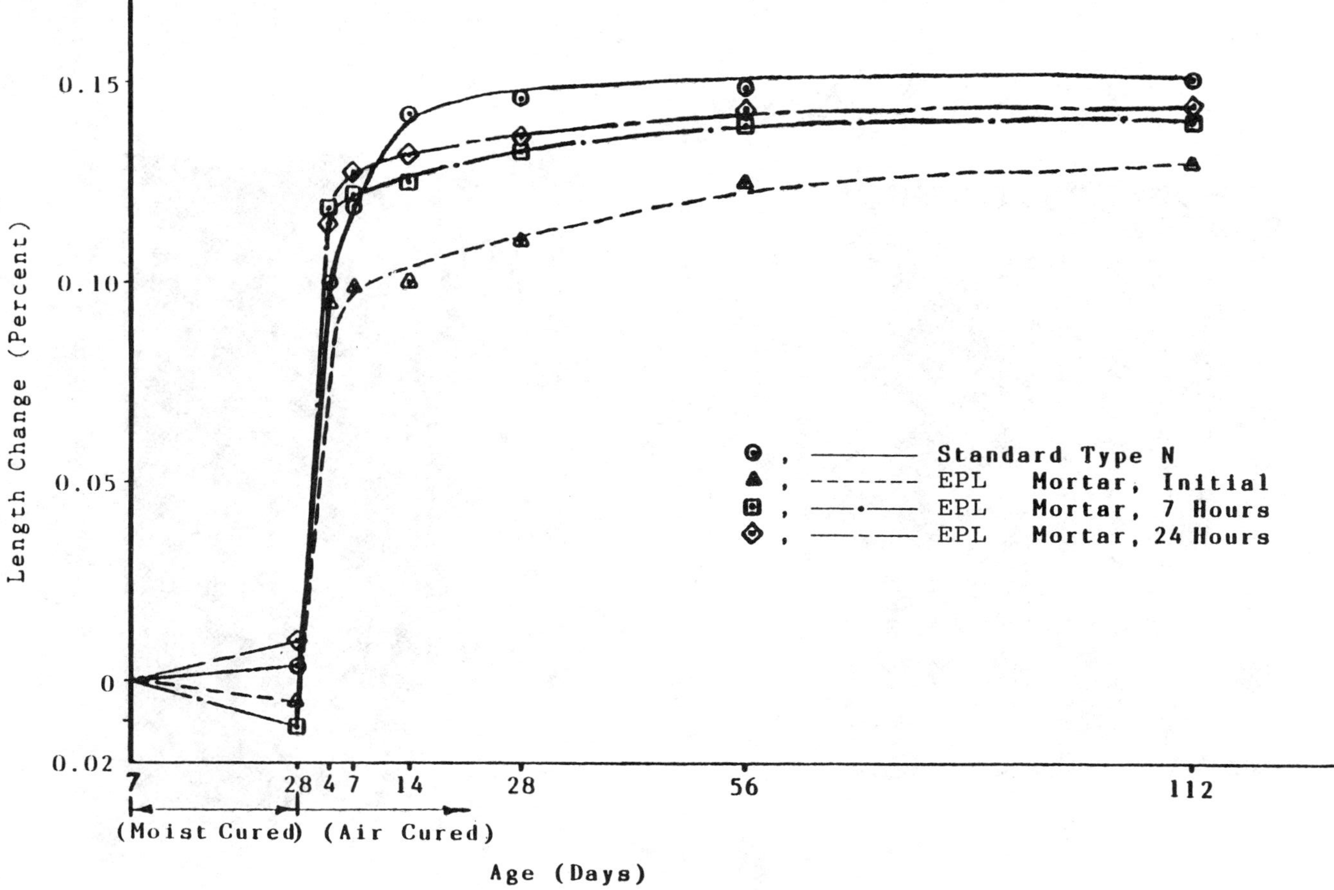

FIG. 27—*Drying shrinkage of mortar with time (ASTM C 157 procedure).*

FIG. 28—*Corrosion test molds.*

FIG. 29—*Corrosion test specimens.*

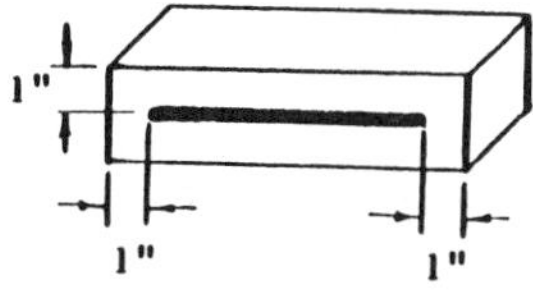

a) Embedded Condition, 2"x2"x12" Mold

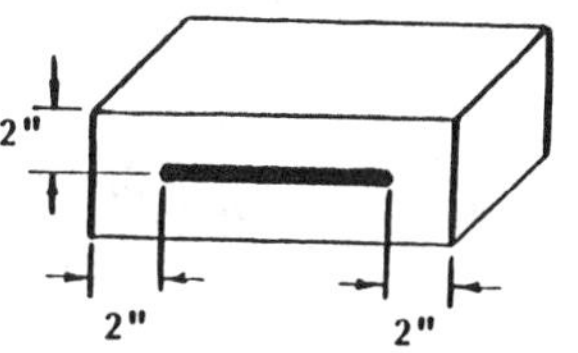

b) Embedded Condition, 4"x4"x12" Mold

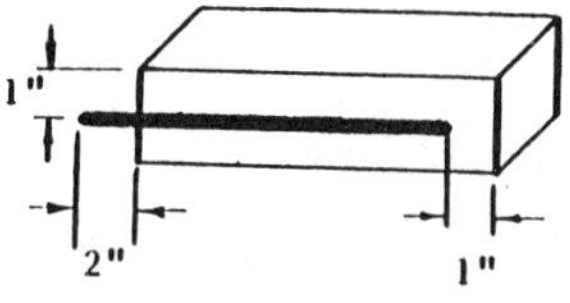

c) Exposed Condition, 2"x2"x12" Mold

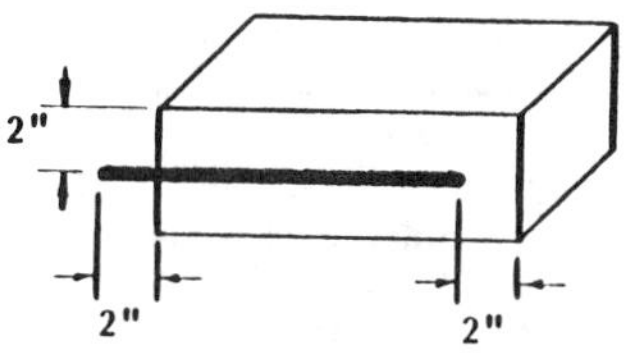

d) Exposed Condition, 4"x4"x12", Mold

FIG. 30—*Placement condition of rods for corrosion.*

Summary—Phase 2

The one extended plastic life mortar in this phase showed equivalent or better performance as compared to the standard ASTM Type N portland cement lime proportion mortar for all the tests conducted on specimens constructed from initial mixing and the 6 to 7-h suspension period. The EPL mortar exhibited sufficient compressive strength and also good bonding to the masonry unit. Shrinkage at all suspension times of EPL mortar was less than Type N mortar. Freeze-thaw resistance of the initially mixed EPL mortar was far superior to that of the construction mortar. The EPL mortar did not appear to promote corrosion.

However, the bond strength and compressive strength of masonry prisms made for the 24-h suspended mortar were about 50% less than those for Type N mortar. This effect is associated with retempering of the mortar. In all cases the extended plastic life mortar had good workability for a minimum period of 20 h. If retempering is needed it is recommended that the flow of EPL mortar examined here be kept in the 80 to 100% range.

William B. Coney[1] *and Jerry G. Stockbridge*[2]

The Effectiveness of Waterproofing Coatings, Surface Grouting, and Tuckpointing on a Specific Project

REFERENCE: Coney, W. B. and Stockbridge, J. G., **"The Effectiveness of Waterproofing Coatings, Surface Grouting, and Tuckpointing on a Specific Project,"** *Masonry: Materials, Design, Construction, and Maintenance, ASTM STP 992,* H. A. Harris, Ed., American Society for Testing and Materials, Philadelphia, 1988, pp. 220–224.

ABSTRACT: Water leakage in a building gave the authors an opportunity to try a variety of widely used procedures to improve the watertightness of masonry walls.

The water permeance of the masonry walls in their original condition ranged from 11 to 60 L/h. Three types of clear waterproofing coatings, two types of surface grout, and conventional tuckpointing were tested. Each procedure was applied to two wall areas. Water permeance tests were performed on each sample area in accordance with a field-modified version of ASTM Test Method for Water Permeance of Masonry (E 514). Sample areas were tested prior to repair, after repair, and in some cases after weathering through one winter.

The tests relate to one type of brick and mortar on one specific building. It would be unwise to extrapolate our findings to other buildings, but the results are interesting, and we believe they contribute to a growing wealth of experience.

KEY WORDS: masonry, water permeability, testing

In recent years, tuckpointing, surface grouting, and waterproof coatings have all been used as methods for decreasing the water permeance of masonry walls, but to our knowledge, their comparative effectiveness has been studied on few actual buildings in the field.

Water leakage in the Science building at Northeastern Illinois University in Chicago and an aggressive University architect gave the authors an opportunity to study a variety of widely used procedures to improve the watertightness of masonry walls. The Science building is four stories high and was completed in 1972. Figure 1 shows a general view of the building from the northwest. The bricks conform to ASTM Specification for Facing Brick (Solid Masonry Units Made From Clay or Shale) (C 216). The mortar was analyzed and found to have Type N proportions in accordance with ASTM Specification for Mortar for Unit Masonry (C 270). The mortar was made with masonry cement containing an air entraining agent. Air contents of the mortar samples tested ranged from 16 to 26%. Petrographic analysis revealed minimal intimacy of contact between the mortar and brick. The contact area on the samples examined was estimated at only about 25%.

When tested in accordance with a field-modified version of ASTM Test Method for Water Permeance of Masonry (E 514), walls were found to have water permeances of 11 to 60 L/h, which are some of the highest values we have seen.

[1]Member of the Evanston Preservation Commission. He is a specialist in rehabilitation and restoration.

[2]Senior architect and vice president, respectively, Wiss, Janney, Elstner Associates, Inc., Northbrook, IL 60062.

FIG. 1—*General view of building.*

Testing Procedures

As mentioned, the water permeance tests were performed based on a field-modified version of ASTM E 514. Two test chambers were used. The test chambers each covered a wall area of 12 ft^2. Test exposure conditions were equivalent to the standard 5.5 in. of rain per h accompanied by a 62.5 mph wind. Our field test varied from the standard laboratory test in that the permeance was determined by measuring the amount of water entering the face of the wall rather than measuring the amount of water exiting the back of the wall. Each test was run for 3 h or until the 15-L capacity of our reservoir was drained. Figure 2 shows one of the test setups.

Test Area

The testing was performed on the masonry walls of the two penthouses on the roof of the building because of easy access. There was one penthouse near the west end of the roof and one near the east end. The two penthouses were each 82 ft long by 46 ft wide by 14 ft high.

The walls of each penthouse were divided into six test areas, creating a total of twelve test areas, as shown in Fig. 3.

Pretreatment Permeance

One E 514 test was run in each of the twelve test areas before any of the masonry treatments were performed. Test areas were selected to miss any cracks in the walls. At some locations where cracks could not be avoided, they were tuckpointed before any testing was performed. Shrinkage separations between brick and mortar were not tuckpointed. Even with visible cracks eliminated from the test, water permeances from 11 to 60 L/h were measured as mentioned. The permeance of each of the twelve tests is presented in Table 1.

Wall Treatments

After pretreatment testing, the twelve test areas were subjected to the following repair procedures.

FIG. 2—*Water permeance test setup.*

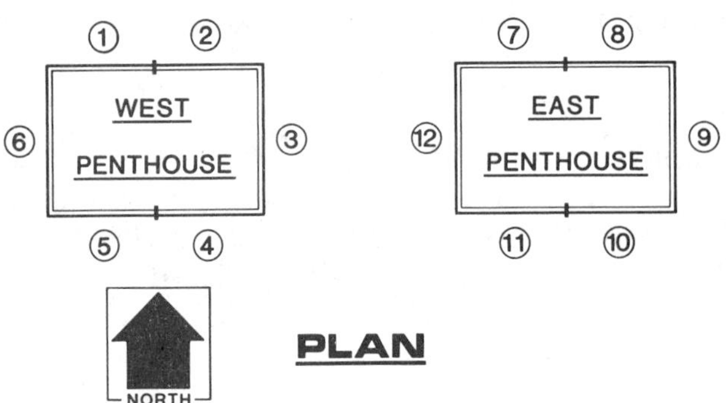

FIG. 3—*Test areas.*

Tuckpointing

Wall Area 3 was tuckpointed in accordance with the recommendation of the Brick Institute of America (BIA 7F). All existing mortar joints were cut out to a depth of at least 1/2 in. Walls were hosed down about 1 h before tuckpointing to remove debris and to wet the brick. A prehydrated mortar mix of one part lime, one part cement, and five and one-half parts sand by volume was used. Portland cement met the requirements of ASTM C Specification for Portland Cement (C 150). Lime met ASTM Specification for Hydrated Lime for Masonry Purposes (C

TABLE 1—*Summary of permeance testing.*

Wall Area	Pretreatment Permeance[a]	Treatment	Post Treatment Permeance[a]	Percent Improvement
1	30	Proprietary grouting of joints	3.2	89
2	11	Acrylic coating	3.7	66
3	30	Tuckpointing with joints tooled	0.5	98
4	45	Alkyl trialkoxy silane coating	15.0	67
5	18	Alkyl trialkoxy silane coating	10.0	44
6	26	Aluminum stearate coating	10.0	62
7	24	Alkyl trialkoxy silane coating	5.5	77
8	22	Aluminum stearate coating	4.5	80
9	60	Proprietary grouting of joints	1.2	98
10	30	Grouting of entire wall surface	0.3	99
11	45	Tuckpointing with joints struck	2.3	95
12	60	Acrylic coating	45.0	25

[a]Units in liters per hour.

207), Type S. Sand met ASTM Specification for Aggregate for Masonry Mortar (C 144). The joints were tooled to form a concave joint.

Wall Area 11 was tuckpointed in substantially the same manner and using the same materials as Wall Area 3, except the joints were simply struck flush rather than being tooled.

Surface Grouting of Entire Wall Surface

Wall Area 10 was surface grouted in accordance with BIA Technical Note 7, July 1961. A grout coating consisting of three quarters part portland cement, one part sand that passed a No. 30 sieve, and one quarter part hydrated lime was used. Shortly before using, all ingredients were mixed with water to obtain a fluid consistency. The joints were thoroughly wetted but were permitted to surface dry before the grout was applied. Stiff fiber brushes were used to work the grout into the joints. No attempt was made to keep the bricks clean, and after all of the joints were treated, the entire wall surface was painted with a diluted coating of the same grout. This method resulted in the joints receiving two coats of grout. The appearance of the wall was changed appreciably.

Proprietary Grouting of Mortar Joints

A proprietary grout was applied to the mortar joints of Wall Areas 1 and 9 with 6-in. nylon masonry brushes. One coat was applied. Templates were used to keep the masonry units clean while applying the grout to the joints. The proprietary grout is a cement-based compound meeting Federal Specifications TT-P-21, Type 11, Class A. Walls were dampened but not saturated before the application.

Aluminum Stearate Clear Coating

Wall Areas 6 and 8 were sprayed with a waterproofing coating of a product containing an aluminum stearate-based compound. It was sprayed on the walls in accordance with manufacturer's recommendations. One coat was applied. It is advertised as a breathable coating.

Alkyl Trialkoxy Silane Clear Coating

Wall Areas 4, 5, and 7 were sprayed with a waterproofing coating product containing 40% (wt/wt) alkyl trialkoxy silane and 60% (wt/wt) ethyl alcohol.

The material was sprayed on the walls according to manufacturer's recommendations. One coat was applied. It is advertised as a breathable coating.

Acrylic Clear Coating

Wall Areas 2 and 12 were sprayed with a waterproofing coating product containing acrylic (a blend of polymeric resins containing mineral spirits and aromatic hydrocarbons). It was sprayed on the walls according to the manufacturer's recommendations. One coat was applied. It is advertised as a breathable coating.

Post Treatment Permeance

After the surface treatments were allowed to cure for at least 15 days, the permeance tests were rerun. The same equipment was used, and it was positioned in the same positions, reusing wall bolts installed during pretreatment testing.

All of the treatments reduced the water permeance of the masonry to some extent. The clear waterproofing coatings were the least effective treatment. Surface grouting the joints in the walls with the proprietary product was a much more effective treatment. Surface coating the entire wall with grout and conventional tuckpointing with tooled joints were the most effective treatments. The surface grouting did, however, change the appearance of the walls. None of the treatments made the walls completely watertight. We have found clear waterproofing coatings more effective on other projects than they were on this project. The permeance of each of the twelve treated wall areas is presented in Table 1.

Permeance After a Winter of Weathering

It was originally planned to retest all of the test areas after they had gone through a winter of service, but because of the limited success of the clear coatings, only the tuckpointed and the surface-grouted areas were retested. These areas showed no significant change in ability to resist water penetration between the fall and the spring testing.

Despite its good performance on the retests, surface grouting provides only a very thin layer of new material over defects in mortar joints and common sense tells us that its longevity will probably be less than tuckpointing where defects are ground out and a half inch of new, sound material is installed.

Observations

1. All of the treatments reduced the water permeance of the masonry to some extent.
2. On this specific project, clear waterproof coatings were the least effective treatment. Surface grouting of joints was a much more effective treatment. Surface grouting the entire wall surface and conventional tuckpointing with tooled joints were the most effective treatments.
3. None of the treatments made the walls completely watertight.
4. Tuckpointing with struck joints was considerably less effective than tuckpointing with tooled joints.

While it would be unwise to extrapolate our findings on this particular project to other projects, the results are interesting, and we believe that they contribute to a growing wealth of experience.

Maintenance

Sven E. Thomasen[1] *and Carolyn L. Searls*[1]

Diagnosis of Terra-Cotta Glaze Spalling

REFERENCE: Thomasen, S. E. and Searls, C. L., **"Diagnosis of Terra-Cotta Glaze Spalling,"** *Masonry: Materials, Design, Construction, and Maintenance, ASTM STP 992,* H. A. Harris, Ed., American Society for Testing and Materials, Philadelphia, 1988, pp. 227–236.

ABSTRACT: One of the most common types of deterioration observed in terra-cotta buildings is glaze spalling. This separation of the hard, fairly impermeable glaze from the underlying clay bisque exposes the clay to water infiltration and accelerated deterioration. The mechanisms responsible for glaze spalling are fabrication defects, incorrect installation, environmental exposure, and improper maintenance. The investigation of glaze spalling consists of a field survey, field tests, and laboratory analysis. Once a diagnosis of the causes is made, appropriate preventive measures can be implemented.

KEY WORDS: terra-cotta, spalling, glaze spalling, restoration, masonry, deterioration, testing

Terra-cotta was a popular architectural cladding material from the 1880s through the 1930s, coinciding with the building of American cities and the rise of the skyscrapers.

The typical terra-cotta block was fabricated by handpressing clay into wood forms. The back of the block was open with internal stiffeners called webs. The glaze, which was brushed or sprayed on the surface, was either a slip (clay wash) or an aqueous solution of mineral and metal salts.

Terra-cotta is relatively durable and permanent because of the excellent weathering properties and the hard surface of the glaze. But, most terra-cotta buildings are now more than 50 years old and many claddings have started to deteriorate. The historical significance of these buildings, the unmatched richness of detailing, and the vivid color of the terra-cotta make preservation of these structures important.

Deterioration of terra-cotta claddings takes many forms, but the one most frequently seen is glaze spalling. Glaze spalling can first appear as a small blister from which the spalling can grow to cover an entire terra-cotta block. Once glaze spalling has started, it allows a larger amount of water to enter the bisque and in many cases further accelerates the breakdown of the terra-cotta cladding.

It is important to properly diagnose the causes of glaze spalling before initiating a rehabilitation program. This paper discusses glaze spalling and presents techniques for the testing and diagnosis of the causes of glaze spalling.

Types of Glaze Spalling

Terra-cotta glaze spalling can be characterized by the depth of the failure plane.

[1]Senior consultant and senior engineer, respectively, Wiss, Janney, Elstner Associates, Inc., Emeryville, CA 94608.

Failure Between Glaze and Bisque

This kind of glaze spalling is characteristic of inherent material failure where the glaze was not properly fused with the clay. Sometimes this failure takes many years to appear.

Failure in Boundary Bisque Layer

In this common type of glaze spalling, delamination occurs in the boundary layer of the bisque, as shown in Fig. 1, with a small amount of clay body adhering to the spalled glaze. On ashlar blocks, this glaze spalling often takes one of these forms:

Edge Spalling Below or Above Horizontal Joint—The spalling originates at the edge of the mortar joint or in some cases 20 to 80 mm below the joint. After the spalling has started, moisture has access to the exposed bisque, and the spalling progresses rapidly upward or downward from the joint.

Edge Spalling Adjacent to Vertical Edge—The spalling appears to start at the very edge of the block. Water enters the bisque, as shown in Fig. 2, through the mortar or from the corners of the block where the bisque is exposed as the mortar is worn away by the environment. This type of spalling occurs more frequently on blocks where the glaze covers only the exposed face and does not wrap around the edges of the block.

Round Spalls in Center of Glaze—A typical spall in the middle of a block is approximately circular. The spall may have started at a pinhole in the glaze and the diameter increases as the spall progresses outward from the center. Sometimes multiple small round spalls will combine into larger spalling.

Causes of Glaze Spalling

The mechanisms generally recognized as responsible for glaze spalling of terra-cotta are fabrication defects, incorrect installation, environmental exposures, and improper maintenance.

Fabrication

Inherent Glaze Faults—Inadequate firing of the terra-cotta can result in undervitrification, which makes the glaze permeable. High permeability can also result from the glaze being too

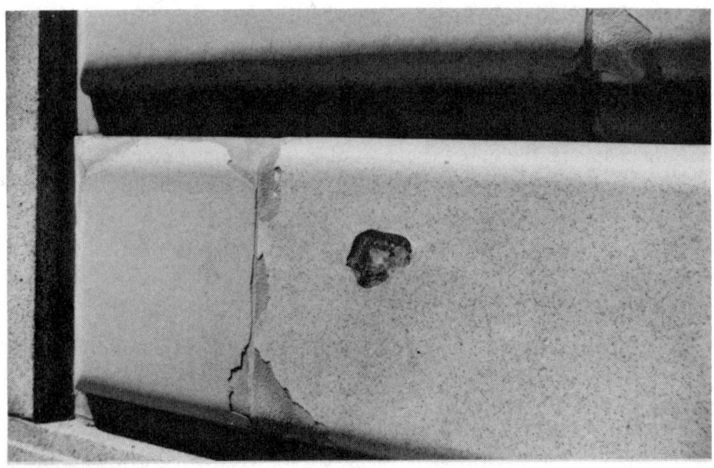

FIG. 1—*Glaze spalling in boundary layer of bisque occurs at mortar joint and in center of block.*

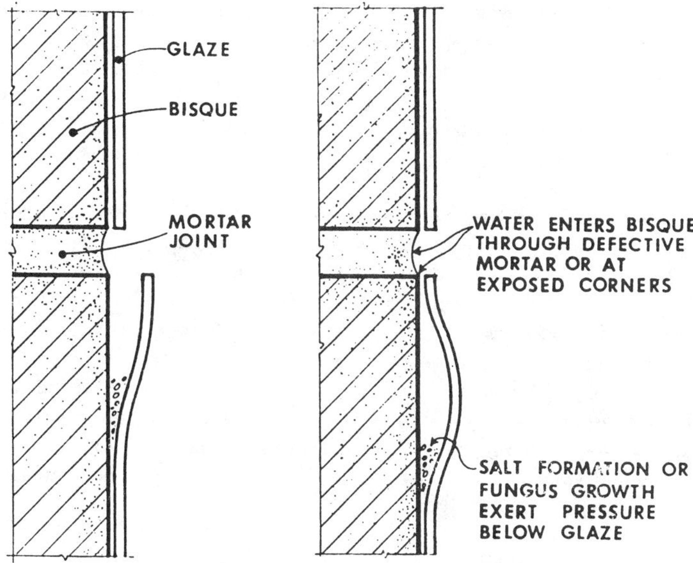

FIG. 2—*Glaze spalling from salt formation or biological growth at mortar joint.*

thin or from insufficient flux in certain glaze formulas. Highly permeable glaze allows water infiltration, which can lead to glaze spalling.

Inherent Bisque Faults—Many terra-cotta blocks were packed by hand into molds. Pressing the clay layers into the mold sometimes produces microcracks and weak planes parallel to the surface of the block immediately below the glaze. Spalling has later occurred along these weak planes.

Crazing—After the firing, the terra-cotta cools and both the glaze and the clay shrink. If the coefficient of thermal expansion for the glaze is larger than the coefficient for the clay, the cooling will create tension in the glaze and cause craze cracks in the glaze. The terra-cotta block has reached its smallest dimension after the firing. As it absorbs moisture from the environment, the bisque expands and the nonabsorbent glaze remains stable. The expansion of the bisque creates tension in the glaze and can result in crazing. The craze cracks are typically microscopic in width and generally do not increase the absorption rate compared to glaze without crazing. The overall effect of crazing on durability is not well understood, but it appears that crazed glazes are less prone to damage from glaze spalling than uncrazed glazes.

Installation

Terra-cotta was sometimes used as a load-bearing masonry, but more often as a cladding anchored to the structural framing system. Normally, no provisions were made in the design for movement, either absolute or differential, in the backup framing or in the cladding. The blocks were typically supported vertically at each floor level by shelf angles. Ties or Z-shaped steel straps were framed into slots or holes in each block and anchored the terra-cotta horizontally to the backup walls of masonry or concrete. Ornamental units generally had multiple anchors. The terra-cotta was installed with solid cement/lime mortar joints about 5 mm wide. The back of the blocks were sometimes fully grouted and sometimes left ungrouted, but whether grouted or ungrouted, the wall cladding had no internal flashings or weepholes.

Strain Buildup—Buildings with concrete frames experience long-term shrinkage of the frames, and differential movement of building components can also cause high stresses, but most strain buildup in terra-cotta claddings results from moisture expansion of the terra-cotta clay body. Wet/dry cycles can cause permanent expansion of the terra-cotta blocks, and this expansion plus the absence of expansion joints in the facade creates high compressive stresses. Failure is usually in the form of crushing or buckling in the bisque body or boundary, as shown in Fig. 3.

Corrosion of Embedded Metals—Most shelf angles and anchors were originally protected with paint but much of this paint has peeled away. As the mortar ages, it carbonates and loses its ability to protect the steel from corrosion. Rust scale expansion is often found at shelf angles and lintels where water is caught at the inside face of blocks. The failure, as shown in Fig. 4, typically is seen at the toe of the shelf angle or lintel steel and usually results in spalling of the clay body, not just the glaze.

Incomplete Anchoring—The terra-cotta cladding is typically supported vertically at the floor levels on shelf angles and anchored horizontally to the backup wall. Incomplete vertical support sometimes results in glaze spalling, but it has usually caused spalling of the bisque and vertical cracking from shear stresses between the supported and the unsupported wall sections. Such

FIG. 3—*Spalling caused by strain buildup.*

FIG. 4—*Spalling at corroding shelf angles.*

incomplete supports can be found at outside building corners, as shown in Fig. 5, where the shelf angles are terminated short of the corner.

Environmental

Terra-cotta is generally durable and permanent because of the excellent weathering properties and the hard surface of the glaze, but the mechanical, chemical, and biological processes resulting from environmental exposure will, with time, alter the durability of the facade. Among the various environmental agents harmful to terra-cotta, moisture is the most important. Water is a carrier for airborne chemicals: it transports salts, supports biological growth, and exerts pressure when it freezes. Water is the single most damaging element to terra-cotta.

Environmental Chemicals—The vitreous glass-like surface of most terra-cotta glazes provides a durable barrier against most environmental chemicals and generally resists the adherence of dirt and the accumulation of airborne gases and salts from acid rain. The exception is some slip glazes, especially those with relatively high permeability or those with rough-textured surfaces. While most glazes are not sensitive to carbon dioxide and sulphur dioxide, the long-term exposure to the acids usually found in polluted air can cause dulling of the glaze. The environmental chemicals will, however, more vigorously attack the mortar in the joints and this will allow water infiltration and accelerated deterioration of the terra-cotta.

Water, Salt Deposits—Water enters the bisque through deteriorated mortar joints, permeable glaze, or damaged terra-cotta blocks, and it travels freely through the porous bisque and masonry backup. Water soluble salts are carried along with the water and these salts are deposited in the boundary layer of the bisque as the water evaporates through permeable glaze. The buildup of salt deposits can exert considerable internal pressure which ultimately fractures the boundary layer of the bisque.

Water, Biological Process—Algae, moss, and lichen thrive in a porous environment when light and moisture are present. Most light-colored glazes permit sufficient sun transmission to support biological growth. The presence of biological organisms prevents the bisque from drying out, and the chemical products of their metabolism disintegrate the mortar. The formation of biological growth below the glaze can exert considerable internal pressure which ultimately fractures the boundary bisque layer, as shown in Fig. 6. Glaze spalling from biological growth is commonly found adjacent to mortar joints or adjacent to previously spalled sections where

FIG. 5—*Cracking and spalling caused by incomplete vertical support around building corners.*

FIG. 6—*Glaze spalling from biological growth.*

moisture and air are readily available; but it can also be found in the middle of an undamaged block, where it can be seen as a glaze blister. As soon as the glaze has spalled, the underlying bisque will dry and the biological growth will die or will move in under the edge of the adjacent glaze.

Water, Freeze/Thaw—Frost damage is dependent on the cycles and speed of temperature variations around the freezing point, the degree of saturation, and the pore distribution and permeability of the bisque. While certain forms of cracking can be attributed to frost damage, it appears that glaze spalling is seldom caused by freeze/thaw action in the bisque. This might be attributed to the large percentage of air voids usually found in the terra-cotta bisque.

Maintenance

Delayed Maintenance—Deteriorated mortar joints, damaged terra-cotta blocks, and spalled glaze are major sources of water infiltration into the bisque. Repairs and maintenance of the terra-cotta should be performed at regular intervals to prevent accelerated deterioration. Water infiltration through cornices, roofs, window sills, and other elements should be reduced by proper maintenance.

Improper Repairs—More damage can be done from incorrect repair than from no repair. Most glaze spalling is water related, and often the water infiltrates through deteriorated mortar joints. While the joints permit water to enter into the bisque, the mortar joints also allow water to escape in the same way. If the joints are covered with sealant, water that enters the block by other sources will tend to evaporate through the glaze rather than the joint, and this will result in glaze spalling.

Improper Cleaning—The hard vitreous surface of the glaze is most often self cleaning, but dirt sometimes collects in crevices or at rough-textured surfaces. Cleaning of dirty surfaces should be done with great care. Improper chemical cleaning agents can dull the gloss of the glaze or attack the mortar, and any abrasive cleaning, especially sandblasting, will damage the glaze surface and the increased permeability will result in future spalling.

Improper Coatings—Epoxy paints have sometimes been used to coat terra-cotta buildings with glaze spalling because the glossy epoxy is a good visual match for the glaze. The epoxy paint traps moisture inside the walls. The water ultimately evaporates through pinholes in the coating and extensive spalling has resulted.

Techniques for Diagnosis of Glaze Spalling

Field Survey

The extent and the nature of the spalling is first evaluated by a visual survey of the building facades. Distress in the terra-cotta is recorded on elevation drawings for analysis and detection of failure trends. An inspection is made of possible sources of water infiltration such as leaking gutters, damaged cornices, or deteriorated window sills. The glaze is examined closely for pinholes, mortar deterioration, and biological growth. In light-colored glazes, biological growth can be seen as a blue-green color showing through the glaze. In advanced stages, the growth can be seen as a buckling of the glaze outward from the clay substrate.

The marked-up field survey is analyzed for typical patterns in the spalling. Glaze spalling in one building consistently started adjacent to the mortar joints and grew in towards the center of the block. Examination revealed that water was being absorbed through the mortar joints and through the corner of the terra-cotta block where the glaze did not wrap around the edge. In another building, spalling occurred at each floor level. It was found that infiltrated moisture collected on top of the shelf angles and caused spalling from corrosion and from salt deposits as the water evaporated.

Field Tests

The deterioration and spalling of terra-cotta is closely related to water infiltration, and two field tests are used to evaluate water tightness. The simplest test is done by attaching an open-ended plastic tube to the terra-cotta with sealant tape, as shown in Fig. 7. The tube is filled with a predetermined head of water, and the decrease in height of water in the tube, representing the amount of water absorbed by the terra cotta, is recorded at intervals of 10 min, 30 min, 1 h, and hourly thereafter. This test is useful for comparing a variety of substrates on a single building, such as crazed glaze, glaze with pinholes, intact glaze, treated glaze, and mortar joints.

A portable aluminum test frame can be attached to the facade to perform a modified version of ASTM Standard Method of Test for Water Permeance of Masonry (E 514). Water and air pressure are applied simultaneously to the wall. The permeance is determined by measuring the loss of water in the recirculating system.

Strain relief testing measures the magnitude and direction of the built-up stresses often caused by expansion of the terra-cotta and absence of expansion joints. Electrical resistance strain gages are attached to the terra-cotta surface and the gages are read. Then the terra-cotta block, with the gages attached, is cut loose from the wall and the gages are read again. The change in gage reading is a measure of the strain in the block. The stress in the block is found by multiplying the measured strain difference by the modulus of elasticity, as determined by laboratory tests. The glaze in the test area should be firmly attached to the clay body and no glaze cracks should occur under the gage. Temperature variations during the day can affect the readings and should be recorded when strain readings are recorded. Compressive vertical stresses as high as 23 MPa have been recorded. This exceeds the maximum compressive strength of some terra-cotta. From multiple strain measurements, a stress map can be made for the exterior elevation. The map is used to evaluate if cutting of expansion joints into the facade will relieve the built-up terra-cotta stresses.

Laboratory Tests

Laboratory tests can be performed on terra-cotta to evaluate the performance of the glaze and to determine the causes of glaze spalling.

Petrographic Analysis—The consistencies of the glaze and the clay body are evaluated

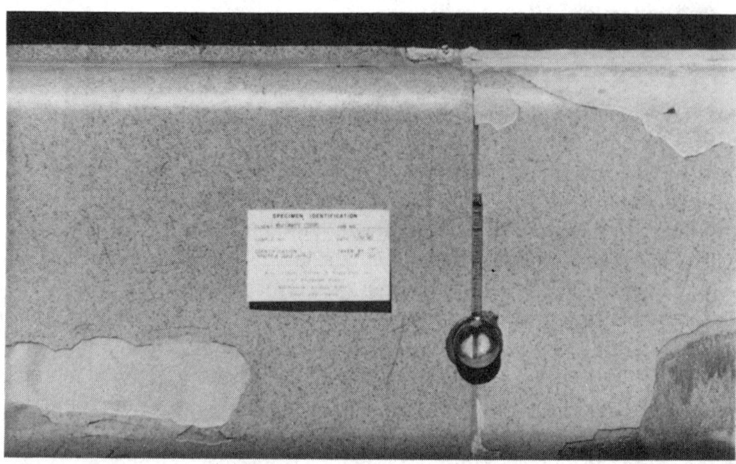

FIG. 7—*Field test for water absorption.*

through a stereomicroscopic examination. The density of the glaze surface, the composition of the material, and the degree of deterioration can be established by an experienced petrographer. The examination of the boundary layer between glaze and clay body is important, and the nature, the magnitude, and the depth of cracks often are an indication of the future performance of the glaze.

Absorption—The absorption test compares the performance of glazed and unglazed specimens. On one terra-cotta sample, the glaze is ground off while on an identical sample the glaze is left intact. The sides of both samples are coated so that water absorption occurs only through the face. The samples are soaked face down in water for 24 h. The weight gain of the glazed compared to the unglazed sample is a measure of the absorption characteristics of the glaze. Ideally, a glazed specimen should produce zero absorption if the glaze is intact, sound and craze-free. New glazes are normally impervious, but tests of 40- to 80-year-old terra-cotta structures found that the glaze at best reduced water absorption about 70% and at worst only 10% compared to unglazed terra-cotta. High water absorption through the glaze is not necessarily indicative of impending failure if the terra-cotta has the ability for water vapor to be transmitted from the clay out through the glaze. Moisture which is trapped behind the glazed surface is the more likely cause of spalling failures.

Thermal Coefficients—Tests are performed both on the complete terra-cotta block and on separate samples of glaze and clay body. Strain gages are mounted on the samples which are then subjected to temperature ranges representing the normal wall exposure. Strain readings are taken at the high, the low, and an intermediate temperature.

Moisture Expansion—During the firing, the free water is removed from the terra-cotta. As it again absorbs moisture, the clay body expands. Some of the expansion is cyclic, but a portion is nonrecoverable. A general magnitude of the moisture expansion can be determined by a reheat test. A terra-cotta sample is measured at 21°C. The sample is then heated to 460°C, again allowed to cool to 21°C, and then measured. The measured shrinkage is an indication of some of the long-term moisture expansion of the terra-cotta. Moisture expansion of the clay body is the cause of much of the cracking, spalling, and buildup of stresses in the terra-cotta walls.

Glaze Adhesion—Glaze spalling may occur because the bond between the glaze and clay has deteriorated or because it was not initially well bonded. The glaze adhesion test is performed by attaching a 25 by 25 by 100 mm test bar to the face of the terra-cotta. The bar is then knocked off and the fracture surface examined. If the glaze is well adhered to the clay, pieces of the clay body will come off with the glaze. A clean separation of glaze and clay indicates poor adhesion.

Preventive Measures

The causes of glaze spalling must be addressed and preventive measures implemented to reduce the ongoing deterioration. The repair of glaze spalling is often part of a larger restoration project. Damaged blocks can be replaced or the spalled section can be patched by applying a surface coating to the exposed bisque.

Since moisture is a common cause of glaze spalling, the elimination of water infiltration is of primary importance. Defective mortar joints should be repointed, and horizontal surfaces should slope to drain or they should be protected against water infiltration. It is generally not recommended to apply coatings to vertical surfaces. Reducing water infiltration will also cut off the supply of moisture for biological growth, but treatment with a biocide may be necessary for some infestations.

Bibliography

Berryman, N. C., "History of Architectural Terra-Cotta," *National Council on Education for the Ceramic Arts (NCECA) Journal,* 1981.

Fidler, J., "The Conservation of Architectural Terra-Cotta and Faience," Association for Studies in the Conservation of Historic Buildings Transactions, Vol. 6, 1981, reprinted in *Friends of Terra Cotta* newsletter, Vol. 3, No. 3, Fall/Winter 1984.

Patterson Tiller, de Teel, "The Preservation of Historic Glazed Architectural Terra-Cotta," Preservation Briefs No. 7, Technical Preservation Services Division, U.S. Department of the Interior, Washington, DC, June 1979.

Prudon, T. H. M., "Architectural Terra Cotta: Analyzing the Deterioration Problems and Restoration Approaches," *Technology and Conservation,* Vol. 3, Autumn 1978.

Robinson, G. S., "The Reversibility of Moisture Expansion," *Ceramic Bulletin,* Vol. 64, No. 5, 1985.

Stockbridge, J. G., "Evaluation of Terra Cotta on Inservice Structures," *Durability of Building Materials and Components, STP 691,* American Society for Testing and Materials, Philadelphia, PA, 1980.

Thomasen, S. E., "Degradation and Rehabilitation of Terra Cotta," Second International Conference on the Durability of Building Materials and Components, National Bureau of Standards, Gaithersburg, MD, September 1981.

Thomasen, S. E. and Ewart, C. S., "Techniques for Testing, Analyzing and Rehabilitation of Terra Cotta," Strengthening of Building Structures—Diagnosis and Therapy, IABSE Symposium, Venezia 1983, final report, International Association for Bridge and Structural Engineering, ETH—Hongger-berg, CH-8093 Zurich, Switzerland.

Tindall, S. M., "Architectural Terra-Cotta Restoration," *National Council on Education for the Ceramic Arts (NCECA) Journal,* 1981.

Dushyant Manmohan,[1] *Robert L. Schwein,*[2] *and Loring A. Wyllie, Jr.*[3]

In-Situ Evaluation of Compressive Stresses

REFERENCE: Manmohan, D., Schwein, R. L., and Wyllie, L. A., Jr., **"In-Situ Evaluation of Compressive Stresses,"** *Masonry: Materials, Design, Construction, and Maintenance, ASTM STP 992,* H. A. Harris, Ed., American Society for Testing and Materials, Philadelphia, 1988, pp. 237–250.

ABSTRACT: The use of terra cotta as a cladding material emerged in the 1880s and began to decline in the late 1930s. During this period, high-rise buildings employing terra cotta cladding included no provisions for differential movement between the cladding and structural framing. Numerous terra cotta clad buildings are showing signs of distress due to weathering and induced stresses from frame shortening under load. Determination of residual compressive stresses within the terra cotta cladding is often necessary for evaluation and repair of the distressed material.

The level of stresses induced in the cladding of a high-rise building in San Francisco was measured by performing strain relief tests. Uniaxial strain gauges were adhered vertically to the face of the terra cotta blocks and initial balance readings were obtained. The mortar bed joints were sawn along the terra cotta units and the strains monitored during and after the cutting. Samples of the terra cotta block were cut, instrumented, and tested in compression to determine elastic properties and ultimate compressive strength.

Having obtained the physical properties of the terra cotta, compressive stresses due to frame shortening in the structure were determined. Strain data obtained indicated stress levels which were low enough to eliminate the need for stress relief. Optimum saw cutting locations were determined had stress levels been high enough to require stress relief.

KEY WORDS: terra cotta facade, residual compressive stresses, in-situ evaluation, strain measurements

Glazed terra cotta was a popular facade material for multi-story buildings constructed between 1880 and 1930. Originally it was employed in the traditional structural load-bearing application in masonry walls in buildings of modest height. Subsequently, its use was widened to high-rise construction where it was used as cladding in purely architectural applications.

Terra cotta, like other building materials, is subject to weathering and deterioration. Most failures like spalling and crazing are due to moisture [*1*]. Vertical cracks, however, that run through several units, stories, or large areas of material are often due to excessive stresses. These stresses could be due to thermal movements or shortening of the structure relative to the terra cotta or both. Stress failures are indicative of inadequate design. An understanding of differential movement and how to compensate for it came after many terra cotta facades were constructed without flexible joints for stress relief.

Reports of terra cotta restoration on the Woolworth Building in New York [*2–4*] indicated very high compressive stresses. Stockbridge [*3*] reported measuring compressive stresses of 4.8

[1]Principal, Applied Materials & Engineering, Inc., Alameda, CA 94501.
[2]Schwein/Christensen Engineering, Lafayette, CA 94549.
[3]Vice president, H. J. Degenkolb Associates, Engineers, San Francisco, CA 94014.

to 23.9 MPa (680 to 3400 psi) in 21 terra cotta clad columns. None of the above reports, however, presented details of experimental techniques used to measure or relieve these stresses.

This paper discusses experimental techniques employed to determine compressive stresses in the terra cotta facade of the Pacific Telesis Headquarters Building in San Francisco.

The study was undertaken with the following objectives in mind:

1. To determine the compressive stresses present in the terra cotta units.
2. If excessive, to determine optimum techniques for stress relief that would be effective in the field.

The Pacific Telesis Headquarters (Fig. 1) is a 27-story building constructed in 1924. The building has a structural steel frame with reinforced concrete floor slabs. The exterior walls consist of glazed terra cotta with unreinforced brick masonry backing. The brick and terra cotta are supported by structural steel ledgers 7.62 by 7.62 cm (3 by 3 in. angles) cantilevered out from the floor framing. Though the terra cotta is supported on every floor, there are no provisions for expansion joints in the cladding.

Figure 2 is a cross-section of a column showing the typical construction. The terra cotta units are 76.2 cm long by 45.7 cm high by 2.5 cm thick (30 in. long by 18 in. high by 1 in. thick) and

FIG. 1—*Photograph of Pacific Telesis Headquarters Building, San Francisco, California.*

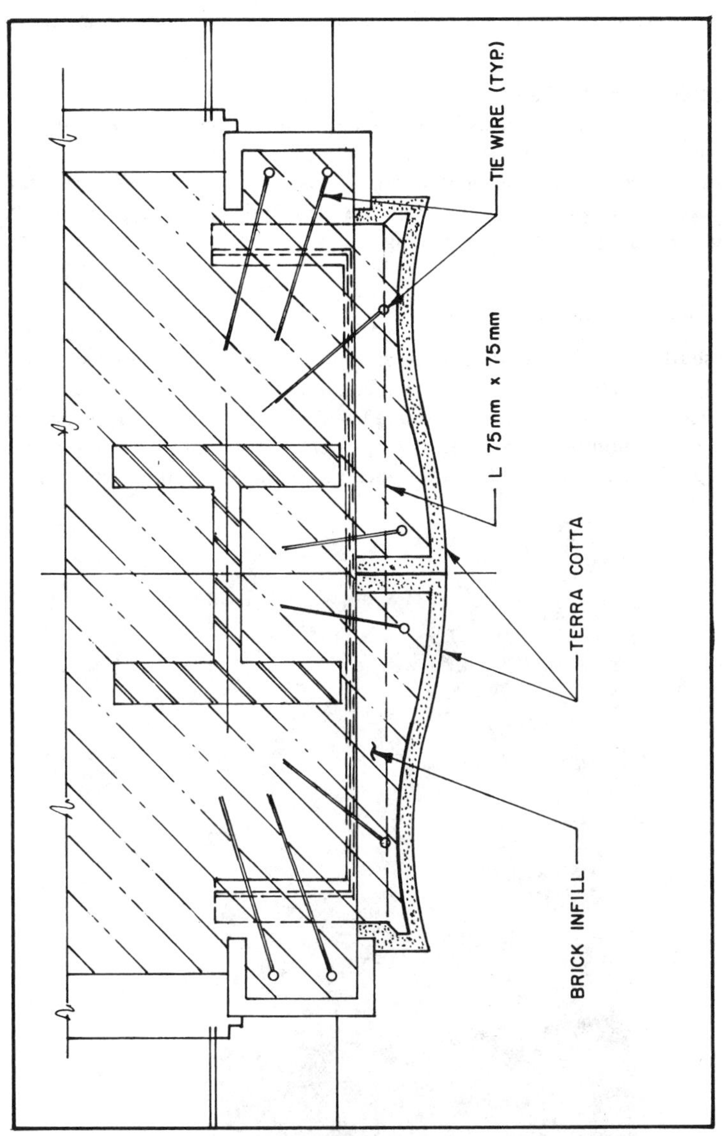

FIG. 2—*Plan of typical terra cotta attachment at a column.*

are attached to the brick by tie wires hooked through attachment holes at the top and bottom of each unit. The tie wires are embedded in the brick mortar.

Distress to the facade was manifested by spalled glaze and fine cracks in some of the terra cotta units.

Procedures

The test procedures to measure stresses due to frame shortening can be summarized as follows:

1. Instrumenting selected terra cotta units with strain gauges.
2. Measuring and establishing the as-instrumented strain as the base line strain.
3. Saw cutting two horizontal mortar joints and remeasuring strain.
4. Determining the compressive strength and modulus of elasticity of the terra cotta to convert strain to stress. The magnitude of the change in strain, and hence stress, was then assumed to be the residual stress due to frame shortening.

Selection of Strain Gauge Locations

A site on the south face of the building at the sixth floor elevation was selected for the study. This particular bay is located just east of the first column line from the southeast corner of the building. The distress in this area was considered typical of that observed in the terra cotta cladding on the entire building. Two vertical strips were instrumented, a column area and a mullion area (Fig. 3). Figure 4 is a schematic which shows the blocks that were instrumented.

FIG. 3—*Photograph of elevation instrumented with gauges. The shadows are across the fifth floor windows.*

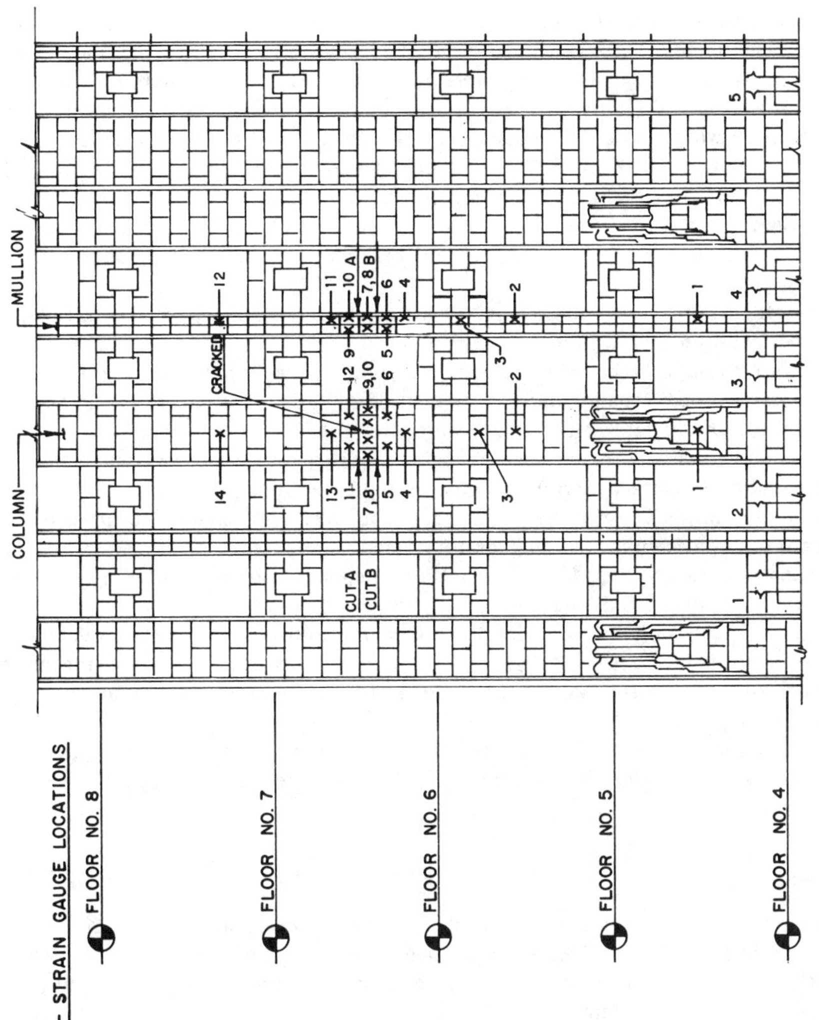

FIG. 4—*Schematic of gauge locations on column and mullion.*

As can be seen in Fig. 4, the main concentration of gauges was on terra cotta blocks on either side of the saw cuts. There was one gauge placed one floor above the saw cuts and three gauges spaced out two floors below. The four locations on the 4th, 5th, and 7th floors were selected to determine the extent of stress relief provided on floors adjacent to the one with saw cuts. A total of 14 gauges were installed on the column units and 12 on the mullion units.

Gauge Installations

Micro-Measurement's CEA-06-500UW-120 electronic strain gauges were selected for making strain measurements. These gauges have an active length of 1.2 cm (1/2 in.). Based on the fact that terra cotta is a relatively fine-grained material, it was decided that an active gauge length of 0.64 to 2.54 cm (1/4 to 1 in.) would provide satisfactory results. The other important selection criterion was that the above gauges are self-temperature compensating [at ambient temperatures of 10 to 38°C (50 to 100°F)] for a coefficient of thermal expansion of 10.8×10^{-6} cm/cm/°C (6×10^{-6} in./in./°F). Prudon et al. [*3*] measured the thermal coefficient of expansion of the glaze and of the terra cotta to be 16.6 cm/cm/°C (9.2×10^{-6} in./in./°F) and 7.92 cm/cm/°C (4.4×10^{-6} in./in./°F), respectively, making these gauges ideal for the application.

The glazed faces of the terra cotta tile units were uneven and contained small surface asperities. A cylindrical drum grinder equipped with 100-grit abrasive was used to smooth and prepare the glazed surfaces. Precautions were taken to prevent the glaze from being penetrated. The surface was finished by hand sanding with 120 and 240-grit emery paper. The gauges were bonded with a cyanoacrylate adhesive in accordance with the manufacturer's instructions, wired, and protected with a urethane coating and rubber pad. The wiring was routed along the face of the building and attached with duct tape. Each of the gauge wires was cut to a common length and routed inside a nearby office. A strain gauge was mounted on a cut piece of terra cotta and was removed from the building for use as a temperature-compensating element which was free of load-induced strains. The compensation gauge was also wired with a length of conductor equal to that of the active gauges.

Each of the 26-gauge circuits was calibrated with shunt resistors which are equivalent to 500 and 1000 microstrain. A Daytronics digital strain instrument was used for measuring strains. Vishay switch and balance units were incorporated into the circuitry to allow rapid scanning of the strain gauges.

Strain measurements were made for two days prior to the stress relief operations. These measurements were made to determine the effect of thermally induced strains due to varying sunlight on the facade.

Strain Measurements

Stress relief operations were carried out in the evenings, several hours after the sun had left the south face of the building. This was done deliberately to minimize the effect of thermal strains. While making strain measurements, the temperature-compensating gauge was placed on the outside face of the building at the sixth floor.

The sequence of saw cutting is reported in Table 1. Strains were measured on the column gauges after each cut on the column mortar joint. Similar measurements were made on the mullion. Final measurements were made two days after the saw cutting operations were complete.

Saw cutting was accomplished dry using a regular Skil-saw and a diamond-tipped blade. The cuts were deep enough to cut through the terra cotta, leaving the brick intact.

Subsequent to saw cutting of the mortar joints, terra cotta Blocks 8 and 9 from the column and Block 7 from the mullion were removed and taken to the laboratory. Cube samples were

TABLE 1—*Schedule of saw cutting and strain measurements.*

Time	Condition
	04/26/86
6:05 p.m.	Initial balance of strain gauges.
6:53 p.m.	Completed cut on Joint B of column. Measure strain on column gauges.
7:02 p.m.	Completed cut on Joint A of column. Measure strain on column gauges.
7:22 p.m.	Completed cut on Joint A of mullion. Measure strain on mullion gauges.
7:33 p.m.	Completed cut on Joint B of mullion. Measure strain on mullion gauges.
	04/27/86
4:31 p.m.	Measure strain on both column and mullion gauges.
	04/28/86
4:36 p.m.	Measure strain on both column and mullion gauges.

prepared from each of the three blocks and instrumented with strain gauges. Two gauges, identical to those used in the field, were applied to opposite faces of the cubes. One gauge was placed on the glazed surface and the other on the fired clay face. The cubes were tested in compression to obtain stress versus strain data to determine the compressive modulus of elasticity. The calculated modulus for the glaze and fired clay were averaged and the average modulus used for determining field frame shortening stresses. In addition, the compressive strength of the terra cotta was also determined.

Results

The stress versus strain values for terra cotta samples tested in compression are shown in Fig. 5. The average modulus of elasticity of the two samples was determined to be 10.92×10^6 kPa (1.56×10^6 psi). This value was used for determining frame shortening stresses in the terra cotta blocks. The average compressive strength of the two terra cotta blocks was measured to be 23.70 MPa (3300 psi).

The strains measured after each saw cut and at 24 and 48 h after saw cutting are tabulated in Tables 2 and 3 for the column and mullion gauges, respectively. Corresponding calculated stress relief has also been reported. Figure 6 is a graphical representation of stresses immediately after the final cut and 24 h later for the column gauges. Figure 7 is a similar representation as above for the mullion gauges.

Immediately after the final cut the maximum strain and stress relief measured in the column were 520 microstrain and 5.6 MPa (800 psi), respectively. The maximum strain relief was measured at Location 8 between saw cuts. The maximum strain relief in the mullion cladding immediately after saw cutting was 125 microstrain, corresponding to a maximum stress of approximately 1.40 MPa (200 psi). This location was also in between saw cuts.

Discussion

On the column, the maximum and average stress relief provided by the saw cutting operation was approximately 5.6 MPa (800 psi) (Location 8) and 3.5 MPa (500 psi) (Locations 7, 8, 9, and

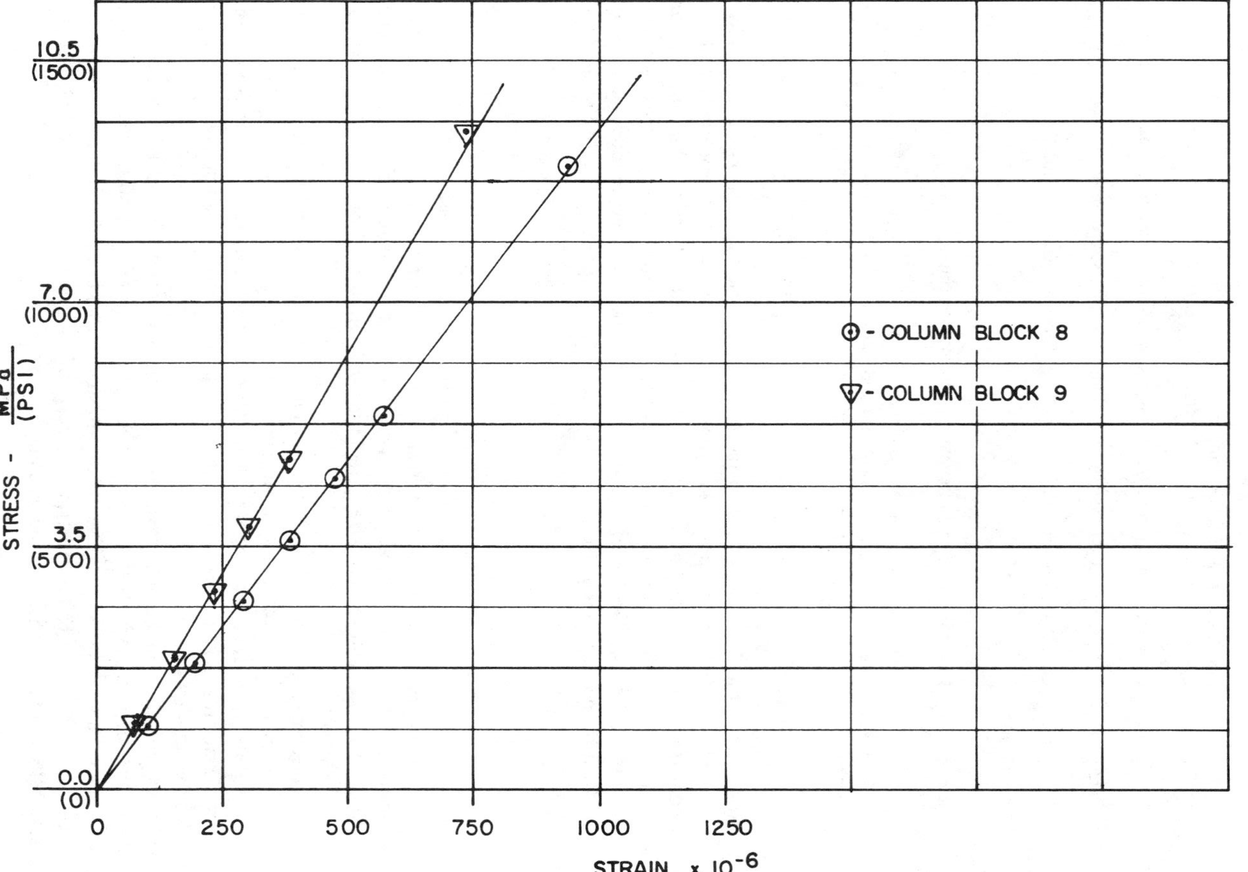

FIG. 5—*Stress-strain data on terra cotta cube samples.*

TABLE 2—*Residual column stress relief. Compressive strain and stress relief.*[a]

Gauge Number	Cut B		Cut A		24 h after Cut B		48 h after Cut A	
	Micro-strain	Stress, MPa (psi)	Micro-strain	Stress, MPa (psi)	Micro-strain	Stress, MPa (psi)	Micro-strain	Stress, MPa (psi)
1	0	0 (0)	40	0.43 (62)	0	0 (0)	20	0.22 (31)
2	−10	−0.11 (−15)	35	0.38 (54)	0	0 (0)	20	0.22 (31)
3	0	0 (0)	30	0.32 (46)	0	0 (0)	15	0.16 (23)
4	50	0.54 (77)	75	0.81 (116)	75	0.81 (116)	75	0.81 (116)
5	200	2.16 (308)	235	2.53 (362)	260	2.80 (400)	260	2.80 (400)
6	210	2.26 (323)	255	2.75 (393)	305	3.29 (470)	310	3.35 (477)
7	130	1.40 (200)	255	2.75 (393)	240	2.59 (370)	245	2.64 (377)
8	395	4.26 (608)	520	5.61 (801)	...[b]	...[b] ...[b]	...[b]	...[b] ...[b]
9	195	2.10 (300)	315	3.40 (485)	...[b]	...[b] ...[b]	...[b]	...[b] ...[b]
10	105	1.13 (162)	230	2.48 (354)	245	2.64 (377)	235	2.53 (362)
11	60	0.64 (92)	175	1.89 (270)	165	1.78 (254)	180	1.94 (277)
12	45	0.48 (68)	165	1.78 (254)	210	2.26 (323)	220	2.37 (339)
13	15	0.16 (23)	50	0.54 (77)	45	0.48 (69)	65	0.70 (100)
14	0	0 (0)	40	0.43 (62)	0	0 (0)	15	0.16 (23)

[a]Stress calculated using $E = 10.92 \times 10^6$ kPa (1.56×10^6 psi).
[b]Wires cut by workmen.

TABLE 3—*Residual mullion stress relief. Compressive strain and stress relief.*[a]

	Cut A		Cut B		24 h after Cut B		48 h after Cut B	
Gauge Number	Micro-strain	Stress, MPa (psi)	Micro-strain	Stress, MPa (psi)	Micro-strain	Stress, MPa (psi)	Micro-strain	Stress, MPa (psi)
1	0	0 (0)	0	0 (0)	−15	−0.22 (−31)	0	0 (0)
2	5	0.06 (8)	10	0.11 (15)	0	0 (0)	15	0.16 (23)
3	25	0.27 (39)	35	0.38 (54)	25	0.27 (39)	35	0.38 (54)
4	35	0.38 (54)	45	0.48 (69)	60	0.64 (92)	50	0.54 (77)
5	35	0.38 (54)	90	0.98 (140)	150	1.62 (231)	150	1.62 (231)
6	30	0.32 (46)	115	1.24 (177)	205	2.21 (316)	205	2.21 (316)
7	75	0.81 (116)	100	1.08 (154)	...[b]	...[b] ...[b]	...[b]	...[b] ...[b]
8	65	0.70 (100)	125	1.35 (193)	...[b]	...[b] ...[b]	...[b]	...[b] ...[b]
9	60	0.64 (92)	45	0.48 (69)	220	2.37 (339)	220	2.37 (339)
10	50	0.54 (77)	140	1.51 (216)	325	3.50 (500)	330	3.56 (508)
11	20	0.22 (31)	40	0.43 (62)	110	1.18 (169)	110	1.18 (169)
12	15	0.16 (23)	20	0.22 (31)	15	0.16 (23)	15	0.16 (23)

[a]Stress calculated using $E = 10.92 \times 10^6$ kPa (1.56×10^6 psi).
[b]Wires cut by workmen.

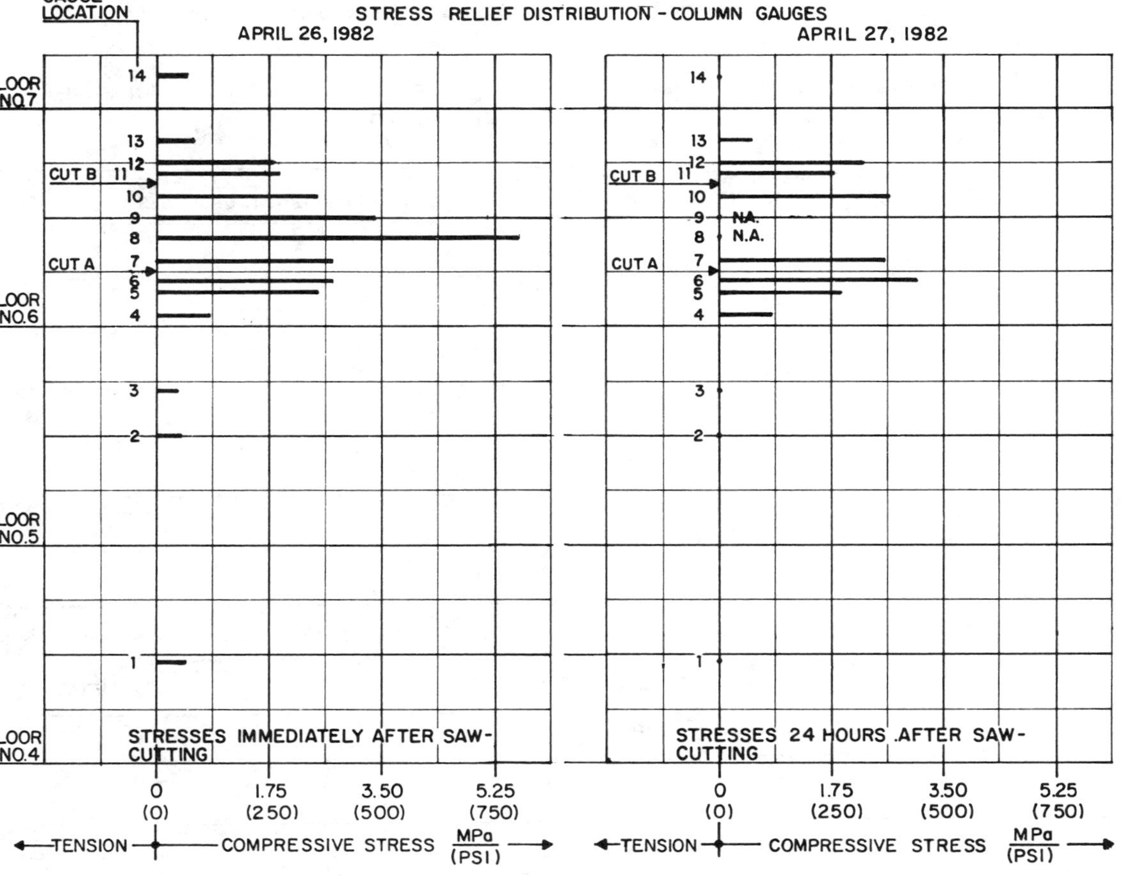

FIG. 6—*Graphical representation of stress relief on column gauges after saw cutting and 24 h later.*

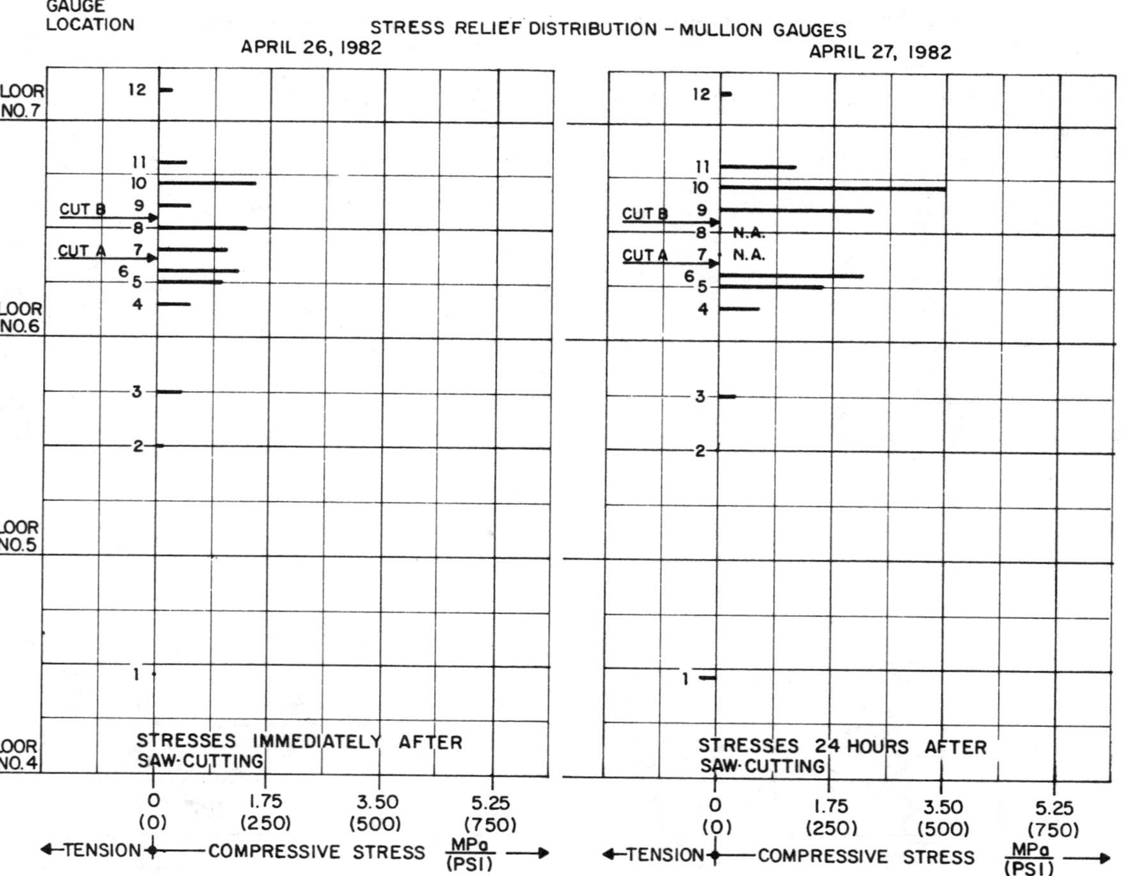

FIG. 7—*Graphical representation of stress relief on mullion gauges after saw cutting and 24 h later.*

10), respectively. The compressive strength of the terra cotta was determined to be 22.4 MPa (3200 psi), which is four times the maximum stress measured. In this case, therefore, it was decided not to proceed with stress-relieving operations on the remainder of the building. The above results were consistent with the distress noted. There was little evidence of large longitudinal or horizontal cracks traversing several pieces of terra cotta, as would be expected if compressive failure had occurred.

Had the stress relief measured been greater than one half the compressive strength of the terra cotta, stress relieving may have been required. Prior to making this decision, however, at least one other area would have been instrumented and relief stresses measured to verify findings of the original data.

As can be seen from Fig. 6, stress relief at Locations 4 and 13 on the column are approximately 35% of the relief at Locations 7 to 10. These locations are only one block away from the saw cuts. In the event that it was necessary to stress relieve the entire building, the recommended saw cutting option may have been every other bed joint on the columns. It is likely that saw cutting every other joint would have relieved the stress uniformly and to acceptable levels. For reasons similar to those presented above, excessive stresses on the mullion terra cotta would also have been relieved uniformly and to acceptable levels by saw cutting out every other joint. In order to determine the optimum spacing of cuts, the following need to be considered:

1. Compressive strength of the terra cotta to establish the maximum acceptable stress level.
2. Decay pattern of stress relief away from cut joint until it reaches an acceptable level.
3. Verification of the stress relief procedures by performing cuts at optimum locations on a new section of instrumented terra cotta.
4. The configuration of the facade. Window openings, block layout, and future maintenance of cut and sealed joints.

It should be pointed out that the stress relief pattern shown in Fig. 6 closely follows the expected behavior of the cladding under residual compressive loading in the structure. The stress relief due to saw cutting should be highest directly adjacent to the horizontal cuts, diminishing with distance due to shear drag along the masonry backing-to-terra cotta joint. Within experimental error, negligible stress relief was obtained on adjacent floors.

No attempt was made to monitor strain after removal of the terra cotta blocks. Thus, we did not measure stress relief due to shear drag from brick backing bonded to the terra cotta through the mortar joint. Compressive stresses due to shear drag, in our opinion, are relatively low and of a magnitude of less than 0.70 MPa (100 psi) depending on the quality of the bond; stresses higher than above would have resulted in shear failure at the mortar-terra cotta interface. Shear drag stresses would not be higher than 0.70 MPa (100 psi) for the following reasons:

1. Current UBC (Uniform Building Code) requirements [Chapter 30, Section 3004 (b)] for minimum shear strength of bonded veneer is 0.35 MPa (50 psi).
2. Experimental data obtained in our laboratory of mortar bonded to brick indicate the highest shear strength obtainable to be approximately 1.05 MPa (150 psi). This value was measured using current materials technology and under laboratory conditions. Considering that the subject project was constructed in 1924, shear strength of bond will definitely be less than 1.05 MPa (150 psi) and most likely less than 0.70 MPa (100 psi).

The maximum stress relief expected due to shear drag [$<$0.70 MPa (100 psi)] is, therefore, substantially less than the maximum stress relief measured on the terra cotta [5.6 MPa (800 psi)].

On the column, strains measured 24 h after saw cutting indicated that there was additional stress relief of approximately 13% at locations adjacent to the saw cuts. Little or no change was measured at locations between saw cuts. The changes in strain after 24 h were insignificant.

On the mullion, however, the additional average stress relief at 24 h on blocks adjacent to saw cuts (Numbers 5, 6, 9, and 10) was 130% of the average relief immediately after saw cutting. This is visually apparent from Fig. 7. We are not sure of the reason for the magnitude of delayed stress relief on the mullion as compared to the column. Nevertheless, it is advisable to wait at least 24 h to obtain final stress measurements. As on the column, there were insignificant changes in strain recorded after 24 h.

During this study, it was assumed that the stress distribution across the full thickness of the terra cotta units and the mechanical property specimens was linear. This is not entirely accurate as the glaze layer is of higher modulus than the fired clay core and, consequently, exhibits higher stress values than the core under a given strain. It should be kept in mind, however, that the fired clay is extremely thick compared to the glaze and will carry most of the load.

Strains measured in the field were converted to stress from modulus values obtained only on terra cotta. This conversion did not take into consideration the brick backing behind the terra cotta.

No significant strains were induced by load or thermal changes in the building during the two-day period when strains were recorded prior to the saw cutting.

Conclusion

The procedures described in this paper can provide valuable information on stresses built up in terra cotta facades due to lack of expansion joints. This information is required during restoration in order to make decisions on the need for and methods of repair.

The relatively low stresses measured resulted in considerable cost savings on the restoration of the Pacific Telesis Building.

References

[1] Wilson, F., *Building Materials Evaluation Handbook*, Van Nostrand Reinhold Co., New York, 1984, pp. 90–99.

[2] Stockbridge, J. G., "Evaluation of Terra Cotta on In-Service Structures," *Durability of Building Materials and Components, ASTM STP 691*, P. J. Sereda and G. G. Litvan, Eds., American Society for Testing and Materials, Philadelphia, 1980, pp. 218–230.

[3] Prudon, T. and Stockbridge, J., "Restoration of the Facade of the Woolworth Building," *Rehabilitation, Renovation and Preservation of Concrete and Masonry Structures, ACI SP-85*, G. Sabnis, Ed., American Concrete Institute, Detroit, 1985, pp. 209–227.

[4] Gaskie, M. F., *Architectural Record*, Vol. 169, No. 11, Mid-August 1981, pp. 90–95.

Tom Sourlis[1]

Restoration of the John J. Glessner House

REFERENCE: Sourlis, T., **"Restoration of the John J. Glessner House,"** *Masonry: Materials, Design, Construction, and Maintenance, ASTM STP 992,* H. A. Harris, Ed., American Society for Testing and Materials, Philadelphia, 1988, pp. 251-256.

ABSTRACT: Glessner House in Chicago was designed in 1886 by H. H. Richardson contemporaneously with the first skyscrapers. The Chicago Architecture Foundation, the present owner of Glessner House, commissioned tuck-pointing and cleaning of exterior granite facades for the building's centennial year. Sourlis Masonry Restoration, Inc. was awarded the contract and proceeded with the project in seven phases: (1) rigging and protection of the building; (2) grinding of mortar joints; (3) preliminary tuck-pointing of mortar joints to prevent penetration of caustic cleaning chemicals; (4) cleaning of the granite; (5) repair of the chimneys; (6) pink beading of the mortar joints; and (7) breakdown of scaffolding and landscaping repairs.

KEY WORDS: granite, masonry, restoration, historic structure, tuck-pointing, mortar joint, scaffolding, alkaline gel

History of Glessner House

The John J. Glessner House, 1800 S. Prairie, Chicago, IL, was designed in 1886 by Henry Hobson Richardson for the family of the Chicago industrialist whose name it bears. Construction had barely begun at the time of Richardson's death in April of 1886. This was a crucial moment in the history of American architecture for the year 1886 also saw the completion of Burnham & Root's Rookery Building in Chicago, one of the first structures to use a metal skeleton to support exterior stone walls. Although Richardson himself never built the skyscrapers which would develop from this technique, his expressive use of stone, both in residences such as Glessner House (Fig. 1) and in commercial buildings such as the Marshall Field Warehouse (Chicago, 1885-1887), influenced later architects of the modern metropolitan skyline.

The particular stone selected by Richardson for Glessner House was Milford granite, so named because it was quarried in Milford, MA. The original construction of the building was carried out by Norcross Brothers, the first general contractor of the nineteenth century. They maintained their own quarries and lumber mill and were responsible for most of Richardson's important buildings. Glessner House has exterior masonry-bearing walls of gray granite arranged in a U-shape around an interior courtyard, thus shutting out the turmoil of the city. Although the facades are simple with little ornamentation, they are composed of beautifully proportioned structural elements including particularly fine arches. The plan for Glessner House was a strikingly modern statement for a period of time when most residences were being designed in French chateau and Victorian style.

[1]President, Sourlis Masonry Restoration, Inc., Munster, IN 46321.

FIG. 1—*East facade (front).*

Historic Preservation

Preserving a historic structure such as Glessner House is difficult and expensive; great care must be taken so that irreplaceable elements of the building are not damaged. Also, the restoration must be historically accurate. Both the scholarly interest and the funding which made possible this type of restoration came about because of a change in attitude toward our architectural heritage in the 1960s. People began to have a deeper regard for older buildings constructed of time-honored materials. The movement gained momentum with the National Historic Preservation Act of 1966. In 1976, the year of America's bicentennial, the Tax Reform Act included legislation encouraging owners of historic property to preserve it. This law disallowed both the deduction of demolition expenses and the accelerated depreciation of any new structure built on the site. As a consequence of these events, restoration of historic buildings such as Glessner House has become more widespread.

The National Historic Preservation Act of 1966 coincided with the purchase of Glessner House by the Chicago School of Architecture Foundation (now the Chicago Architecture Foundation). Glessner House was designated a Chicago landmark in October of 1970. After focusing early restoration work on the interior of the house, the Foundation decided to prepare for the building's 100th birthday by commissioning complete tuck-pointing and cleaning of the exterior granite facades. This was a major undertaking since the building had weathered two of America's severe pollution eras: the smoke from early factories and locomotives and the exhaust fumes from the ever more numerous automobiles. The granite walls of Glessner House had gone from a sparkling light gray to soot black.

Bidding the Project

Sourlis Masonry Restoration, Inc. initially bid the project at one price which included all tuck-pointing and cleaning of the granite facades and the restoration of five of the seven chimneys. For the project layout, we defined the order of work in seven phases: (1) rigging and protection of the building; (2) grinding of mortar joints; (3) preliminary tuck-pointing of mortar joints to prevent penetration of caustic cleaning chemicals; (4) cleaning of the granite; (5) repair of the chimneys; (6) pink beading of the mortar joints; and (7) breakdown of scaffolding and landscaping repairs.

The original spec called for complete pipe scaffolding of the wall perimeter in order to insure a minimum amount of contact with the delicate terra-cotta tiles of the roof and with the granite stones of the facades. We devised a system of rigging the building with pipe scaffold towers at 20 and 12-ft intervals and hanging swing stages in between the towers. Because the swing stage is the most efficient working platform for a tuck-pointer, this innovation enabled us to bid the project competitively. As a result, we were awarded the contract.

Rigging and Protection

Eleven towers of scaffolding including one double section were attached to the mortar joints of the house using a minimum number of masonry bolts. These were red plastic pressure anchors installed in 1/4-in.-diameter holes drilled 1 in. deep with two anchors used per tower. A temporary chain link fence was installed in order to keep visiting architecture buffs and the general public at a safe distance during the project.

Before the actual cleaning could commence, certain components of this historic structure had to be protected. The window panes of French antique glass (considered irreplaceable) were coated with a strippable acrylic mask to protect the glass from the caustic alkaline gel which would be used. Glazing tape was then applied around the entire perimeter of each window frame. Finally, particle board was nailed over the windows to further protect them from the chemical spray and from any falling tools or debris. Where the window frames meet the stone return, we left a 1/2-in. reveal to allow for complete access in cleaning the granite.

After the building was completely scaffolded and protected, John Vinci, the architect, and Elaine Harrington, the curator of Glessner House, conducted an inspection to examine some previously inaccessible areas and to determine the exact extent of the work to be done. During this survey, several interesting discoveries came to light. It became apparent that pieces of slate had been used as shims to help stabilize the granite blocks while they were being leveled. This was a standard nineteenth century masonry technique. Another discovery was that the rosy-red toned mortar in use throughout the rest of the house had not been used in the arch over the entry. Gray mortar was used in a row of decorative ball and bead carved granite. Gray was also used to blend in and unite the separate stones of the inner arch. All these mortar colors would be carefully matched prior to the tuck-pointing application.

Grinding of the Mortar Joints

With the scope of the project now defined, we began the actual work. Initially we agreed to chisel out all the mortar joints and then give the building a preliminary tuck-pointing with standard Type N mortar and then finish tuck-pointing with a pink beaded joint of Type N mortar. The mortar joints were to be chiseled out to a 2-in. depth and then tuck-pointed to within about 1/2 in. of the surface. This preliminary tuck-pointing would stabilize the masonry and insure that none of the chemicals would be water blasted into the interior of the building during the cleaning process. As the chiseling began, we found that the process was painstakingly slow and not economically feasible. Therefore it was agreed that we would perform a sample cutting of the joints with electric grinders and, upon acceptance of this procedure, grind rather than chisel all the mortar joints. This procedure was in fact permitted, although it did not fall within the strictest interpretation of historical restoration. The grinding proceeded on schedule.

As the tuck-pointers worked over the building at close range, they carefully saved iron ivy nails presumably installed under Mrs. Glessner's direction for training ivy on the facade of the house. They also collected bits of original building materials found near the foundation below grade.

Preliminary Tuck-pointing of the Mortar Joints

After the joints were ground out to a 2-in. depth, they were then tuck-pointed to within 3/8 in. of the surface with Type N natural color mortar. Sample mortar mix had been made and sample pointing of the deep mortar had been done for inspection by the architect. Type N mortar was selected to match the original specification which stated:

> The mortar used in the construction of the building above the ground level, unless otherwise specified, will be composed of the best lime mixed with the best Utica cement, and clean sharp sand in the following proportions: one barrel cement, one barrel lime, and the due proportion of sand. Below the ground level the mortar used will be composed of two barrels cement, one barrel lime, and the due proportion of clean, sharp sand. None but the best quality of lime equal to the best Stearns lime, and Utica cement, and clean, sharp sand, free from extraneous matter, shall be used in the work.

The joints were struck with a key down the center in order to assure a good bond with the pink finish mortar which would be applied after cleaning.

Cleaning of the Granite

Upon completion of the preliminary tuck-pointing, the building was completely sealed so that the cleaning process could begin. Prior to the start of the project, even before the scaffolding had been erected, we had applied various cleaning chemicals to sample areas of the stone and had chosen alkaline gel manufactured by Prosoco because it works slowly on granite and poses no threat to the limestone sidewalk surrounding the house, the terra-cotta roof, or the brick inner courtyard and walls where they abut the granite.

The sample applications also determined that the method of application would involve two steps. The alkaline gel would be scrubbed onto the building and allowed to set for 24 h then pressure washed with 600-lb water pressure. After the gel was rinsed off, an acid-based afterwash was applied to the building in order to remove any residue of the alkaline prewash, the actual cleaning agent. The afterwash required no setting time and was immediately rinsed from the building. This two-step process was done twice to thoroughly clean the granite. The alkaline gel was removed by pressure washing starting from the bottom of the building and working upward toward the roof.

Although the dwell time for the alkaline gel was increased from 24 to 48 h on some heavily soiled areas because of the temperature difference between the August test samples and the actual application in October, the cleaning process resulted in virtually 100% removal of soot and grime from the surface of the granite. Split granite samples were used to check the efficacy of the cleaning process. When the cleaning was completed, what had been a dark stone building having the appearance of a fortress (Fig. 2) was transformed into a light gray residence with glittering mica chips and particles of pink stone dispersed throughout the granite facades (Fig. 3).

Repair of the Chimneys

The chimneys had deteriorated to the point where plants were growing from some of the chimney tops and in some sections the mortar was completely missing, allowing our crew members to peer through the joints. In one chimney approximately a dozen stones were removed and replaced; then the entire chimney was tuck-pointed as were the others which we repaired.

During this phase of the operation, we were allowed to walk on the roof as long as we accepted

FIG. 2—*Entry door/decorative arch.*

FIG. 3—*Entry arch after cleaning.*

responsibility for any damage and took every precaution to prevent damage. The terra-cotta tiles are thin, and some have come loose with age. In order to protect these fragile tiles, special ladders that were padded with foam and supported with wood struts from the topmost structural tile of the building were designed. The peak tiles were just strong enough to bear this weight. Chimney repair involved many less man-hours than cleaning but was completed over a longer time schedule because of the required phases and difficult access to the work area.

Pink Beading of the Mortar Joints

After the cleaning and restoration of the chimneys, we were to begin the bead tuck-pointing of the joints in the original pink color for the majority of the building and in gray over the front entry. Two areas of the house had been cleaned but not tuck-pointed: the two servants' porches, one on the first floor, and the loggia on the second floor. The mortar in these areas had been protected from the weather and so was left intact for historical purposes. Accurate mortar color was determined by checking the joints on these porches.

The pink mortar was applied with a 3/16-in. bead. Tools for this size beading had to be handmade by a metalsmith as the standard beading tool of today creates too large a bead. The beading process was delicate and time-consuming since the joints had to be tuck-pointed and then struck several times in order to bring the cement content of the mortar to the surface. This procedure gave a very smooth and uniformly colored joint.

Breakdown of Scaffolding and Landscape Repairs

All of the window protection and swing scaffolding were removed from the building upon completion of the final tuck-pointing. The pipe scaffolding was then removed, and the lawns were resodded. As a final touch, some of the mortar joints required spot tuck-pointing where the scaffolds had been anchored to the building. This was easily done by removing the plastic anchors, grouting in the holes, striking the fresh mortar with a beading tool to blend in with the finished joints. The project which we had begun on August 24 was now completed just before Thanksgiving.

Glessner House, the last surviving example of H. H. Richardson's four Chicago buildings, now celebrates its 100th year with granite facades free from the last century's grime and fortified by this century's restoration techniques.

Acknowledgments

I would like to gratefully acknowledge the editorial assistance of Jacquelyn Scruggs in preparing this paper.

Bibliography

"John J. Glessner House," pamphlet, Commission on Chicago Historical and Architectural Landmarks, Chicago, IL, 1984.

Hitchcock, H. R., *The Architecture of H. H. Richardson and His Times*, Archon Books, Hamden, CT, 1936 revised 1961, pp. 277–278, 290–294, Figs. 105–106.

New Life for Old Buildings, M. F. Schmertz, Ed., McGraw-Hill, New York, 1982, pp. vi–vii.

Van Rensselaer, M. G., *Henry Hobson Richardson and His Works*, Houghton Mifflin, 1888, reprinted Dover Publications, New York, 1969, pp. 109–110.

Clayford T. Grimm, [1]

Masonry Cracks: A Review of the Literature

REFERENCE: Grimm, C. T., "**Masonry Cracks: A Review of the Literature,**" *Masonry: Materials, Design, Construction, and Maintenance, ASTM STP 992,* H. A. Harris, Ed., American Society for Testing and Materials, Philadelphia, 1988, pp. 257–280.

ABSTRACT: Masonry surface cracks are objectionable because they are the primary source of water permeance and may be aesthetically displeasing or indicative of structural distress. Cracks are the most frequent cause for masonry's failure to perform as intended. The types, locations, patterns, sizes, and causes of cracks are discussed. Methods are described for their prevention and repair.

KEY WORDS: block (concrete), bricks, corrosion, cracks, expansion, failure, inspection, masonry, mortar, movement (structural), repair, sealants, shrinkage, strain

Cracking is probably the most frequent cause of masonry performance failure [*30*] and has been an engineering concern for at least the last 150 years [*3, 7, 10, 26, 36, 54, 58, 70, 71, 88, 91, 92, 97, 111, 112*]. More recently, cracked masonry has generated litigation [*107–110*]. Cracks are caused by movement (strain), which can not be prevented but can be accommodated. Thus, cracks can be eliminated or made so small as to be unobjectionable.

A crack is here defined as a break, split, fracture, fissure, separation, cleavage, or elongated narrow opening visible to the normal human eye and extending from the surface and into a masonry unit, mortar joint, interface between a masonry unit and adjacent mortar joint, or into the joint between masonry and an adjacent construction element.

Crack Classification

Masonry cracks may be classified by: (1) structure type; (2) masonry type; (3) location; (4) pattern; (5) width; and (6) cause. For a given type of structure and material, the location, pattern, and width often provide clues to the cause. Cracks result from strain which induces stress in excess of strength in compression, tension, or shear. Strain may be induced by the imposition of loads or by restraint of volume changes in the masonry materials. Volume changes include those induced by change in temperature, moisture, water or salt crystallization, or corrosion. Loads may be imposed by movements of foundations, structural frames, shelf angles, roof slabs, spreading of pitched roofs, wood expansion, or retaining wall deflection. Cracks may also be caused by vibration, blasting, and fire.

Types of masonry structures include: (1) arches and shells; (2) fireplaces and chimneys; (3) floors and pavements; (4) revetments and channels; (5) beams and slabs; (6) bearing walls and columns, and (7) nonbearing walls. Types of cracked masonry units include: (1) brick; (2) concrete masonry units (CMU); (3) stone; and (4) terra cotta. Cracks are located: (1) in masonry

[1]Consulting architectural engineer and senior lecturer in architectural engineering, University of Texas at Austin, Austin, TX 78712.

units; (2) in mortar joints; (3) between units and mortar; and (4) continuously through units and mortar. Crack patterns are: (1) horizontal, (2) vertical, or (3) diagonal. If vertical, they may be straight or cogged, and, if diagonal, they may be straight or stepped.

Cracks smaller in width than 0.1 mm are insignificant to water permeance because that is the least width through which wind-driven rain will enter [*12*]. The maximum width of a crack that will neither impact appearance nor cause alarm is said to be 0.25 to 0.38 mm [*14*]. Bidwell [*11*] classified cracks as fine (up to 1.5 mm), medium (1.5 to 10 mm), and wide (above 10 mm). Rainer [*81*] classified crack width as very slight (less than 1 mm), slight (1 to 5 mm), moderate (5 to 15 mm), and severe (more than 15 mm). Kaminetzky [*53*] classified crack widths as negligible (less than 0.1 mm), very light (0.4 mm), light (0.8 to 3.2 mm), moderate (3.2 to 12.7 mm), extensive (12.7 to 25.4 mm), or very extensive (more than 25.4 mm). The term "hairline crack" often appears in engineering literature to describe vaguely a narrow fracture. The thickness of human hair varies with race, age, body location, circularity (measurement axis), fiber length, relative humidity, and tensile stress [*111*]. Among teen-aged caucasoids for hair from the occipital area of the scalp, the mean diameter is 70 μm with a standard deviation of 20 μm ($\bar{X} = 0.07$ mm, $s = 0.02$ mm). If the data are normally distributed, probably only 1 of 15 "hairline" cracks would not admit wind driven rain, that is, diameter <0.1 mm.

Cracking Strain

The theory of fracture mechanics has been applied to masonry by Shrive [*94*]. He observed that for a crack to be visible, the surfaces of the newly formed crack must separate, indicating the previous existence of tensile stresses. Thus, it is tension which causes cracks, whether the loads are compressive, shear, or tensile.

Compression induces tension transverse to the axial force, which may cause splitting (see Fig. 1). The strain required to cause brick masonry to crack in compression occurs at about half the ultimate strength [*39*]. Frequent application and withdrawal of load may cause fatigue and strength reduction and, therefore, increased cracking probability. As few as 40 cycles of compressive load is said to cause a 30% reduction in static strength [*1*].

Shear induces diagonal tension. The shear load at first visible crack in concrete masonry walls was measured at about 64% of failure load with a coefficient of variation of 25% for 15 walls [*69,89,90*]. Mayes et al. [*63*] found visible cracks in concrete masonry shear panels at stresses exceeding 50 psi (345 kPa). Schneider [*87*] found first crack at an average of 64% of the ultimate shear strength for 29 concrete masonry piers. Meli [*66*] found the ratio of shear load at first crack to ultimate load on various types of masonry walls to be 0.83 with a coefficient of variation of 20% for 19 specimens.

The first crack in brick masonry in flexure occurs at about 80% of failure strength [*46*]. Lawrence and Morgan [*59*] found that the flexural stress in masonry at first crack may be estimated at 30% of the ultimate strength plus 29 psi (200 kPa). The tensile strain at rupture for concrete masonry walls is about 0.021% [*67*]. The effect of specimen size and test method on flexural strength is described in Ref *41*.

Cement Shrinkage

Mortar, grout, concrete masonry units, and reinforced concrete shrink upon drying. When excessive shrinkage is restrained, cracks result. Shrinkage of materials made with portland cement is caused by water loss and by carbonation. Water loss shrinkage is reversible. Carbonation shrinkage is not. As CMU dry the recession of water in capillaries creates surface tension, which places the material in compression and thus reduces volume. Because sand and gravel are stiffer than lightweight aggregates, CMU made with such aggregates have greater shrinkage. Carbonation is primarily a reaction between calcium hydroxides [$Ca(OH)_2$], released by hydration of cement, and carbon dioxide (CO_2) from the air to produce calcium carbonate [$CaCO_3$]

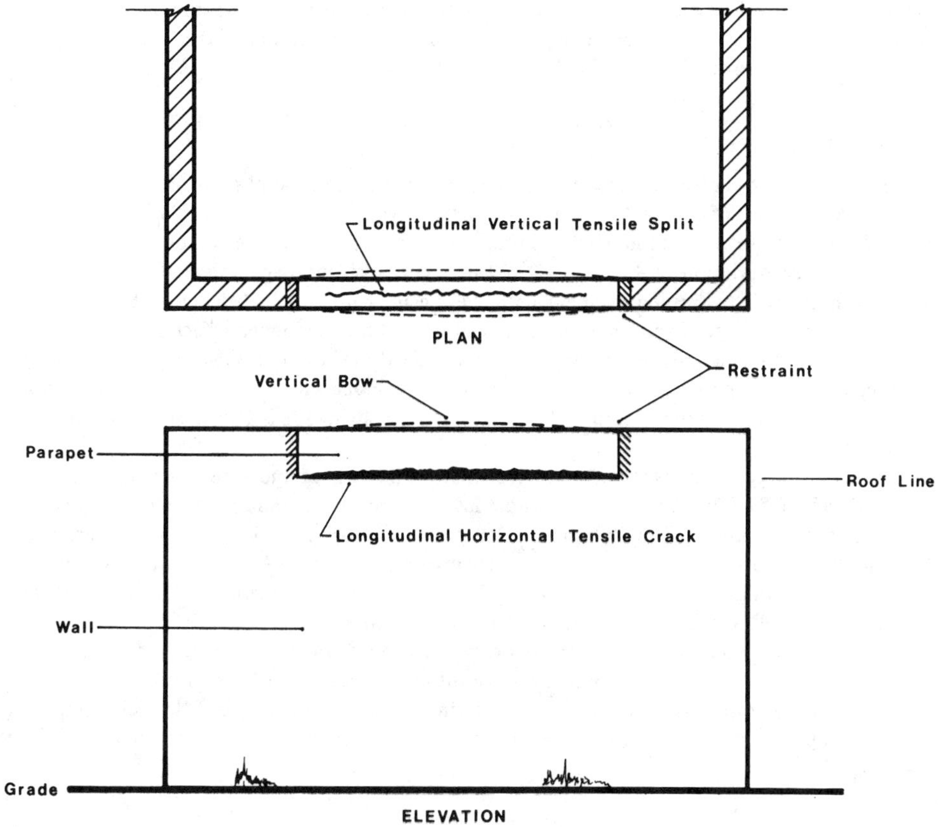

Longitudinal Vertical Tensile Split

PLAN

Restraint

Vertical Bow

Parapet

Roof Line

Longitudinal Horizontal Tensile Crack

Wall

Grade

ELEVATION

FIG. 1—*Restrained parapet: vertical bow or longitudinal split.*

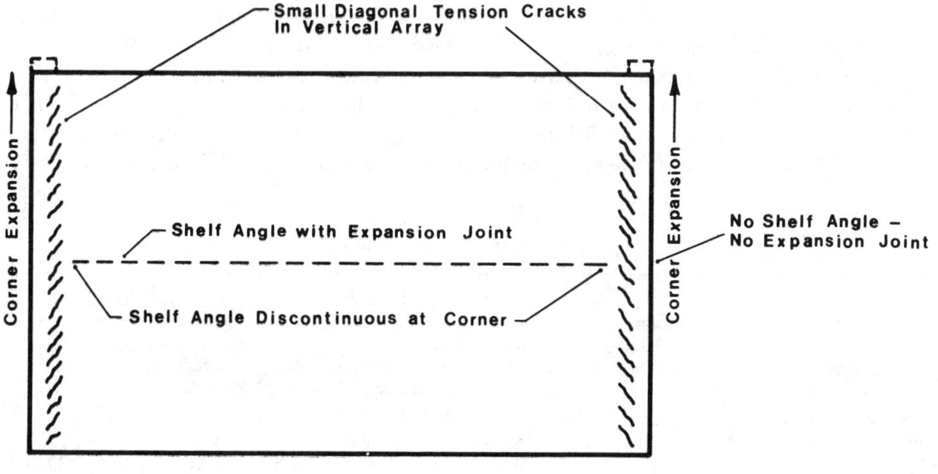

FIG. 2—*Differential vertical movement at corners due to discontinuity of shelf angle.*

and water [*93*]. Carbon dioxide may also react with other cement paste components. Lime also carbonates and therefore shrinks. The physiochemistry of carbonation is discussed by Kroone and Blakey [*57*], Powers [*78*], Verbeck [*103*], and Kamimura [*51*].

Mortar Shrinkage

Unrestrained, 28-day shrinkage of mortar specimens cast in metal molds is said to have a mean value of about $650 \times 10^{-4}\%$ with a standard deviation of $150 \times 10^{-4}\%$ [*42*]. The ultimate (23 year), unrestrained, mortar shrinkage is estimated to have a mean of $1800 \times 10^{-4}\%$ with a standard deviation of $420 \times 10^{-4}\%$ [*40*]. The horizontal shrinkage of mortar in bed joints in masonry is restrained by shear with the masonry units, which Ritchie [*83*] found reduces mortar shrinkage. From those data it is estimated that the mean effect of restraint is to reduce mortar shrinkage by about 30% with a standard deviation of 22%, when adjusted for sample size. Anderegg [*5*] found that mortar left in contact with brick had shrinkage about 50% less than that of mortar cast in metal molds. Mortar shrinkage is also discussed in Refs *44*, *55*, and *58*.

Mortar shrinkage increases with water-cement ratio, which increases with lime content [*28,33,49,65,73,83,106*]. Voss [*104*] found that some mortars made with dolomitic hydrate shrink more than some made with high calcium putty. As mortar sand fineness increases, water demand to provide workability increases and shrinkage increases [*13*]. Thus, Anderegg [*6*] and Conner [*18*] found that mortar joint cracking increased with sand fineness. Masonry sands are also discussed in Ref *95*. Increased air content also increases shrinkage [*13*].

The use of calcium chloride as a mortar additive can cause cracking by accelerating corrosion of metals in contact with mortar and by substantially increasing drying shrinkage of mortar [*19*]. Mortar cracks may also be due to weathering or sulfate attack [*32,84*]. McBurney [*64*] describes expansion due to delayed hydration of magnesium oxide in mortar as causing severe cracking of masonry.

Shrinkage of Concrete Masonry Units

Baker and Jessop [*9*] reviewed the literature on CMU shrinkage. Shrinkage of CMU increases as density is reduced and water absorption of the concrete is increased [*50*]. Shrinkage of low pressure cured CMU is about 85% greater than for high pressure cured units [*16*]. As with all products, availability of autoclaved CMU should be determined prior to specification. Table 1 provides statistical data on CMU shrinkage based on data given in Ref. *16*.

When CMU, which had been previously dried to equilibrium with low humidity, were subsequently sprayed on one face for 2 h to simulate rain or painted with water cement paint, they reexpanded approximately one third of their original shrinkage from a saturated condition [*86*]. ASTM specifications provide for maximum moisture content of CMU ranging from 25 to

TABLE 1—*CMU shrinkage, $10^{-4}\%$.*

Statistic	Probability of Being Exceeded, %	All CMU	Low Pressure			High Pressure		
			Sand & Gravel	Light Weight	Pumice	Sand & Gravel	Light Weight	Pumice
Low characteristic	95	130	162	254	400	138	131	...
Mode	...	260	269	400	596	192	215	373
Mean	...	330	270	420	630	190	230	390
High characteristic	5	600	338	610	838	238	323	481

45% depending on CMU shrinkage and average annual relative humidity at the construction site [*8*].

Concrete Masonry Wall Shrinkage

Total prevention of cracks in concrete masonry is said to be technically and economically unfeasible [*20*]. Shrinkage of mortar and of concrete masonry units in concrete masonry walls results in wall shrinkage which is greater than CMU shrinkage, perhaps 30% greater [*86*]. Total concrete masonry wall shrinkage may range from 0.01% to 0.1% [*14*]. More precise data is given in Table 2.

A rationale for recommending a maximum ratio of wall length to height was provided by Copeland [*21*]. Wall shrinkage is restrained at the wall base by bond and friction but may not be restrained at the top of the wall. For walls of excessive length to height ratio, this phenomenon results in a vertical crack near the center of the wall. Such cracks are wider at the top and tapered in width toward the wall base. Depending on the relative tensile strength of CMU and mortar, the vertical crack may be cogged or relatively straight.

In the absence of excessive wall height, shrinkage cracks are of even width and occur at the weakest part of the wall. With long walls, vertical cracks occur at the midpoint or at rather even intervals. Cracks may be vertical or cogged and sometimes stepped, especially near wall ends. Vertical shrinkage cracks are common at reentrant corners [*22,25,96,100*].

Rainer [*81*] reports that wall cracks are most likely to occur at changes in wall dimension; at corners, openings, pilasters, or other wall stiffener; and in areas of greater exposure as in parapets, wing walls, or fences.

The shrinkage cracking of concrete masonry walls is reduced by using two-core rather than three-core CMU and the use of mortars having higher bond strength [*67*]. An increase of 10% in crack resistance is reported. Concrete masonry units should be kept covered and dry during transportation and job site storage until installed in the wall to minimize in-the-wall shrinkage [*79*]. Weaker mortars, because of their greater extensibility (lower modulus of elasticity and greater creep in tension), accommodate CMU shrinkage to a greater extent than stronger mortars [*86*]. Expansion joints are usually not required in CMU walls because shrinkage normally exceeds expansion (see Ref *2*).

Sealant Joints

Sealant joints are sometimes called "movement joints." Three types of such joints are used for crack control in masonry: expansion joints close to accommodate expansion of brick or stone masonry; control joints open to accommodate shrinkage of concrete masonry; construction joints seal the crack between masonry and other materials, such as windows and doors. To avoid considerable confusion, the terms *expansion joint, control joint,* and *construction joint*

TABLE 2—*Unrestrained shrinkage of concrete masonry, 10^{-4}%.*

Aggregate Type	Cure	
	High Pressure	Low Pressure
Sand and gravel	90 to 130	230
Cinders	120	400 to 450
Expanded slag	. . .	340
Expanded shale	. . .	310

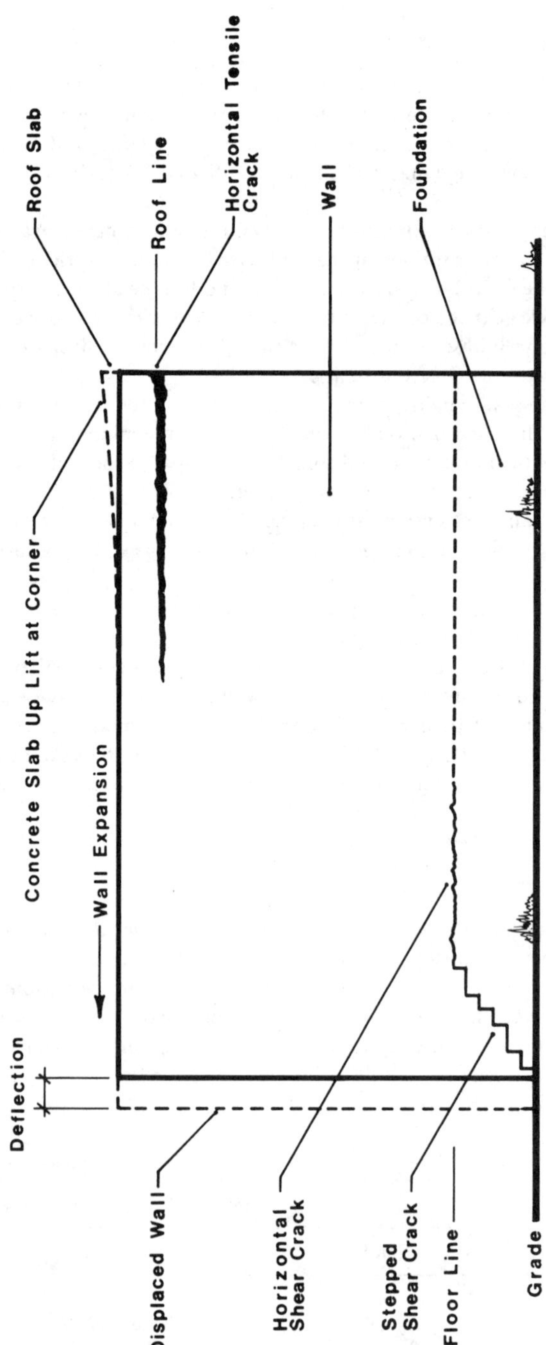

FIG. 3—*Slab curl on right; wall expansion on left.*

should not be used interchangeably. Some types of control joints are filled with mortar, can not close, and can not be substituted for expansion joints without causing significant cracking. The design of sealant joints is discussed in Ref *61* and *75*.

Crack Control in CMU Walls

Methods for controlling cracks in concrete masonry walls include: (1) limitation on length to height ratio of walls; (2) limitations on horizontal distance between vertical control joints; (3) installation of bed joint reinforcement or bond beams; (4) location of control joints at points of stress concentration; (5) control of moisture content in CMU at time of construction; and (6) installation of slip joints [*29,62*].

The maximum spacing of vertical control joints is a function of type of CMU, wall height, and spacing of bed joint reinforcement. The following recommendations for control joint spacings [*62*] apply to walls built of ASTM Specification for Hollow Load-Bearing Concrete Masonry Units (C 90), Type I (moisture controlled) CMU [*8*]. Bed joint reinforcement is two No. 9 cold drawn, steel wires, one in each shell bed. With no bed joint reinforcement, the maximum control joint spacing is 18 ft (5.5 m) in exterior walls in climates where the average annual relative humidity is between 50 and 75%. That control joint spacing is increased 33% to 24 ft (7.3 m) for 16-in. (400-mm) spacing of bed joint reinforcement and increased 67% to 30 ft (9.1 m) for 8-in. (200-mm) spacing of bed joint reinforcement. Those control joint spacings are increased 25% for interior walls. All of those control joint spacings are increased 6 ft (1.8 m) in dry climates, where average annual relative humidity is less than 50% and reduced by 6 ft (1.8 m) in humid climates, where average annual relative humidity exceeds 75%.

In addition, control joints are required at critical points of high stress concentration, that is, at changes in wall height or thickness, above joints in floors or foundations, and below joints in slab roofs bearing on the wall, at one or both sides of wall openings, at a distance from wall intersections or corners not greater than one half the allowable spacing of control joints, and in composite walls at the same location as expansion joints in the brick masonry. In lieu of a control joint at each jamb of a wall opening, bed joint reinforcement may be placed in the first and second joints immediately above and below wall openings, extending at least 2 ft beyond the opening.

In large wall expanses, bond beams may be used in lieu of bed joint reinforcement. Trough or U-shaped CMU are used in a continuous horizontal course, which is filled with concrete and in which one or more reinforcing bars are placed. For two No. 9 bars the vertical spacing of bond beams is four times the spacing of the replaced bed joint reinforcement. If used, bond beams should be placed at the base and top of a wall and below windows [*62*].

Slip joints are horizontal planes of weakness formed by breaking the bond of mortar bed joints with the CMU. Slip joints are placed at the top exterior corners of walls that support cast-in-place concrete roofs or floor slabs and at CMU lintel bearings, where a control joint is located above the jamb at a wall opening.

Cracks in Brick

The plane of cracks in the face of unglazed, extruded brick may be perpendicular to the face and parallel to the direction of extrusion (for example, vertical for brick layed as a stretcher) and are usually located at a core in the brick. They may penetrate only slightly or may extend to the core. Such cracks are typically 2 mm (0.08 or 5/64 in.) or less in width. They are caused by inadequate quality control in the brick manufacture process. ASTM Specification for Facing Brick (Solid Masonry Units Made from Clay or Shale) (C 216) limits facial cracks in face brick to those which may be seen at a distance of 15 ft (4.6 m) or 20 ft (6.1 m), that is, crack widths of about 1.5 mm (0.06 in.). Cracks in brick that are parallel to the face and to the direction of extrusion are called laminations and are not visible on the surface. All extruded brick are laminated to some extent or other. Although laminations are of some concern to ceramic engineers, there is no evidence that they affect brick performance [*85*]. Weathering or excessive compressive stress applied at the face of brick may cause it to spall [*40*] (see Fig. 4).

Brick-Mortar Compatibility

Cracks in mortar may be due to differential movement between brick and mortar. For example, if the coefficient of thermal expansion of brick is greater than that of mortar, vertical cracks may occur in horizontal bed joints.

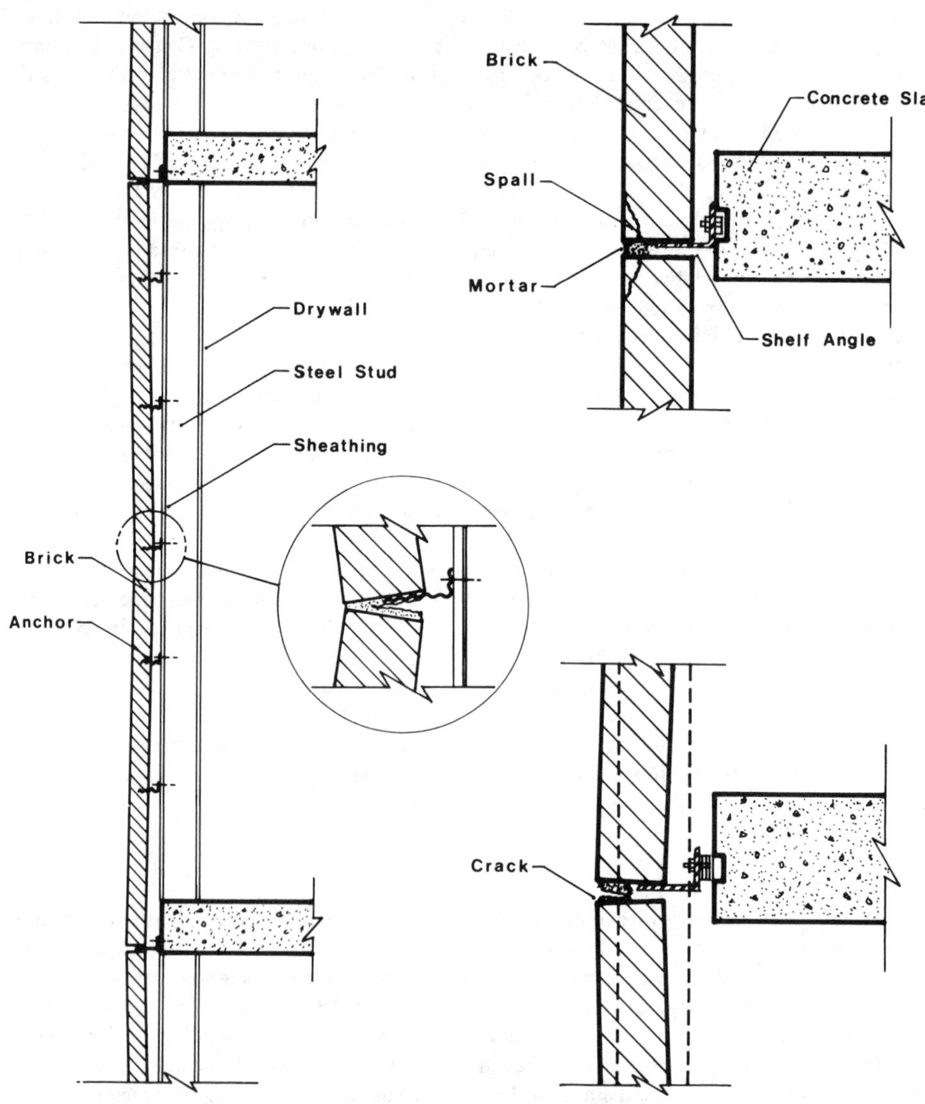

FIG. 4—*Nonbearing wall.*

Hedstrom et al. [*45*] measured the modulus of elasticity of several mortars at 90% of tensile strength. The mean value was 2.87×10^6 psi. If the tensile bond strength of mortar to brick is about 75 psi, the maximum unit strain in the mortar is $75/2.87 \times 10^6$ or 26×10^{-4}%, which is about one eighth of that found by Menzel [*67*]. In any event, if the 28-day shrinkage of restrained mortar is 230×10^{-4}%, to avoid a cracked head joint the differential strain must be compensated for by brick expansion. Irreversible moisture expansion of brick at 28 days has a mean value of 63×10^{-4}% [*42*]. A 7.63-in. (190-mm)-long brick expanding at that rate would produce a strain in a 3/8-in. (10-mm) mortar head joint of 1280×10^{-4}%, that is, some 5.6 times greater than that required to compensate for mortar shrinkage.

Since the most likely mortar shrinkage transverse to the plane of a mortar head joint is much less than the most likely brick moisture expansion, a shrinkage crack in the mortar head joint most likely will not occur in brick masonry. Palmer and Parsons [*74*] found no evidence that volume changes in mortar subsequent to hardening destroyed or weakened mortar bond when extent of bond was good. When water retentivity of mortar was not compatible with the suction rate of brick, mortar volume change was disruptive. High shrinkage mortar could be combined with low moisture expansion brick, in which case the bond strength and extensibility of the mortar becomes important to crack avoidance.

Mortar shrinkage and brick expansion are additive in bed joints. If that differential strain is 290×10^{-4}% and the mortar modulus of rigidity is $0.4 \times 2.87 \times 10^6$ psi, the estimated shear stress is 332 psi. For ASTM C 270 Type N mortars, Nuss, Noland, and Chinn [*72*] found a mean 28-day shear strength of brick masonry to be 394 psi. When the IRA of brick was reduced by wetting, shear strengths were much higher. It was also higher when Types M and S mortar were used. Mortar shrinkage should not cause cracking of well-bonded mortar bed joints in brick masonry.

Facial Separation Cracks

Facial separation cracks are openings in the wall face between brick and mortar, usually 1 mm (0.04 in.) or less in width. They are most frequently caused by inadequate tooling of mortar joints during construction but may also be caused by thermal contraction of masonry units and mortar and less frequently by mortar shrinkage. The effect of facial separation cracks was discussed by Conner [*17*]. He found that the average cracking in the brick masonry wall of 44 buildings which had no wall leaks was 14.7% (that is, 14.7 ft of crack/100 ft of mortar joint), which compared with 36.3% in 34 buildings which did leak.

Flexure induces cracks between brick and mortar rather than in mortar, because bond strength is inevitably lower than the tensile strength of mortar. Horizontal cracks between brick and mortar bed joints may be induced by shear (Figs. 3 and 5) or flexure (Fig. 4).

Brick Masonry Expansion

Expansion of masonry may be caused by heat, moisture, or freezing. Such elongation may cause oversailing of upper portions of a wall over lower portions (see Fig. 5), diagonal shear cracks (see Figs. 3 and 8), bowing of walls (See Fig. 1, 4, 9, and 13), and flexural cracking at corners in a vertical straight or cogged pattern (Fig. 11). Restrained longitudinal expansion may cause delamination and buckling of pavements and bowing of parapets (Figs. 1 and 13).

Cracking due to expansion can be controlled in curtain walls by placing horizontal expansion joints under shelf angles and vertically at appropriate horizontal intervals. For 3/8-in. (10-mm)-wide joints and sealants having 50% compressibility, expansion joints in brick masonry should be spaced at horizontal intervals of about 20 ft (6.1 m) [*42*]. In addition, expansion joints should be located at the same critical points of high stress concentration as in CMU walls. Expansion joints in parapets should occur twice as frequently as those in enclosing walls, unless the parapet is reinforced [*30*].

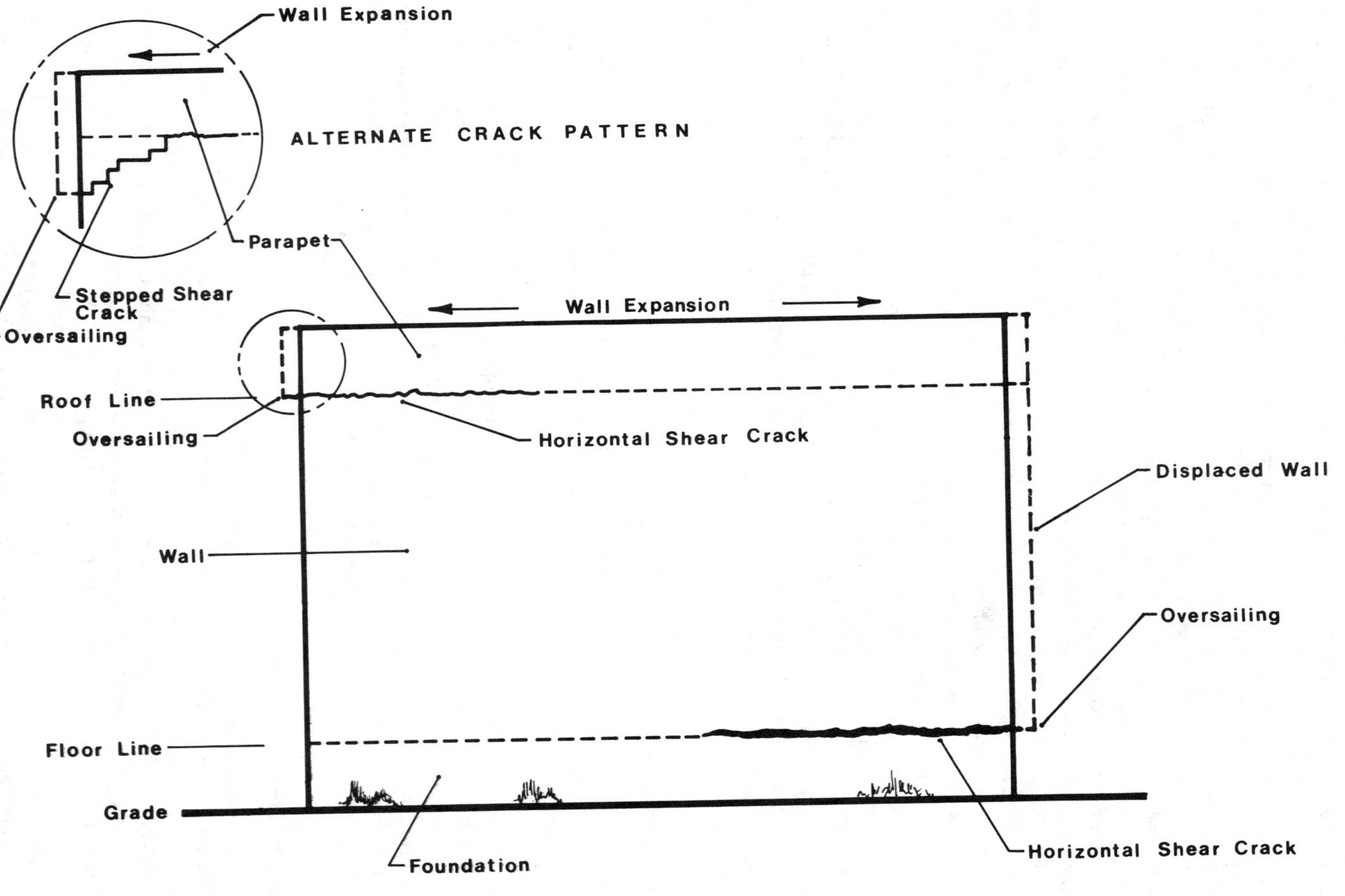

FIG. 5—*Longitudinal wall expansion with alternate crack patterns at parapet.*

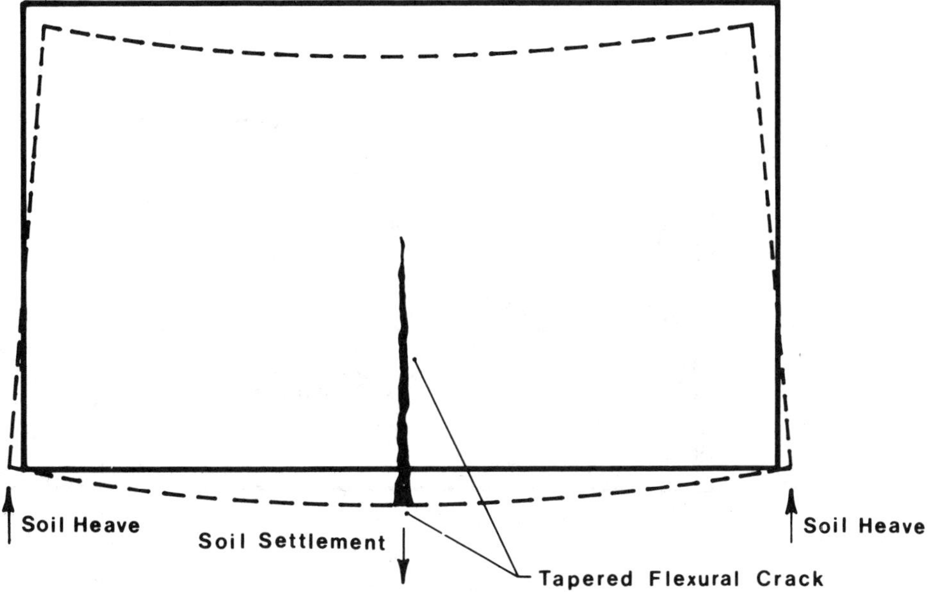

FIG. 6—*Deep beam flexure—corner heave and/or midwall settlement or supporting beam deflection.*

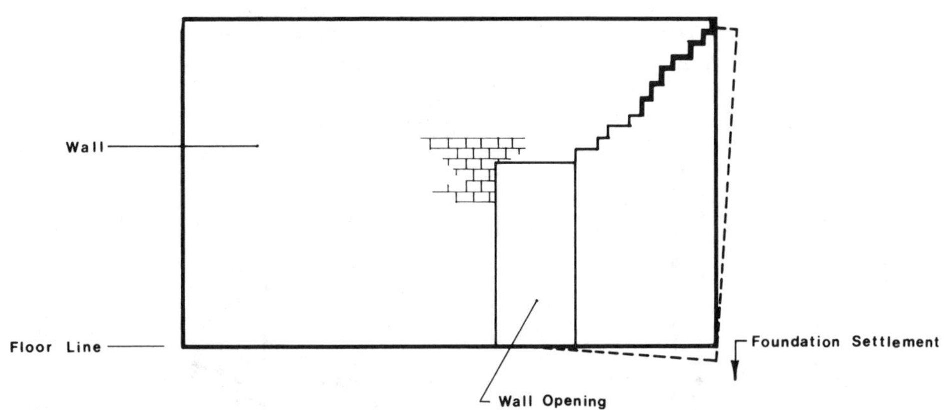

FIG. 7—*Foundation settlement.*

A wall above grade is more prone to volume change than a foundation wall. If the two are rigidly anchored together, the restraint often induces cracks (Fig. 5).

Differential movement will also occur between dissimilar materials in a wall, for example, face brick bonded to CMU in a composite wall [*35*]. If the brick expansion is 0.03% and the concrete masonry contraction 0.03%, the differential is 0.06%. If the modulus of rigidity of the masonry is 500 000 psi, the first approximation of the shear stress in the masonry is 300 psi. Differential vertical movement between dissimilar materials can cause parapets to lean in (see Fig. 12).

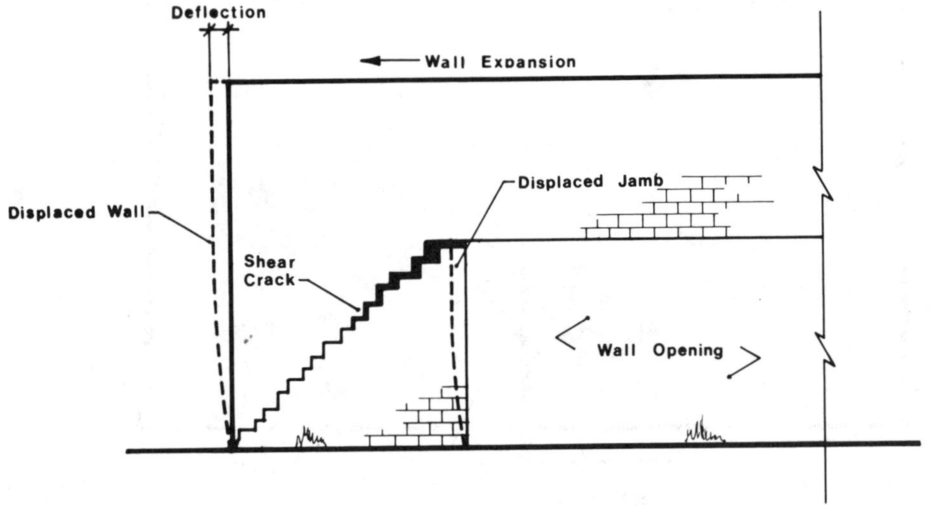

FIG. 8—*Shear crack.*

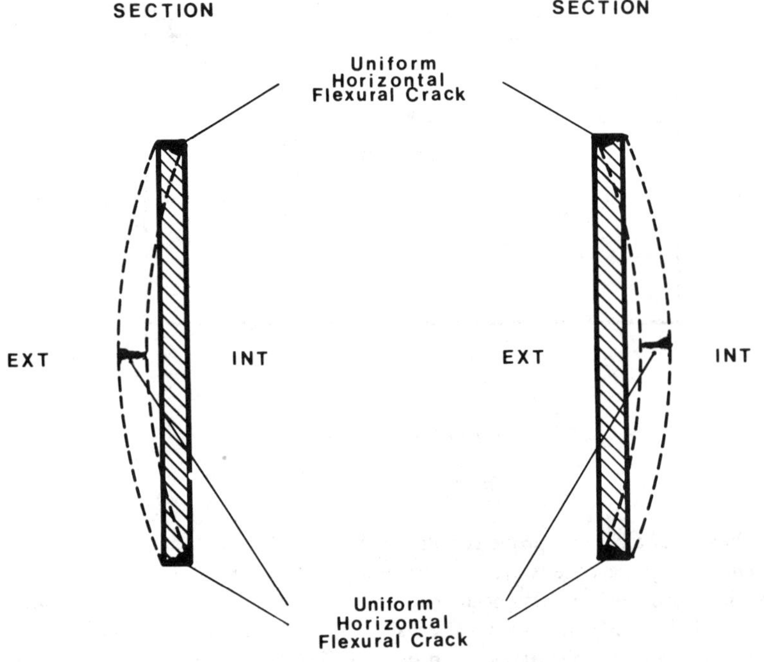

FIG. 9—*Buckled or bowed wall.*

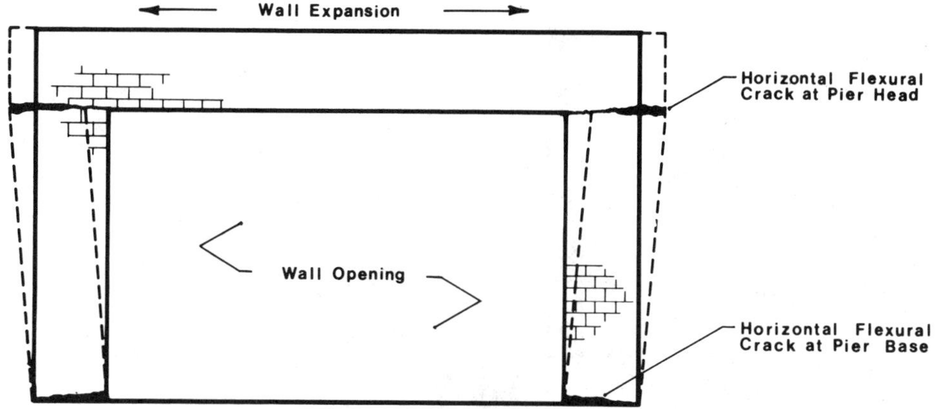

FIG. 10—*Flexural crack at pier head and base.*

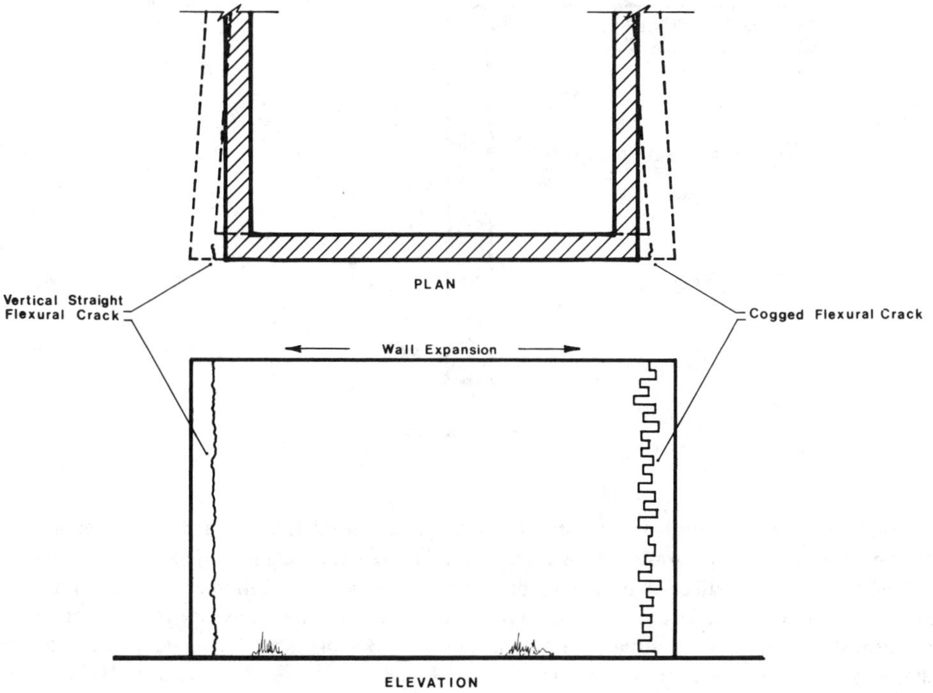

PLAN

Vertical Straight Flexural Crack

Cogged Flexural Crack

Wall Expansion

ELEVATION

FIG. 11—*Flexural crack at corner.*

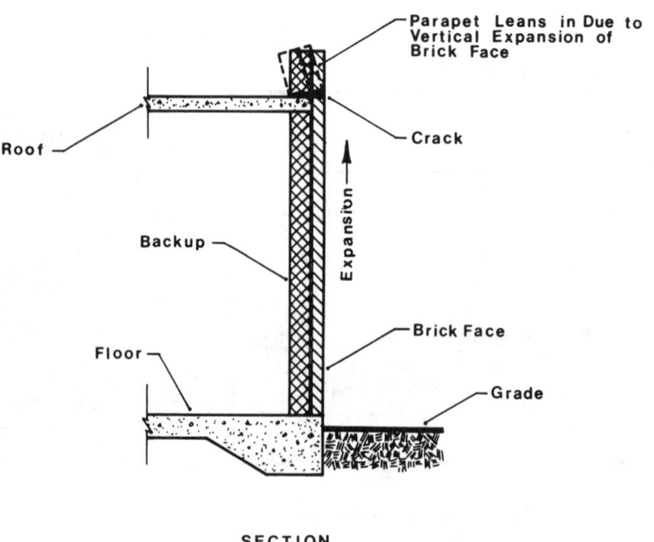

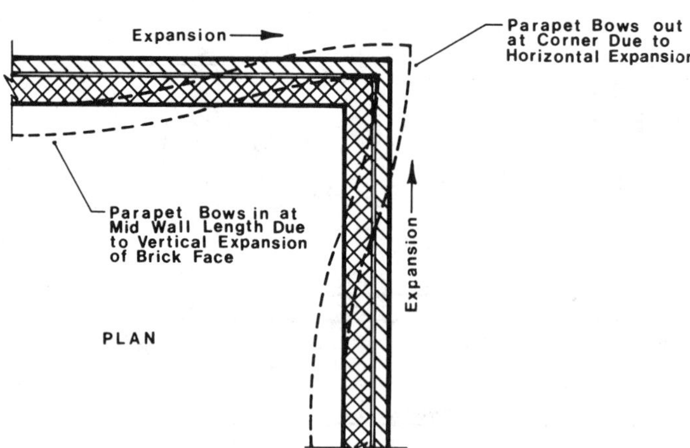

FIG. 12—*Leaning parapet.*

Cracks can occur in bearing wall structures due to differential strain at the intersection of bearing and nonbearing walls. Horizontal cracks at wall midheight may be due to flexure caused by excessive deflection or bowing due to unaccommodated, differential, vertical movement (Fig. 4). Such cracks are also caused by excessive deflection of masonry veneer over flexible steel studs designed for a maximum deflection of L/360 or L/600. A single wythe of brick masonry cracks at a flexural deflection of about L/2000 [*15*]. The horizontal expansions of wood floors has caused bulging and cracks in brick veneer on wood frame buildings.

Masonry having a lower modulus of elasticity has higher strain capacity and is, therefore, less likely to crack. Therefore, the lowest adequate strength of masonry should be used. The lowest strength mortar which is strong enough should always be used.

Foundations

Foundation movement may be caused by uneven settlement, moisture movement in plastic soils, or downhill creep of surface layers. Settlement is caused by soil consolidation, shear failure, and variable soil types. Clays and silts increase in volume with increased moisture content and decrease in volume with reduced moisture. Water content of soil changes with season, trees and shrubs, localized watering and heat, and moisture migration. When ground water reduces the shear strength of sloped soil, downward slides can result [*96*]. Masonry pavements on chalky or fine sandy soils may be subject to upheaval due to ice lensing during severe winters [*23*]. Building on permafrost is a special case with peculiar problems. In coal mining regions ground subsidence may cause a surface wave 2 or 3 ft (0.61 or 0.91 m) high to pass slowly through entire communities [*27, 76*]. The influence of trees on house foundations in clay soils is discussed in a publication of the British Building Research Station [*101*]. The settlement of foundation for masonry walls is discussed by Komornik and Mayurik [*56*].

Cracks that result from uneven settlement of foundations may take any form, but they are most often diagonal or vertical and are usually tapered [*96*] (see Figures 6, 7, 14, 15, 16, and

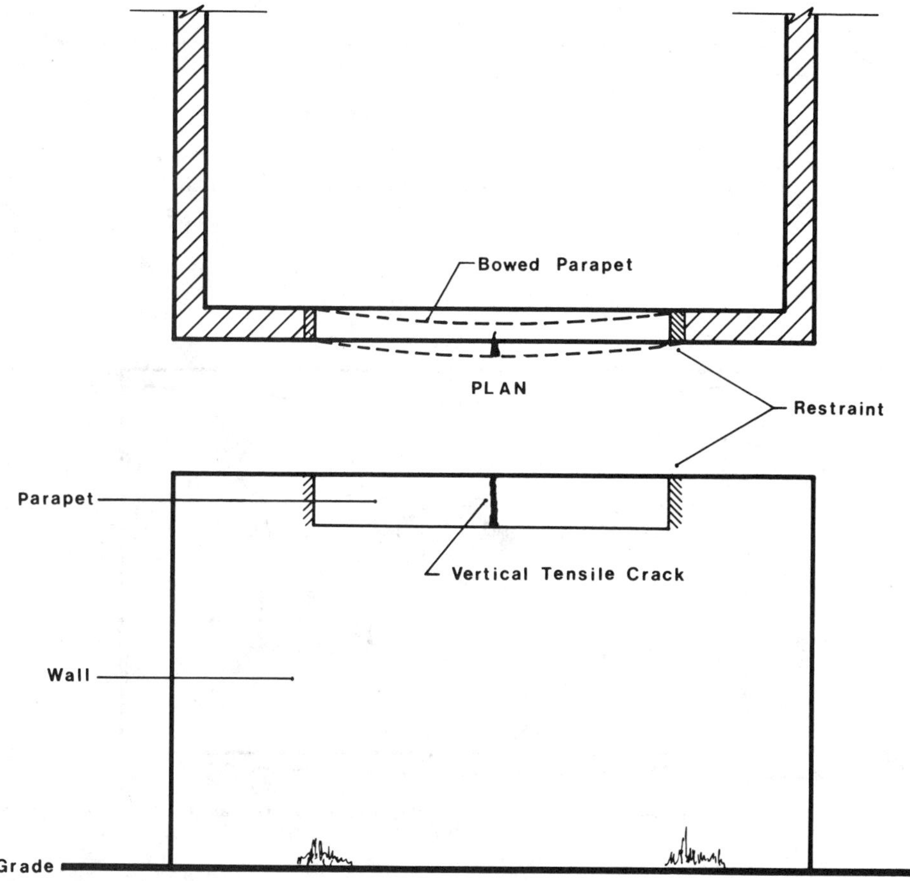

FIG. 13—*Restrained parapet buckled laterally.*

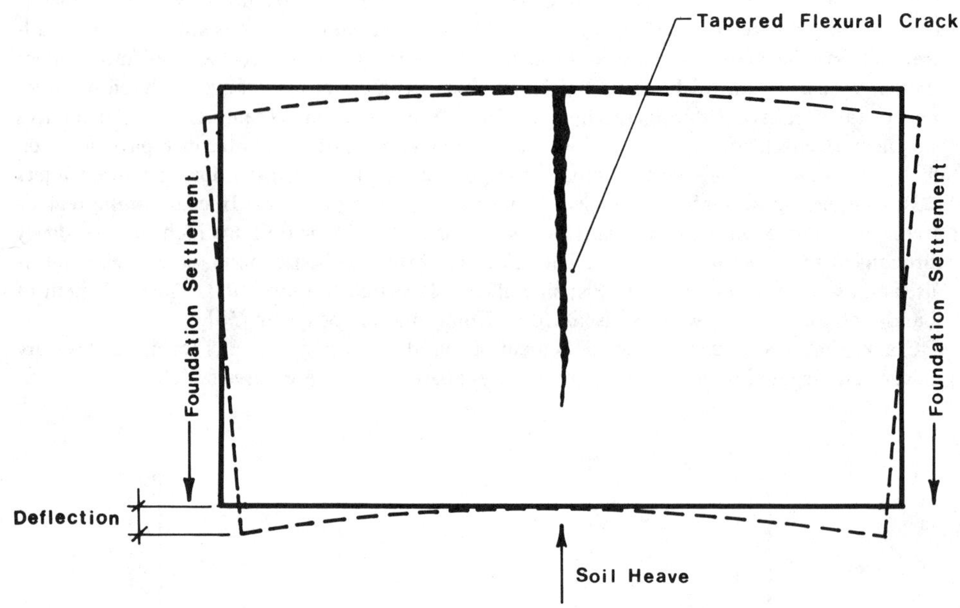

FIG. 14—*Deep beam flexure—corner settlement and/or midwall heave.*

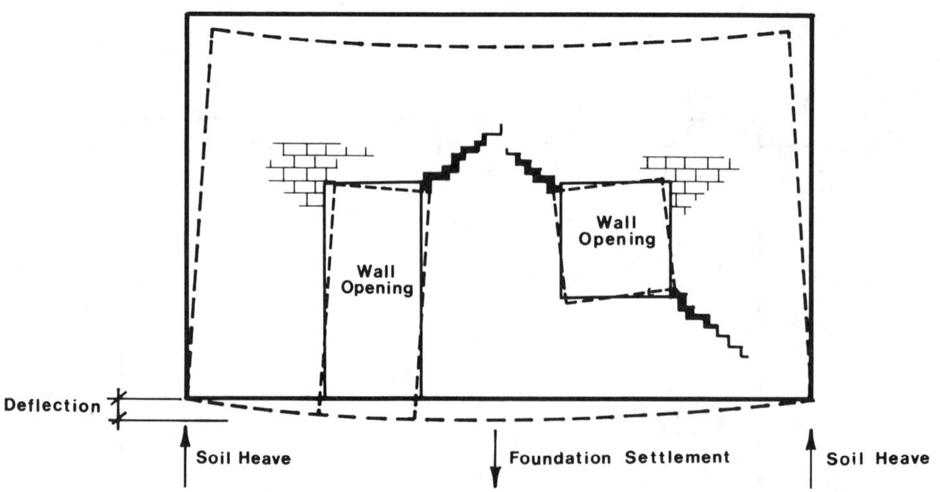

FIG. 15—*Deep beam flexure with wall openings—corner heave and/or midwall settlement or supporting beam deflection.*

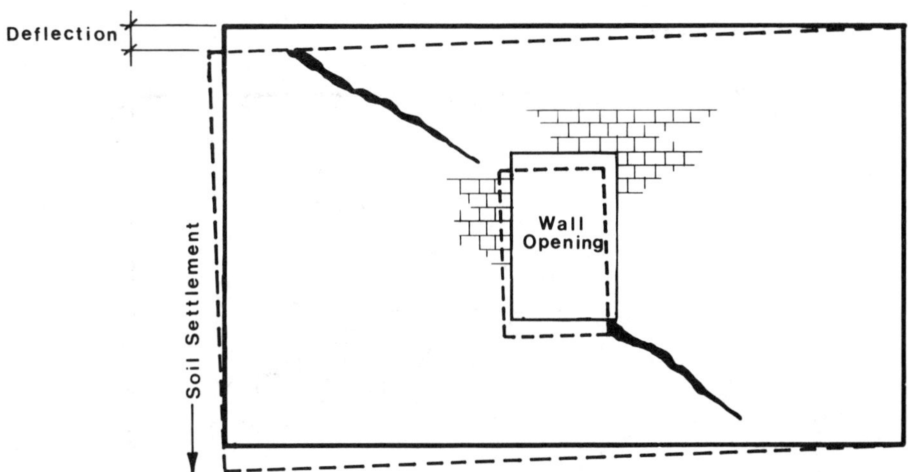

FIG. 16—*Deep beam flexure with wall opening—corner settlement or differential column contraction.*

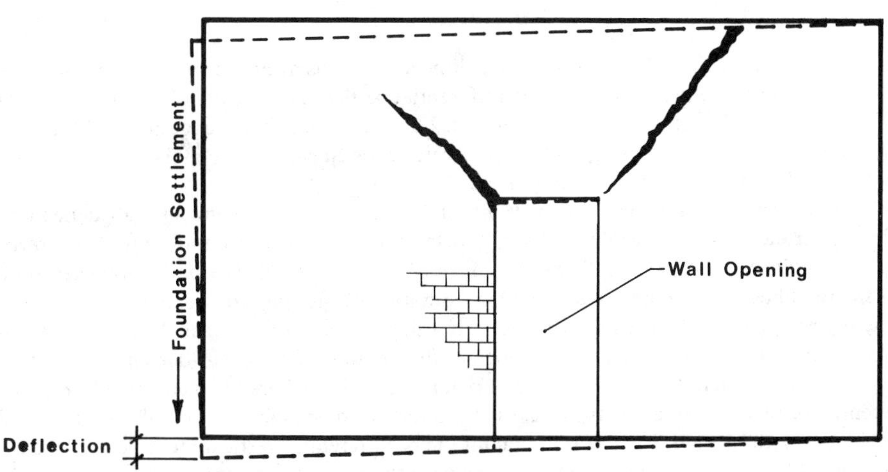

FIG. 17—*Deep beam flexure with wall opening—corner settlement or differential column contraction.*

17). Vertical cracks wider at the top than at the base indicate flexure, sometimes due to foundation movement (see Fig. 14).

Frame Movement

Wind loads [*4*] cause structural frames to sway. Frame drift may induce racking of masonry walls supported on frames (Fig. 18). Concrete columns are subject to thermal, creep, shrinkage, and elastic deformations, which usually are different for each column in the same building. The average contractions of concrete columns in highrise buildings is reported by the Portland Cement Association to be about 0.032% for shrinkage, 0.04% for creep, and 0.042% for elastic deformation, but extreme values may be 65% greater [*14*]. Unaccommodated differential

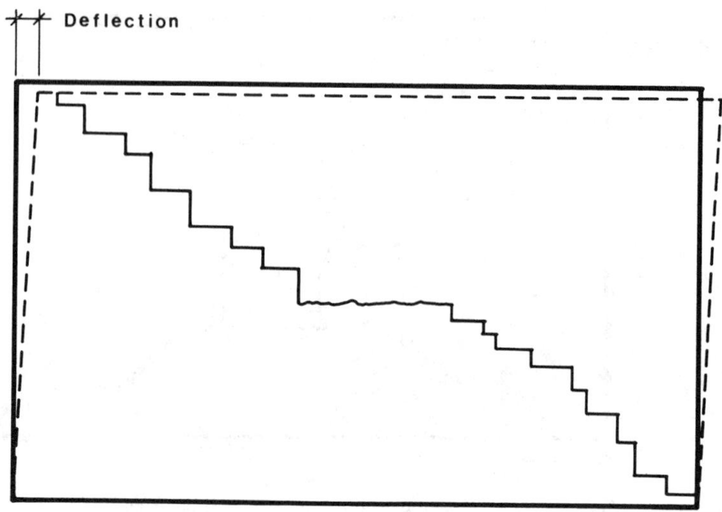

FIG. 18—*Shear in nonbearing wall due to frame side sway (drift).*

movement between adjacent columns will crack masonry walls attached to the frame (see Fig. 16). Thompson and Johnson [*102*] report differential vertical movements of 1/2 in. in 15 ft of wall length as insufficient to cause cracking (0.0028L), but others have suggested limiting such movement to 0.0014L. Such movement, when excessive, often causes diagonal cracks extending to and from the corners of wall openings [*88*].

Horizontal cracks near floor level may indicate excessive floor, beam, or slab deflection. Walls supported on beams having excessive deflection may crack horizontally anywhere, vertically near midspan, and diagonally near span third points [*14,96*]. Masonry supported on a steel spandrel beam may crack due to torsional rotation of the spandrel beam [*52*].

The net expansion of brick masonry due to freezing, moisture, and heat, less restraint and mortar shrinkage, is estimated to have a mean value of about 0.03% but one chance in 20 of being as much as about 0.07% [*42*]. When brick masonry is anchored to a concrete frame, the differential vertical movement may average 0.14% and could be as much as 0.26%, that is, 3.12 in. (79 mm) in 100 ft. (30.5 m). If that magnitude of differential vertical movement is not accommodated by horizontal expansion joints between the masonry and the frame, cracking will result. A restrained movement of only 0.01 in. (0.25 mm) in 20 ft (6.1 m) can produce a bulge of 1 in. (25 mm) [*22*].

Cracks at Shelf Angles

In the absence of a horizontal expansion joint under the shelf angle, which supports masonry on a structural frame, differential vertical movement between the masonry and the frame produces a load on the shelf angle, which may be sufficient to lift the angle from its wedge insert anchor, to yield the angle [*43*], or to cause spalling of brick [*77,98,99*]. Mortar in the joint at the toe of a shelf angle can cause spalling of brick at the shelf angle [*77,98,99*] (see Fig. 4). The deflection of shelf angles should be carefully controlled to insure that the expansion joint under the shelf angle does not close excessively [*43*]. Total shelf angle deflection should not exceed 1/16 in. [*30*].

Vertical cracks frequently occur at corners of walls supported on shelf angles, when the shelf angle is not continuous around the corner, leaving the masonry at the corner continuous vertically past the shelf angle (see Fig. 2).

Cracks at Roofs

Horizontal cracks at corners near concrete roof slabs may be due to slab curl caused by differential shrinkage between the top and bottom of the slab [*24,48*] (see Fig. 3). Roof movement can cause diagonal cracks in masonry walls parallel to the roof movement (see Fig. 5) and horizontal cracks in masonry perpendicular to the roof movement [*96*]. Roof movement may be due to concrete shrinkage or thermal movement in steel roofs. Horizontal cracks near eaves may indicate lateral movement of pitched roofs, vaults, or shells. Dimensional change of wood plates rigidly anchored to masonry walls may cause masonry cracks. The holes in such plates through which the anchor bolts pass should be larger than the bolts, and the anchor nut should be tightened only by hand.

Vibration-Induced Cracks

Most building vibrations generated internally are caused by machines (cranes, elevators, fans, pumps, and punching presses) or by people (walking, jumping, running, dancing). Externally generated vibrations are commonly caused by road or rail traffic, subways, sonic booms, strong wind, earthquake, blasting, excavation, soil compaction, or pile driving [*80*]. Relatively small vibration may add to built-up stress concentrations and lead to unexpected masonry cracks even when vibration levels are within recommended limits [*80*]. In tall buildings wind-induced vibrations can lead to cladding cracks. Dowding and Corser [*31*] describe cracks caused by blasting due to: (1) vibration of the structure or its foundation; (2) impact of flying rock; (3) permanent ground distortion; and (4) air blast [*31*].

Other Crack Causes

Cracks in chimneys may be caused by sudden and wide temperature changes or by the freezing of condensate from the combination of natural gas. Severe fire causes cracking and bulging of masonry as well as surface spalling or possibly vitrification of clay brick. Although severe damage to masonry may be caused by earthquakes, well designed and built masonry may be crack free after imposition of significant seismic loads.

When steel corrodes, the ferric oxide occupies more than twice the volume of steel from which it was formed [*43*]. Corrosion of imbedded reinforcing steel may cause a crack at the wall surface along the length of the steel. In walls, horizontal cracks at regularly spaced vertical intervals may be due to corrosion of bed joint reinforcement or wall ties.

Crack Inspection

Although no absolute determination as to the cause of masonry cracking can be made solely on the basis of visual observation, cause clues are readily obtained. What to notice about cracks [*32*]: (1) direction (pattern); (2) extent (where it begins and ends); (3) width (uniform or tapered and if so how); (4) depth (through the paint, the plaster, and the wall); (5) alignment (in plane or laterally offset); (6) edge sharpness (rough, rounded, or broken edges may be indicative of compression failure); (7) cleanliness (new cracks have clean sides, not coated with paint, dirt, or algae); and (8) crack dynamics (static or changing in size, shape, or direction).

Information on the date of crack occurrence is suspect because a crack is very seldom noticed at first unless its formation is accompanied by a loud noise. Hearing a noise and then finding a crack is not uncommon, but a cause-effect relationship is seldom justified [*32*]. Crack width may be gauged by use of the Avongard Calibrated Crack Monitor (2836 Osage, Waukegan, Il).

Under about 15 foot candles of illumination visual acuity is about 0.5 min of arc, that is, under that illumination, a crack can be seen at a distance up to about 6900 times its width [*47*], for example, a crack width of 0.1 mm can be seen at a distance of about 2 ft-3 in. (960 mm).

The frequency with which masonry should be inspected for cracks varies from one to five years [*105*]. The legal liability assumed by architects and engineers who inspect building facades has caused considerable concern [*60*].

Repair

Tests made at the Building Research Station in England have shown that the capacity of 9-in. (229-mm)-thick brick walls to carry vertical loads is reduced no more than 30% by a stepped or slanted crack up to 1 in. (25 mm) wide, provided that the damage is not accompanied by considerable transverse movement [*82*]. If a wall is out of plumb not more than 1 in. (25 mm) or bulges no more than 1/2 in. (12 mm) in a normal story height, no repair would usually be needed on structural grounds alone [*82*].

Crack repair methods may be classified as those which do not significantly change wall appearance and those which do. Fine cracks [less than 1/16 in. (1.5 mm)] are not very conspicuous and in brick masonry would often be made more unsightly by repointing [*82*]. Such cracks can be filled by surface grouting, which will prevent objectionable water permeance and not greatly change wall appearance, if the masonry surface texture is relatively smooth. Clear coatings for masonry typically do not bridge cracks and, therefore, do not prevent water permeance. Crack repair methods for masonry are discussed in Ref. *34, 37,* and *38*.

References

[*1*] Abrams, D. P., Noland, J. L., and Atkinson, R. H., "Response of Clay-Unit Masonry to Repeated Compressive Forces," *Proceedings of the 7th International Bricks Masonry Conference,* Brick Development Research Institute, University of Melbourne, Melbourne, Australia, February 1985, p. 565.

[*2*] Allan, W. D. M., "Shrinkage Measurements of Concrete Masonry," *Journal of the American Concrete Institute,* Detroit, MI, Vol. 26, No. 6, 1930, pp. 699–713.

[*3*] Ameny, P. and Jessop, E. L., "Masonry Cladding: A Report On Causes and Effects of Failures," *Proceedings of the Seventh International Brick Masonry Conference,* University of Melbourne, Melbourne, Australia, February 1985, p. 261.

[*4*] *American National Standard Minimum Design Loads for Buildings and Other Structures,* ANSI A58.1—1982, American National Standards Institute, Inc., New York, NY, 1982.

[*5*] Anderegg, F. O., "Some Properties of Mortars in Masonry," *Proceedings,* American Society for Testing and Materials, Philadelphia, Vol. 40, 1940, p. 1134.

[*6*] Anderegg, F. O. and Anderegg, J. A., "Some Volume Changes in Mortar and Concrete," *ASTM Bulletin,* American Society for Testing and Materials, Philadelphia, December 1955, pp. 60–63.

[*7*] Anderson, G. W., "The Design of Brickwork for Differential Movement," *Techniques,* No. 6, Brick Development Research Institute, Parkville, Victoria, Australia, January 1979.

[*8*] *1985 Annual Book of Standards,* American Society for Testing and Materials, Philadelphia, Vol. 4.05, 1985.

[*9*] Baker, L. R. and Jessop, E. L., "Moisture Movement in Concrete Masonry," *International Journal of Masonry Construction,* London, England, Vol. 2, No. 2, 1982, pp. 75–80.

[*10*] Baker, M. C., "Introduction to the Problem of Cracks, Movement, and Joints in Buildings," *Cracks, Movement, and Joints in Buildings,* Division of Building Research, National Research Council of Canada, Ottawa, Ontario, August 1972.

[*11*] Bidwell, T. G., *The Conservation of Brick Buildings, The Repair, Alteration, and Restoration of Old Brickwork,* Brick Development Association, London, England, August 1977, p. 6.

[*12*] Birkeland, O. and Sevendsen, S. D., "Norwegian Test Methods for Rain Penetration through Masonry Walls," *Symposium on Masonry Testing, STP No. 320,* American Society for Testing and Materials, Philadelphia, February 1963, pp. 3–15.

[*13*] Bloem, D., "Effects of Aggregate Grading on Properties of Masonry Mortar," *Symposium on Ma-*

sonry Testing, ASTM STP 320, American Society for Testing and Materials, Philadelphia, June 1962, pp. 67-91.

[*14*] *Building Movement Joints,* Portland Cement Association, Skokie, IL, 1982, p. 30.

[*15*] *Compressive, Transverse, and Racking Strength Tests of Four-inch Brick Walls,* Structural Clay Products Research Foundations Research Report No. 9, Brick Institute of America, Reston, VA, August 1965, p. 17.

[*16*] *Concrete Masonry Shrinkage,* National Concrete Masonry Association, Herndon, VA, 1961.

[*17*] Conner, C. C., "Factors in the Resistance of Brick Masonry Walls to Moisture Penetration," *Proceedings of the American Society for Testing and Materials,* ASTM, Philadelphia, Vol. 48, 1948, pp. 1-35.

[*18*] Conner, C. C., "Some Effects of the Grading of Sand on Masonry Mortar," *Proceedings of the American Society for Testing and Materials,* ASTM, Philadelphia, Vol. 53, 1953, pp. 933-945.

[*19*] *Control by Cracking in Concrete Structures,* ACI 224 R-80, American Concrete Institute, Detroit, MI, July 1985, p. 13.

[*20*] Copeland, R. E., "Procedures for Controlling Cracking in Concrete Masonry," *Concrete Products,* Chicago, IL, Vol. 67, No. 9, September 1964, pp. 48-52.

[*21*] Copeland, R. E., "Shrinkage and Temperature Stresses in Masonry," *Journal of the American Concrete Institute,* ACI, Detroit, MI, Vol. 53, 1957, pp. 769-780.

[*22*] "Cracking of Concrete Masonry—Causes and Suggested Remedies," *Concrete,* Vol. 56, No. 4, Chicago, IL, April 1948, p. 6ff.

[*23*] "Cracking in Buildings," *Building Research Digest,* No. 75, Building Research Station, Garston, Waterford, England, October 1966.

[*24*] "Cracking Tendencies in Brick or Stone Masonry Walls at the Structural Slab," *Journal of The American Concrete Institute,* Detroit, MI, January 1947, Vol. 18, No. 5, pp. 606-608.

[*25*] *Cracks Control in Concrete Masonry Unit Construction,* Federal Construction Council Technical Report No. 48, National Academy of Sciences, Washington, DC, 1964.

[*26*] *Cracks, Movements, and Joints in Buildings,* Division of Building Research, National Research Council of Canada, Ottawa, September 1976.

[*27*] Crawford, C. B., "Deformation Due to Foundation Movements," *Cracks, Movement, and Joints in Buildings,* Division of Building Research Council of Canada, Ottawa, Ontario, August 1972.

[*28*] Davis, R. E. and Troxell, G. E., "Volumetric Changes in Portland Cement Mortars and Concrete," *Proceedings of the American Society of Civil Engineers,* Philadelphia, Vol. 25, 1929, pp. 210-260.

[*29*] "Design of Concrete Masonry for Crack Control," *NCMA-TEK No. 53,* National Concrete Masonry Association, Herndon, VA, 1973.

[*30*] "Differential Movement," *Technical Notes on Brick Construction,* No. 18, Bricks Institute of America, Reston, VA, April 1963.

[*31*] Dowding, C. H. and Corser, P. G., "Cracking and Construction Blasting," *Journal of The Construction Division,* American Society of Civil Engineers, New York, NY, March 1981, pp. 89ff.

[*32*] Eldridge, H. J., *Common Defects in Buildings,* Her Majesty's Stationery Office, London, England, 1974, p. 85ff.

[*33*] Evans, D. N. et al., "Properties of Some Masonry Cement," *Journal of Research of the National Bureau of Standards,* Research Paper 2427, Washington, DC, Vol. 51, No. 1, July 1953, pp. 11-16.

[*34*] Filler, J. D. and Kriegh, K. D., "*A Guide to Pressure Grouting Cracked Concrete and Masonry Structures with Epoxy Resins,* National Technical Information Service, Springfield, VA, February 1973, AD-755-926.

[*35*] Fowler, D. W. and Grimm, C. T., "Differential Movement in Composite Load Bearing Masonry Walls," *Journal of the Structural Division,* Proceedings of the American Society of Civil Engineers, Vol. 105, No. ST 7, New York, NY, July 1979, pp. 1277-1288.

[*36*] Fricki, K. E. et al., "Problems in Masonry Walls—A Case Study," *Proceedings of the First North American Masonry Conference,* University of Colorado, Boulder, CO, August 1978, p. 113-1.

[*37*] Grimm, C. T., "Masonry Maintenance and Restoration—A Guide to the Literature," *Structural Renovation and Rehabilitation of Buildings,* Boston Society of Civil Engineers, Section/ASCE, Boston, MA, Nov. 1979, pp. 71-90.

[*38*] Grimm, C. T., "Water Permeance of Masonry Walls—A Review of the Literature," *Masonry: Materials, Properties, and Performance, ASTM STP 778,* American Society for Testing and Materials, Philadelphia, 1982, pp. 178-199.

[*39*] Grimm, C. T. and Fok, C.-P., "Brick Masonry Compressive Strength at First Crack," *Masonry International,* University of Edinburgh, Edinburgh, Scotland, Vol. 1, No. 2, July 1984, pp. 18-23.

[*40*] Grimm, C. T., "Durability of Brick Masonry—A Review of the Literature," *Masonry: Research, Application, and Problems, ASTM STP 871,* American Society for Testing and Materials, Philadelphia, PA, 1985, pp. 202-234.

[*41*] Grimm, C. T., "Flexural Strength of Masonry Prisms vs. Wall Panels," *Journal of the Structural Division,* American Society of Civil Engineers, New York, NY, September 1985, pp. 2021-2032.

[*42*] Grimm, C. T., "Probabilistic Design of Expansion Joints in Brick Masonry," *Proceedings of the 4th Canadian Masonry Symposium,* University of New Brunswick, Fredericton, NB, June 1986.
[*43*] Grimm, C. T. and Yura, J. A., "Shelf Angles for Masonry Veneer," *Journal of the Structural Division,* American Society of Civil Engineers, New York, NY, in review.
[*44*] Hansen, T. C., "Effect of Wind on Creep and Drying Shrinkage of Hardened Cement Mortar and Concrete," *Materials Research and Standards,* American Society for Testing and Materials, Philadelphia, January 1966, pp. 16-19.
[*45*] Hedstrom, R. O., Litvin, A., and Hanson, J. A., "Influence of Mortar and Block Properties on Shrinkage Cracking of Masonry Walls," *Journal of the PCA Research and Development Laboratories,* Portland Cement Association, Skokie, IL, January 1968.
[*46*] Hendry, A. W. and Kheir, A. M. A., "The Lateral Strength of Certain Brickwork Panels," *Proceeding of the Fourth International Brick Masonry Conference,* Groupement National de l'Industrie da la Terre Cuite, Brussels, Belgium, April 1976, p. 4. a. 3.1-4.
[*47*] *IES Lighting Handbook,* Illuminating Engineering Society, New York, NY, 1959, p. 2-9.
[*48*] "Job Problems and Practice," *Journal of The American Concrete Institute,* Detroit, MI, Vol. 43, January 1947, pp. 606-608.
[*49*] Johnson, H. V., "Cement-Lime Mortars," *Technologic Papers of the Bureau of Standards,* Superintendent of Documents, U.S. Government Printing Office, Washington, DC, Jan. 29, 1926, Vol. 20, No. 308, pp. 256.
[*50*] Kalouseb, G. L., *Relation of Shrinkage to Moisture Content in Concrete Masonry Units,* Housing Research Paper No. 25, Housing and Home Finance Agency, Superintendent of Documents, U.S. Government Printing Office, Washington, DC, March 1953, p. 4.
[*51*] Kamimura, K., et al., "Changes in Weight and Dimensions in the Drying and Carbonization of Portland Cement Mortars," *Magazine of Concrete Research,* Vol. 17, No. 10, March 1965, pp. 5-14.
[*52*] Kaminetzky, D., "Preventing Cracks in Masonry Walls," *Architectural Record,* Vol. 136, November 1964, pp. 210-214.
[*53*] Kaminetzky, D., "Verification of Structural Adequacy," *Rehabilitation, Renovation, and Preservation of Concrete and Masonry Structures,* American Concrete Institute, Detroit, MI, 1985, p. 141.
[*54*] Keller, H. and Suter, G. T., "Concrete Masonry Veneer Distress A Case Study," *Proceedings of the Third North American Masonry Conference,* University of Texas at Arlington, Arlington, TX, June 1985, p. 3.
[*55*] Kessler, D. W. and Anderson, R. E., *Studies in Stone Setting Mortar,* BMS Report 139, National Bureau of Standards, Superintendent of Documents, U.S. Government Printing Office, Washington, DC, Nov. 23, 1953, p. 23.
[*56*] Komornik, A. and Mayurik, A., "Restrained Settlement of Masonry Buildings," *Proceedings of the International Conference of Soil Mechanics and Foundation Engineering,* Japanese Society of Soil Mechanics and Foundation Engineers, Tokyo, 1977 and 1978, VI, pp. 613-618.
[*57*] Kroone, B. and Blakey, F. A., "Reaction Between Carbon Dioxide Gas and Mortar," *ACI Proceedings,* American Concrete Institute, Detroit, MI, Vol. 56, 1960, pp. 497-510.
[*58*] "Lack of Design Data, New Building Techniques Cause Facade Failures," *Engineering News Record,* 2 Feb. 1978, p. 9.
[*59*] Lawrence, S. S. and Morgan, J. W., "*Investigations of the Properties of Small Brick works Panels in Lateral Bending,* Experimental Building Station, North Ryde, N.S.W., Australia, TR 52/75/418, January 1975.
[*60*] Le Patner, B. B., "Caveat Architectus: Facade Inspections and the Design Professional," *Architectural Record,* July 1981, p. 57.
[*61*] Mansfield, G. A., Sirrine, C. A., and Wilk, B., "Control Joints Regulate Effects of Volume Change in Concrete Masonry," *Journal of the American Concrete Institute,* Detroit, MI, July 1957, Vol. 54, pp. 59-70.
[*62*] *Masonry Structural Design for Buildings,* TM 5-809-3/NAVFAC DM—2.9/AFM 88-3, Chap. 3, Department of the Army, Navy, and the Air Force, Washington, DC, 16 Jan. 1985.
[*63*] Mayes, R. L., Yutaro Omote, and Clough, R. W., *Cyclic Shear Tests of Masonry Piers, Vol. 1., Test Results,* National Technical Information Service, PB-264-424, Springfield, VA, May 1976, p. 77.
[*64*] McBurney, J. W., "Cracking in Masonry Caused by Expansion of Mortar," *Proceedings of the American Society for Testing and Materials,* ASTM, Philadelphia, Vol. 52, 1952, pp. 1-20.
[*65*] Mears, A. R. and Hobbs, D. W., "The Effect of Mix Proportions Upon the Ultimate Air-Drying Shrinkage of Mortars," *Magazine of Concrete Research,* Slough, England, Vol. 24, No. 79, June 1972, pp. 77-84.
[*66*] Meli, R., "Behavior of Masonry Walls Under Lateral Loads," *Proceedings of the Fifth World Conference on Earthquake Engineering,* Rome, Earthquake Engineering Research Institute, El Cerrito, CA, 1972.
[*67*] Menzel, C. A., "General Considerations of Cracking in Concrete Masonry Walls and Means for

Minimizing It," *Development Department Bulletin D 20,* Portland Cement Research and Development Laboratories, Skokie, IL, September 1958, pp. 1-15.

[*68*] Monk, C. B., "Testing High-Bond Clay Masonry Assemblages," *Symposium on Masonry Testing, ASTM STP 320,* American Society for Testing and Materials, Philadelphia, February 1963, pp. 31-66.

[*69*] Moss, P. J. and Scrivener, J. C., "Concrete Masonry Wall Panel Tests—The Effect of Cavity Filling on Shear Behavior," *New Zealand Concrete Construction,* Concrete Publications, Private Bag, Porirua Bay, NZ, April 1968.

[*70*] Myren, B. J., "Cracks in Housing Project Walls," *Engineering News Record,* 17 Mar. 1938, p. 390.

[*71*] Niermann, T. H., "Cracks in Brick Walls," *Engineering News Record,* 5 May 1938, pp. 640-641.

[*72*] Nuss, Larry K., Noland, J. L., and Chinn, J., "The Parameters Influencing Shear Strength Between Clay Masonry Units and Mortar," *Proceedings of the (First) North American Masonry Conference,* University of Colorado, Boulder, CO, August 1978, p. 13.

[*73*] Palmer, L. A., "Volume Changes in Brick Masonry Materials," Research Paper No. 321, *Journal of Research,* National Bureau of Standards, Superintendent of Documents, U.S. Government Printing Office, Washington, DC, Vol. 6, June 1931, pp. 1003-1026.

[*74*] Palmer, L. A. and Parsons, D. A., "A Study of the Properties of Mortars and Bricks and Their Relation to Bond," R.P. 683, *Journal of Research,* National Bureau of Standards, May 1934, Vol. 12, pp. 609-644.

[*75*] Panek, J. R. and Cook, J. P., *Construction Sealants and Adhesives,* Wiley-Interscience, New York, NY, 1984.

[*76*] Parate, N. S., "Some Observations on Masonry Structure Behavior Due to Ground Movement," *Proceedings of the Second North American Masonry Conference,* University of Maryland, College Park, MD, August 1982, p. 31.

[*77*] Plewes, W. G., "Failure of Brick Facing on High-Rise Buildings," *Canadian Building Digest,* No. 185, April 1977, p. 3.

[*78*] Powers, T. C., "A Hypothesis on Carbonization Shrinkage," *Journal of Portland Cement Research and Development,* Portland Cement Association, Skokie, IL, May 1962, pp. 41-49.

[*79*] "Prevention of Cracking," *Rock Products,* March 1942, Vol. 45, No. 3, March 1942, p. 79.

[*80*] Rainger, J. H., "Vibrations in Buildings," *Canadian Building Digest,* No. 232, National Research Council of Canada, Ottawa, Ontario, Canada, May 1984.

[*81*] Rainer, P., *Movement Control in the Fabric of Buildings,* Nichols Publishing Co., New York, NY, 1983.

[*82*] "Repairing Brickwork," *Building Research Station Digest,* No. 4, Garston, Waterford, Hurts, England, 1960.

[*83*] Ritchie, T., "Effect of Restraint on the Shrinkage of Masonry Mortars," *Materials Research and Standards,* American Society for Testing and Materials, Philadelphia, Vol. 6, No. 1, January 1966, pp. 13-16.

[*84*] Ritchie, T., "Influence of Lime in Mortar on the Expansion of Brick Masonry," *Reaction Parameters of Lime, ASTM STP 472,* American Society for Testing and Materials, Philadelphia, 1969, pp. 67-81.

[*85*] Ritchie, T., "Measurement of Laminations in Brick," *Bulletin,* American Ceramic Society, Columbus, OH, No. 9, Vol. 54, 1975, pp. 725-726.

[*86*] Russell, W. A., *Shrinkage Characteristics of Concrete Masonry Walls,* Housing and Home Finance Agency, Housing Research Paper 34, Superintendent of Documents, U.S. Government Printing Office, April 1954.

[*87*] Schneider, R. R., *Shear in Concrete Masonry Piers,* Masonry Research of Los Angeles, Los Angeles, CA, undated.

[*88*] Schubert, P. and Glitza, H., "Resistance to Cracking of Masonry Subjected to Vertical Deformation," *Proceedings of the Fourth International Brick Masonry Conference,* Bruge, Belgium, April 1976, p. 4.6.9.

[*89*] Scrivener, J. C., "Concrete Masonry Wall Panel Tests—Static Racking Tests with Predominant Flexural Effect," *New Zealand Concrete Construction,* Concrete Publications, Private Bag, Porirua Bay, Porirua Bay, New Zealand, July 1966.

[*90*] Scrivener, P. J., "Static Racking Tests on Concrete Masonry Walls," *Proceeding of the First International Masonry Conference,* Gulf Publishing Co., Houston, TX, May 1969, p. 185.

[*91*] Severud, F. N., "Avoiding Cracks in Building Walls," *Engineering News Record,* 6 July 1939, pp. 45-46.

[*92*] Shakir, A., "Failure of Masonry Structures," *Proceedings of the Third North American Masonry Conference,* University of Texas at Arlington, Arlington, TX, June 1985, p. 21.

[*93*] Shideler, J. J., "Carbonation Shrinkage of Concrete Masonry Units," *Journal of the Portland Cement Association Research and Development Laboratories,* Skokie, IL, Vol. 5, No. 3, September 1963, pp. 36-51.

[*94*] Shrive, N. G., "A Fundamental Approach to the Fracture of Masonry," *Proceedings of the Third Canadian Masonry Symposium,* University of Alberta, Edmonton, Alberta, Canada, June 1983, pp. 4-1.

[*95*] Sinha, B. P. and Currie, D. W., "Survey of Scottish Sands and Their Characteristics Which Effect the Mortar Strength," *International Journal of Masonry Construction,* United Trade Press, London, England, Vol. 2, No. 1, 1981, pp. 2-12.

[*96*] Sorensen, C. P. and Tasker, H. E., *Cracking in Brick and Block Masonry,* Experimental Building Station, Technical Study No. 43, Australian Government Publishing Service, 1976.

[*97*] Stevenson, J. H., "Cracking Walls," *Engineering News Record,* 31 Aug. 1939, p. 38.

[*98*] Stockbridge, J. G., "Cladding Failures—Lack of Professional Interface," *Journal of The Technical Councils of ASCE,* American Society of Civil Engineers, New York, NY, December 1979.

[*99*] Suter, G. T. and Hall, J. S., "How Safe Are Our Cladding Connections," *Proceedings of The First Canadian Masonry Symposium,* University of Calgary, Calgary, Alberta, 1976, pp. 95-109.

[*100*] "The Avoidance of Cracking in Masonry Construction of Concrete and Sand-Lime Brick," *Building Research Digest,* No. 6, Building Research Station, Gaston, Waterford, Herts, England, 1949, revised 1957.

[*101*] "The Influence of Trees on House Foundations in Clay Soils," *Digest,* No. 298, Building Research Station, Garston, Waterford, England, June 1985.

[*102*] Thompson, J. N. and Johnson, F. B., "Design for Crack Prevention", *Insulated Masonry Cavity Walls,* National Academy of Sciences, Publication No. 793, Washington, DC, 1960.

[*103*] Verbeck, G., "Carbonization of Hydrated Portland Cement," *Cement and Concrete, ASTM STP 205,* American Society for Testing and Materials, Philadelphia, 1958, pp. 17-36.

[*104*] Voss, W. C., "Lime Characteristics and Their Effect on Construction," *Symposium on Lime,* American Society for Testing and Materials, Philadelphia, 1939, p. 14.

[*105*] "Wall Cladding Defects and Their Diagnosis," *Building Research Establishment Digest,* No. 217, Building Research Station, Garston, Waterford, England, September 1978, p. 2.

[*106*] Watstein, D. and Seese, N. A., "Properties of Masonry Mortars of Several Compositions," *ASTM Bulletin,* No. 147, American Society for Testing and Materials, Philadelphia, Aug. 1974, pp. 77-81.

[*107*] Society of Mt. Carmel v Fox, 31 IlAp3d 1060, 335 NEZd 588 (1975).

[*108*] Schreiner v Miller, 67 Ia 91, 24 NW 738 (1885).

[*109*] President and Directors of Georgetown College v Madden, 660 FZd 91 (4th Cir 1981).

[*110*] Society of Mt. Carmel v Fox & Fox, 90 IlAp3d 537, 413 NEZd 480 (1980).

[*111*] Robbins, C. R., *Chemical and Physical Behavior of Human Hair,* Van Nostrand Reinhold Co., New York, NY, 1979, p. 179.

[*112*] Bartlett, W. H. C., "Experiments on the Expansion and Contraction of Building Stones by Variations of Temperature," *American Journal of Science,* first series, Vol. 22, 1830, pp. 136-140.

Author Index

Subject Index

A

B

D

E

F

G